Natural Products and Cardiovascular Health

Clinical Pharmacognosy Series

Series Editors

Navindra P. Seeram and Luigi Antonio Morrone

Herbal Medicines
Development and Validation of Plant-Derived Medicines for Human Health
Giacinto Bagetta, Marco Cosentino, Marie Tizianna Corasaniti, and Shinobu Sakurada

Natural Products Interactions on Genomes
Siva G. Somasundaram

Aromatherapy
Basic Mechanisms and Evidence Based Clinical Use
Giacinto Bagetta, Marco Cosentino, and Tsukasa Sakurada

Natural Products and Cardiovascular Health
Catherina Caballero-George

Natural Products and Cardiovascular Health

Edited by
Catherina Caballero-George

CRC Press is an imprint of the
Taylor & Francis Group, an **informa** business

CRC Press
Taylor & Francis Group
6000 Broken Sound Parkway NW, Suite 300
Boca Raton, FL 33487-2742

Printed on acid-free paper

International Standard Book Number-13: 978-1-4987-8900-4 (Hardback)

Library of Congress Cataloging in Publication Data

Names: Caballero-George, Catherina, editor.
Title: Natural products and cardiovascular health / [edited by] Catherina C. Caballero-George.
Description: Boca Raton : Taylor & Francis, 2018. | Includes bibliographical references.
Identifiers: LCCN 2018035552 | ISBN 9781498789004
Subjects: | MESH: Cardiovascular Diseases—therapy | Cardiovascular Diseases—drug therapy | Phytotherapy—methods | Biological Products—chemistry
Classification: LCC RC671 | NLM WG 120 | DDC 616.1/06—dc23
LC record available at https://lccn.loc.gov/2018035552

Visit the Taylor & Francis Web site at
http://www.taylorandfrancis.com

and the CRC Press Web site at
http://www.crcpress.com

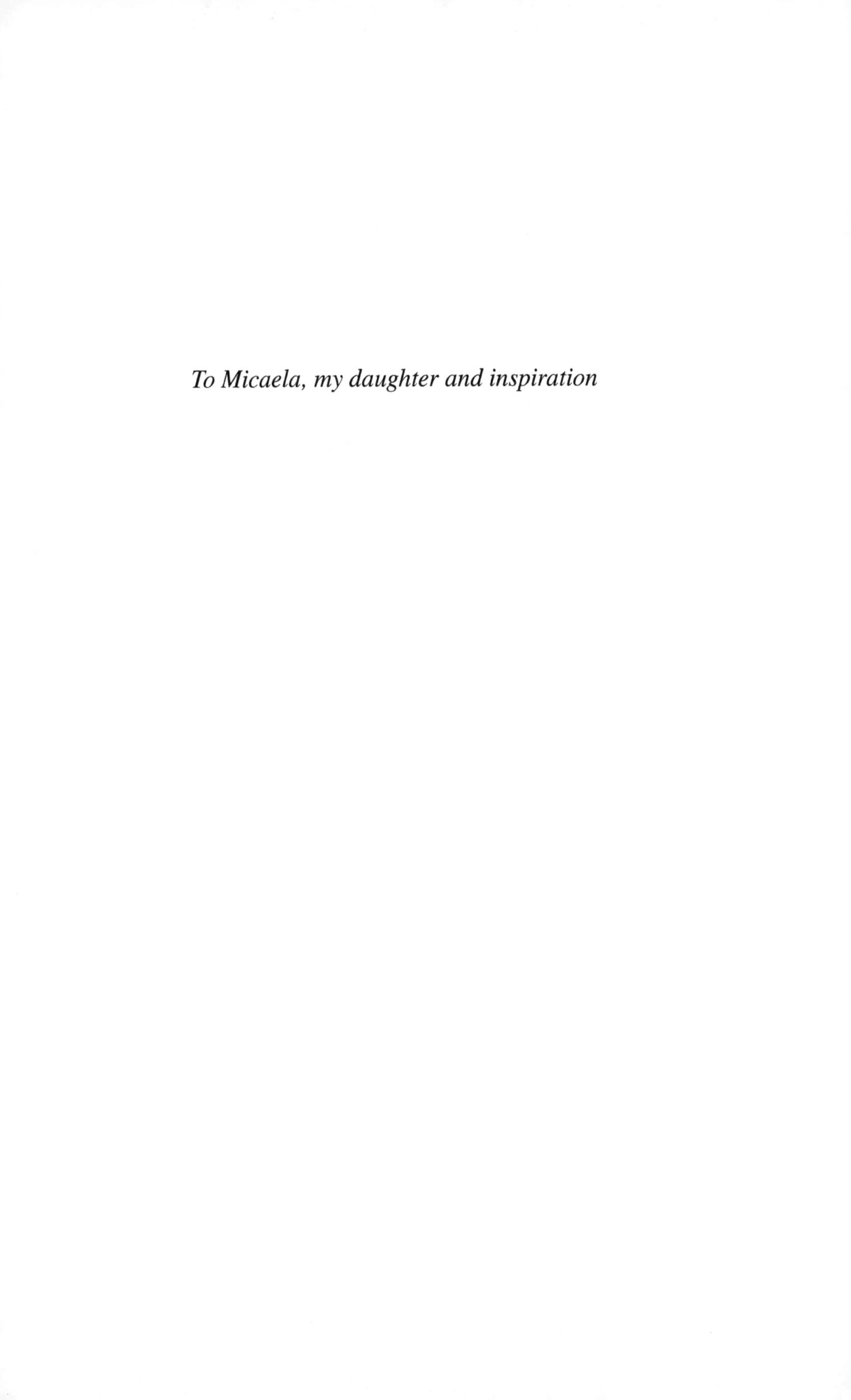

To Micaela, my daughter and inspiration

Contents

Foreword

Cardiovascular diseases (CVDs), including coronary heart disease, cerebrovascular disease and peripheral arterial disease, are one of the leading causes of premature death in most Western countries. Although the classical risk factors such as smoking and obesity are more and more becoming managed, a continuous increase of a cluster of disorders, including dyslipidemia, hypertension, abdominal obesity and hyperglycemia, since the beginning of this century, referred to as the metabolic syndrome, has become more a primary risk factor for CVDs and diabetes.

Many bioactive compounds in foods and herbal medicinal products provide potential health benefits in cardiovascular diseases. Compiling data from experimental, epidemiological and clinical studies shows that these natural products have important cardioprotective effects in the primary as well as secondary prevention of cardiovascular diseases based on the following two conditions: that the amount consumed is high enough and the bioavailability is guaranteed. These bioactive products vary widely in chemical structure and function and are classified accordingly. In this book, several of these classes of bioactive natural compounds are discussed in nine chapters.

In Chapter 1, the beneficial effects of omega-3-fatty acids and extra virgin olive oil are discussed. Omega-3-fatty oils largely contain polyunsaturated fatty acids (PUFAs). The types of PUFAs involved in human physiology are alpha-linolenic acid (ALA), found in plant oils, and eicosapentaenoic acid (EPA) and docosahexaenoic acid (DHA), both commonly found in fish oils. It is suggested that supplementation with PUFAs modestly lowers blood pressure, stimulates blood circulation and increases the breakdown of fibrin involved in blood clotting.

Olive oil is the primary source of fat in the Mediterranean diet and is associated with a low mortality for CVDs. Besides the classical benefits on the lipid profile provided by olive oil consumption compared to that of saturated fat, a broad spectrum of benefits on cardiovascular risk factors is now emerging that is associated with extra virgin olive oil. In Chapter 7, the effects on blood pressure are the main focus.

A systematic review of the efficiency of natural polyketides (statins) for the prevention of coronary events are presented in Chapter 2.

Polyphenols are abundant micronutrients in our diet, and evidence for their role in the prevention of degenerative diseases such as CVDs and cancer is emerging daily. Polyphenolics such as flavonols, anthocyanins and proanthocyanidins are present in many fruits, cereals, legumes, vegetables, tea and red wine. There is some evidence to suggest that these bioactive compounds are effective in reducing CVD risk factors with respect to anti-hypertensive effects, inhibition of platelet aggregation and increasing endothelial-dependent vasodilatation. It is believed that the potential health benefits arise mainly from their antioxidant activity, namely, free radical scavenging activity, chelation of transition metals and inhibition of enzymes.

Chapters 3, 5, 6 and 8 deal with the protective effects of polyphenolics on cardiovascular health.

Chapter 9 is devoted to the study of the binding capability of serum albumin to different ligands, which may interfere with specific molecular mechanisms modifying the pharmacological profile of these chemicals. This is relevant to those studying natural products for their capacity to interact with molecular targets involved in CVDs, since serum albumins are present in all mammals and are used in a wide array of cellular and biochemical assays in drug discovery programs.

This book has been written by authors who are all experts in their fields and who give a concise and comprehensive overview of the current knowledge of bioactive natural products and their benefits in the prevention of cardiovascular diseases. It has been designed as a reference and resource for all who are interested in this discipline. It is our hope that the future of science and product development in this area will be an exciting adventure for years to come.

Professor Dr. Emeritus Arnold J. Vlietinck
University of Antwerp, Belgium

Preface

It is a fact that there has been since time immemorial and there still continues to be a popular predilection for using herbal medicines to treat many diseases, from a simple cold to the most complex ailment. Patients trust these medicines because they have been used for centuries, and all the knowledge acquired from their existence and medicinal properties have been passed on from generation to generation. However, in spite of the presumption that anything natural *is safe*, the use of herbal medicines has drawbacks that call into question the correct identification and manipulation of the plant material. Misidentification of the plant and the use of adulterants in commercially available products have resulted in deaths and raised awareness of the importance of having a complete understanding of the botany, chemical composition, pharmacology and toxicology of medicinal plants.

Nowadays, the number of patients with cardiovascular diseases who prefer herbal medicines over conventional drugs is increasing. Therefore, the search for new and more effective treatments that integrate natural products as phytomedicines or functional foods are a contemporary trend in treating these patients

This book provides updated information on secondary metabolites obtained from selected organisms (plants or microorganisms) with beneficial effects in the treatment of cardiovascular diseases or in the prevention of their progression. The aim of this book is also to provide, whenever possible, a detailed description of the source, chemistry, mechanism of action and clinical studies relating to such metabolites.

In order to use natural products to efficiently treat any disease, a clear connection should be made between chemical structures, pre-clinical studies and their clinical significance.

Even though the subject is extensive, the important topics discussed in this book will be useful to students and professionals in the area of natural product research, pharmacognosy, analytical chemistry, pharmacology, pharmacy, biochemistry, molecular biology, alternative medicine and integrative health programs and food science, as a guide for a better understanding of the cardio- and vasculoprotective potential of different groups of secondary metabolites.

Dr. Catherina Caballero-George
Institute of Scientific Research and High Technology Services (INDICASAT AIP)
July 2018

Acknowledgments

I am highly grateful to Professor Navindra Seeram for trusting me to be the editor of this important and timely publication and to all contributing authors for their generosity in sharing their knowledge with the scientific community.

I am especially indebted to Professor Mahabir Gupta for going above and beyond by revising critical parts of this book, to Professor Arnold Vlietinck for accepting the invitation to be an important part in this endeavor and to Dr. Rae Matsumoto for her constant valuable recommendations.

I wish to thank Hilary LaFoe, Natasha Hallard, James Murray and Danielle Zarfati for their excellent editorial support.

Thanks are also due to the National System of Investigation (SNI) from the National Secretariat of Science, Technology and Innovation of the Republic of Panama (SENACYT) for financial support of my research.

I am deeply grateful to my husband, Roberto, for his patience and understanding, and to my parents, Jilma and Arcenio, for their never-ending support when I needed it the most.

Contributors

Arpita Basu
Department of Kinesiology and Nutrition Sciences
University of Nevada, Las Vegas
Las Vegas, Nevada

Paramita Basu
Department of Biology
Texas Woman's University
Denton, Texas

Nancy Betts
Department of Nutritional Sciences
Oklahoma State University
Stillwater, Oklahoma

Catherina Caballero-George
Center for Innovation and Technology Transfer
Institute of Scientific Research and High Technology Services (INDICASAT-AIP)
Panama, Republic of Panama

Angela I. Calderón
Department of Drug Discovery and Development, Harrison School of Pharmacy
Auburn University
Auburn, Alabama

Tess De Bruyne
Laboratory of Natural Products and Food–Research and Analysis (NatuRA)
University of Antwerp
Antwerp, Belgium

Estela Guerrero De León
Department of Pharmacology, School of Medicine
University of Panama
Panama, Republic of Panama

Guido De Meyer
Laboratory of Physiopharmacology
University of Antwerp
Antwerp, Belgium

Germán Domínguez-Vías
Unit of Physiology, School of Medicine
University of Cádiz
Cádiz, Spain

Sarah Engelbeen
Bioengineering Sciences Department
Vrije Universiteit Brussel
Brussels, Belgium

Mahabir Prashad Gupta
Center for Pharmacognostic Research on Panamanian Flora, College of Pharmacy
University of Panama
Panama, Republic of Panama

Mark T. Hamann
Charles and Carol Cooper/SmartState Endowed Chair
Drug Discovery, Biomedical Sciences and Public Health
The Hollings Cancer Center
Charleston, South Carolina

George Hanna
Drug Discovery, Biomedical Sciences and Public Health
The Hollings Cancer Center
Charleston, South Carolina

Nina Hermans
Laboratory of Natural Products and Food–Research and Analysis (NatuRA)
University of Antwerp
Antwerp, Belgium

Marina Hidalgo
Unit of Microbiology, Department of Health Sciences
University of Jaén
Jaén, Spain.

Haroon Khan
Department of Pharmacy
Abdul Wali Khan University
Mardan, Pakistan

Graham C. Llivina
Department of Drug Discovery and Development, Harrison School of Pharmacy
Auburn University
Auburn, Alabama

Timothy J. Lyons
Division of Endocrinology
Medical University of Charleston
Charleston, SC, USA

Magdalena Martínez-Cañamero
Unit of Microbiology, Department of Health Sciences
University of Jaén
Jaén, Spain

Marya
Department of Pharmacy
Abdul Wali Khan University
Mardan, Pakistan

Juan Antonio Morán-Pinzón
Department of Pharmacology, School of Medicine
University of Panama
Panama, Republic of Panama

Orlando O. Ortiz
Herbario PMA
Universidad de Panamá, Estafeta Universitaria
Panama City, Republic of Panama

Luc Pieters
Laboratory of Natural Products and Food–Research and Analysis (NatuRA)
University of Antwerp
Antwerp, Belgium

Ada Popolo
Dipartimento di Farmacia
University of Salerno
Fisciano, Italy

Isabel Prieto
Unit of Physiology, Department of Health Sciences
University of Jaén
Jaén, Spain

Manuel Ramírez-Sánchez
Unit of Physiology, Department of Health Sciences
University of Jaén
Jaén, Spain

Luca Rastrelli
Dipartimento di Farmacia
University of Salerno
Fisciano, Italy

Andrés Rivera-Mondragón
Natural Products & Food Research and Analysis (NatuRA), Department of Pharmaceutical Sciences
University of Antwerp
Antwerp, Belgium

Harry Robberecht
Laboratory of Natural Products and Food–Research and Analysis (NatuRA)
University of Antwerp
Antwerp, Belgium

Lynn Roth
Laboratory of Physiopharmacology
University of Antwerp
Antwerp, Belgium

Ana Belén Segarra
Unit of Physiology, Department of Health Sciences
University of Jaén
Jaén, Spain

Patrick M. L. Vanderheyden
Bioengineering Sciences Department
Vrije Universiteit Brussel
Brussels, Belgium

Megan M. Waguespack
Department of Drug Discovery and Development
Harrison School of Pharmacy
Auburn University
Auburn, Alabama

1 Beneficial Effects of Omega-3 Fatty Acids on Cardiovascular Disease

Estela Guerrero De León, Mahabir Prashad Gupta and Juan Antonio Morán-Pinzón

CONTENTS

1.1 INTRODUCTION

Cardiovascular diseases (CVDs) are the principal cause of mortality in many economically developed countries, and its incidence is increasing rapidly in emerging economies. Since an appropriate diet ensures the supply of essential nutrients for the regulation of diverse physiological functions, a nutritional imbalance increases the risk of cardiovascular and metabolic diseases. Despite some countries having established healthy lifestyle policies, implementation is not always easy, as people are not always sufficiently motivated to make recommended modifications. Parallel strategies are directed to exploration of the use of nutraceuticals to promote cardiovascular health, as some specific components of diet have been described as possessing therapeutic roles in human health. Specifically, a higher consumption of fish as source of essential fatty acids (EFAs) is promoted because they could prevent the development of CVDs such as dyslipidemias, arrhythmias, arterial hypertension and atherosclerosis.

1.1.1 Chemistry and Metabolism of Essential Fatty Acids

EFAs are a group of organic compounds that have a carboxylic group (-COOH) at one end, and methyl (H_3C-) at another end. The nature of the rest of the molecule, a chain of hydrocarbons, determines the chemical and biological characteristics of the different fatty acids. Structural differences in the hydrocarbon chain are fundamentally based on the number of carbon atoms, absence or presence of double bonds in saturated and unsaturated fatty acids and location of these bonds *cis* or *trans* in their configuration. Fatty acids are classified by the length of the carbon chain (long chain, n-20 to 22; intermediate chain, n-18) and the number of double bonds (saturated, monounsaturated, polyunsaturated) (DeFilippis and Sperling, 2006).

Long-chain polyunsaturated fatty acids, known as PUFAs, are present in all mammalian tissues. However, mammals cannot directly synthesize these fatty acids because they lack the enzymes to make double bonds at some position in the fatty acid chain. Therefore, long-chain PUFAs should be consumed with diet and are, therefore, "essential" fatty acids (EFAs) (Dimitrow and Jawien, 2009).

PUFAs comprise two main classes: ω-6 PUFA and ω-3 PUFA, also known as ω-6 and ω-3, respectively. These two families of PUFAs distinguish themselves in the position of the first double bond, counted from the end methyl group of the molecule. Major sources of ω-6 PUFAs are vegetable oils such as corn, safflower and soybean oil, and they are derived from linoleic acid (LA; C18:2 ω-6), while ω-3 PUFAs are derived from α-linolenic acid (ALA; C18:3 ω-3), and their main sources are fish such as salmon, trout and tuna (De Caterina et al., 2003). Unsaturated fatty acids also include the ω-9 series, derived from oleic acid (OA, C18:1 ω-9) and the ω-7 series, derived from palmitoleic acid (C16:1 ω-7), which are not essential (Das, 2006; Wallis et al., 2002).

Once consumed, these EFAs are further metabolized within mammalian cells by the same set of enzymes to their respective long-chain metabolites (Figure 1.1). The main limiting step in the biosynthesis of ω-6 and ω-3 PUFAs is the microsomal Δ^6-desaturation of LA and ALA (Bernert and Sprecher, 1975; Sprecher et al., 1999). As a result of this metabolic reaction, the λ-linolenic (GLA; 18:3n-6) and stearidonic (SDA; 18:4n-3) acids are obtained, which are elongated, respectively, to dihomo-λ-linolenic (DGLA; 20:3n-6) and eicosatetraenoic (ETA; 20:4n-3) acids. Both PUFAs, Δ^5-desaturase substrates, generate arachidonic (AA; 20:4n-6) and eicosapentaenoic acids (EPA; 20:5n-3), which undergo two consecutive elongation processes to their respective products 24:4n-6 and 24:5n-3. A second microsomic Δ^6-desaturation takes place again in the biosynthesis of PUFAs (Baker et al., 2016). The products of this desaturation, 24:5n-6 and 24:6n-3, are then converted by enzymatic β-oxidation to 22:5n-6 and docosahexaenoic acid (DHA; 22:6n-3), respectively.

As can be seen, both EPA and DHA, the two ω-3 PUFAs with greater biological activity, are obtained from the same precursor, ALA (Robinson and Stone, 2006). However, the effectiveness of conversion of ALA into EPA and DHA is limited in humans. Thus, in humans ALA can be converted in EPA up to 8%, while the

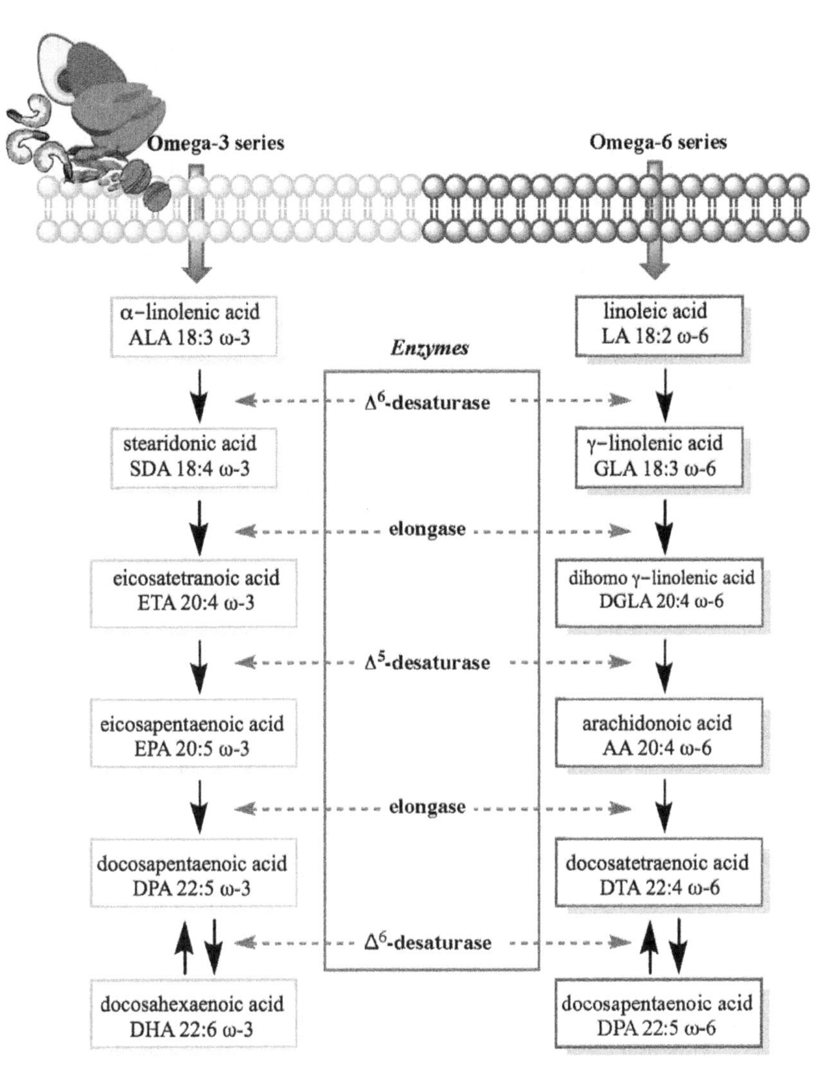

FIGURE 1.1 Biosynthesis of the principal polyunsaturated fatty acids and their metabolism.

ALA-to-DHA conversion rate is estimated to be less than 4% (Burdge et al., 2002; DeFilippis and Sperling, 2006). Although the regulation of this process is poorly understood, EPA and DHA have been reported to play a vital role in brain development and cardiovascular health, and have anti-inflammatory and other effects (Das, 2006; Wang et al., 2006).

The biochemical processes described above regulate a delicate balance between ω-6 and ω-3 fatty acids. These PUFAs are powerful regulators of cellular functions and constitute the main lipid mediators with biological activity. It should be noted that these two types of PUFAs are not interconvertible, metabolic and functionally distinct, and often have opposite physiological functions (Simopoulos, 1991). Thus, the pathways described for ω-3 PUFAs and the biosynthesis of LA-derived metabolites are obtained through the concerted action of elongases and desaturases. Evidence from several *in vivo* and *in vitro* studies indicates that these two PUFAs families not only share these enzymes but also establish a system of competition. It is known that the affinity of LA and ALA for Δ^6-desaturase is different; in fact, in order to inhibit up to 50% of GLA formation, a concentration of ALA approximately 10 times greater than LA is required (Mohrhauer et al., 1967). These findings suggest that in the presence of higher LA concentration, as occurs in a living system, the pathway leading to the synthesis of AA is preferred. For this reason, the LA/ALA ratio of dietary components is important, because mammalian cells lack the ω-3 enzyme desaturase and are unable to convert ω-6 into ω-3 fatty acids.

Unhealthy eating is one of the factors that can be attributed to the pandemic of CVDs affecting most countries in the world. Compared to our Paleolithic ancestors, for whom consumption of products rich in ω-3 was higher, the modern Western diet is considered to be up to 20 times richer in ω-6. In relation to the consumption rate, excessive amounts of ω-6 PUFAs and a very low proportion of ω-3/ω-6 has been described as promoting the pathogenesis of many diseases, including CVDs, cancer and inflammatory and autoimmune diseases.

1.1.2 Beneficial Effects of ω–3 PUFAs

The beneficial effects that the ω–3 PUFAs, in particular EPA and DHA, provide—protection against CVDs—have been supported by an important number of clinical, experimental and epidemiological studies (Mozaffarian and Wu, 2011). A universally known fact is that, in the primary prevention of CVD, a ratio of 4/1 (ω–3/ω–6) is associated with a 70% reduction in total mortality (de Lorgeril et al., 1994; Simopoulos, 2006). It is important to consider that this relationship may vary when the objective is to prevent other diseases.

In general, ω–6 fatty acids are described as proinflammatory and capable of triggering thrombosis and contributing to the formation of atherosclerotic plaque, whereas ω–3 fatty acids are described to have anti-inflammatory and cardioprotective effects. Although this is generally accepted, the scientific community continues to generate data to establish the variables associated with the biological potential of PUFAs (Szostak-Wegierek et al., 2013).

In 2002, the American Heart Association (AHA) published a scientific statement "Fish Consumption, Fish Oil, Omega-3 Fatty Acids, and Cardiovascular Disease" (Kris-Etherton et al., 2003). The main observation is summarized as the capacity of EPA and DHA to reduce fatal cardiac events. Further studies are suggested to confirm these findings and to define the health benefits of ω–3 PUFAs supplementation in primary and secondary prevention of CVDs. Since then, there have been numerous studies in humans and experimental models to define the impact of ω–3 PUFAs on reducing cardiovascular risk. Some studies have focused on understanding their effects and possible mechanisms of action. The aim of this chapter is to review the evidence for the beneficial effects of fatty acids, particularly the effects of ω–3 PUFAs on cardiovascular risk factors.

1.2 EFFECTS ON PLASMA LIPIDS AND LIPOPROTEINS

Even though risk management of coronary heart disease (CHD) has historically focused primarily on low-density lipoproteins (LDL) targets, greater emphasis recently has been placed on non-high-density lipoproteins (HDL) and apolipoprotein B (*apo* B) as more robust predictors of CHD death. Also, high plasma triglyceride (TG) levels represent a risk factor for atherosclerosis and related cardiovascular disease (Labreuche et al., 2009), and treatment with ω–3 PUFAs is an established intervention to reduce TGs (Harris, 2013; Mori, 2014). The role of TGs in the development of CHD is a function of several pathophysiological factors such as: **(1)** In normal conditions, TGs are transported mainly by TG-rich lipoproteins, such as very low-density lipoproteins (VLDL) derived from the liver and chylomicrons (CMs) derived from the intestine. After lipoprotein lipase (LPL) mediated triglyceride hydrolysis to free fatty acids (FFAs), the remaining lipoproteins, including LDL, are formed. Thus, it is considered that hypertriglyceridemia can generate an excessive amount of highly atherogenic LDL (Goldberg et al., 2011; Schwartz and Reaven, 2012) **(2)**. In hypertriglyceridemic conditions, the metabolism of VLDL shifts from a system dominated by apolipoprotein E (*apo* E), and characterized by its rapid elimination from vascular circulation, to a system dominated by apolipoprotein C III (apo C III), which is characterized by a preferential conversion of TG-rich lipoproteins into small amounts of highly dense atherogenic LDL.

Oral administration of ω-3 PUFAs from both marine and plant sources have been shown to reduce the risk of death from CHD. It is believed that this beneficial cardiovascular effect is due in part to its anti-atherosclerotic properties, which are associated with a reduction in TG levels. It has been proposed that ω-3 PUFAs exert these TG-lowering effects via a number of mechanisms (Backes et al., 2016): (1) by suppression of the expression of protein-1c, and thus a reduction of liver lipolysis, (2) by inhibition of phosphatidic acid phosphatase (PAP) and diacylglycerol acyltransferase (DGAT), key enzymes in the hepatic synthesis of TGs, (3) by an increase in the oxidation of fatty acids, resulting in a reduction of the available substrate required for TG and VLDL synthesis, (4) by increments in LPL expression, a key component of the TG-rich lipoproteins' biosynthetic pathways, leading to greater removal of TG from circulating VLDL and CM particles.

Many of the mechanisms involved in PUFAs-mediated reduction of TGs consist in regulating the gene expression of key proteins in lipid metabolism (Figure 1.2). PUFAs suppress the nuclear abundance of several transcription factors involved in lipid and carbohydrate metabolism, including sterol regulatory element binding proteins (SREBPs) (Horton et al., 2002) and carbohydrate responsive element binding protein (ChREBP) (Postic et al., 2007).

There are three isoforms of SREBPs: SREBP-1a and -1c, which originate from the same gene, and SREBP-2. All of them are activated in response to a decrease in free cholesterol content in the endoplasmic reticulum (ER). SREBP-2 stimulates primarily transcription of genes related to cholesterol biosynthesis and LDL receptor (rLDL). SREBP-1c is the predominant form of SREBP-1 in most tissues, and together with SREBP-1a, regulates the synthesis of FAs, TGs and phospholipids. Most of the lipogenic effects of insulin depend upon the induced expression of SREBP-1c and

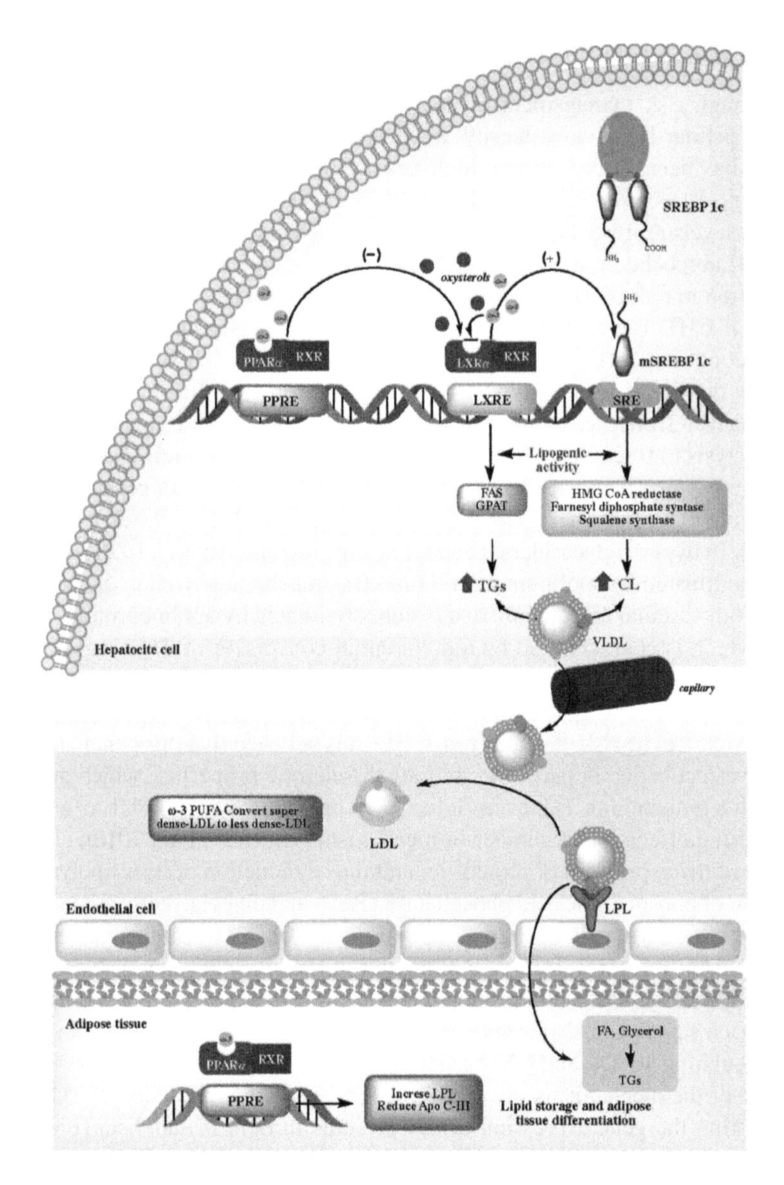

FIGURE 1.2 Mechanisms involved in the regulation of genes by non-esterified fatty acids (ω-3 PUFAs) (1). PUFAs in hepatocytes inhibit the binding of LXR-RXR to LXRE, thus suppressing the gene expression of SREBP-1c, which is a transcription factor that activates genes that encode enzymes involved in the synthesis of fatty acids that lead to the formation of triglycerides and phospholipids (2). In adipocytes, through the activation of PPARα-RXR binding, ω-3 PUFAs regulate the expression of crucial genes in lipid degradation, such as LPL and apo C III. (3) Greater catabolism of VLDLs increases LDL levels; however, the characteristics of less dense lipoproteins, resulting in reduced atherogenic capacity. LDL, low-density lipoprotein; LPL, lipoprotein lipase; PPAR, peroxisome proliferator-activated receptor; PPRE, peroxisome proliferator hormone response element; RXR, retinoid X receptor; SREBP, sterol regulatory element binding proteins; VLDL, very low-density lipoprotein.

the subsequent stimulation of the FA synthesis pathway. The expression of SREBP-1c is also stimulated by the liver X receptor (LXR), through two LXR binding sites (LXRE) present in the SREBP-1c promoter (Desvergne et al., 2006) (Figure 1.2). In contrast to these signaling events, high levels of PUFAs suppress the expression of SREBP-1c and ChREBP, the latter being a sensitive factor for glucose concentration that promotes lipogenesis from carbohydrates. However, these transcription factors are not regulated by the direct binding of fatty acids but rather respond to changes in gene transcription. For example, PUFAs decrease the transcription of SREBP-1c and ChREBP secondary to competition with oxysterols, a positive regulator of the SREBP-1 and ChREBP genes, which bind to the LXR (Ou et al., 2001).

In addition to the LXR, PUFAs bind to the ligand binding domain of several nuclear receptors expressed in the liver, including peroxisome proliferator-activated receptors (PPAR α, β, γ1 y γ2) (Xu et al., 1999), farnesoid X receptor (FXR) (de Urquiza et al., 2000), and hepatic nuclear factor-4α (HNF4α) (Wisely et al., 2002; Yuan et al., 2009). All receptors, except HNF4α, form heterodimers with the nuclear retinoid X receptor (RXR) to regulate gene expression. Fatty acids are hydrophobics, which function as hormones (steroids and thyroid) to control nuclear receptor function (Davidson, 2006).

As mentioned above, the expression of the SREBP-1c gene is partly regulated by the LXR/RXR receptor complex. Data suggest that PUFAs inhibit binding of the LXR/RXR heterodimer to the LXR response elements (LXREs) in the SREBP-1c promoter, a crucial process for the expression of SREBP-1c. The inhibitory magnitude order of each long-chain fatty acid over the expression of SREBP-1c is as follows: AA> EPA> DHA> LA> SA = 0 (Yoshikawa et al., 2002). Other mechanisms have been proposed for explaining the SREBP-1c downregulation induced by PUFAs. For example, in rat hepatocytes, decay of transcription SREBP-1c can accelerate, and PUFAs can also interfere with the maturation process of the SREBP-1c protein, decreasing its activity (Yoshikawa et al., 2002).

The lipogenic activity of LXR/RXR is due to the positive regulation it exerts on the above-mentioned SREBP-1c and to the direct regulation of several genes in the SREBP-1c downstream, including those that encode the synthesis of DGAT, fatty acid synthase (FAS) and acetyl CoA carboxylase (Mozaffarian and Wu, 2011). Through this pathway, synthesized FAs can be used in the synthesis of TGs, which get integrated with cholesterol particles in the VLDLs. This mechanism could explain how LXR agonists produce increased plasma concentration of TGs. Thus, PUFAs cause their hypotriglyceridemic effects by the coordinated suppression of liver lipogenesis through the blockage of LXR activation and suppression in the processing of SREBP-1c and other lipogenesis-related proteins.

PUFAs and products derived from these so-called natural ligands for peroxisome proliferator-activated receptors (PPARs) are also transcription factors that play a crucial role in fatty acid oxidation. Although a possible influence on the concentration of circulating TGs has been described for all PPARs, it is only with PPARα that a hypotriglyceridemic effect is regularly observed. PPARα receptors reduce the availability of FAs for TGs synthesis by regulating specific genes, including LPL (Michaud and Renier, 2001), and apolipoprotein A I (*apo* A I) (Vu-Dac et al., 1994), a key structural element in HDL. In addition, PPARα has been shown to increase the liver expression of apolipoprotein A V (*apo* A V), which stimulates LPL-mediated lipolysis (Prieur et al., 2003). As a result, ω–3 PUFAs stimulate PPARα pathways and increase signaling that lowers TG levels and increases HDL-C levels.

The binding of fatty acid to HNF-4 (α and γ) was documented by X-ray crystallographic analysis of HNF4 binding domains expressed in bacteria (Dhe-Paganon et al., 2002). *In vitro* studies report that fatty acids are irreversibly bound to HNF-4 (Wisely et al., 2002), while others report that binding of LA to HNF-4α is reversible (Yuan et al., 2009). It has also been observed that, while saturated fatty acids stimulate the transcription of HNF-4α, PUFAs inhibit the effects of HNF-4α on gene transcription (Clarke, 2001; Pegorier et al., 2004). Proteins involved in lipoprotein metabolism, including *apo* C III, A I, A IV, are encoded by this receptor (Jump and Clarke, 1999). However, in contrast to the direct regulation of FAs over other nuclear receptors, the occupation of HNF-4α does not appear to significantly affect their transcriptional activity, and are considered indirect mechanisms. For example, PUFA-activated PPARs compete with HNF-4α for binding to the *apo* C III promoter.

According to these mechanisms, in samples of fasting subjects showing moderate to severe hypertriglyceridemia (TGs of 500 to 2000 mg/dL), a 12-week treatment with ω–3 PUFAs revealed a significant reduction in plasma *apo* C III levels (Morton et al., 2016). Another randomized, double-blind, crossover study with eight-week treatment periods using high-dose EPA and DHA, demonstrated potential mechanisms by which ω–3 PUFAs may decrease the risk of coronary heart disease by reducing atherogenic apolipoproteins in individuals with elevated TGs. Treatment with 3.4 g/d of EPA and DHA significantly reduced the concentrations of both *apo* C III and *apo* B, and tended to a modest reduction of lipoprotein associated phospholipase A2 (Lp-PLA2) (Skulas-Ray et al., 2015). The effect of ω–3 PUFAs on *apo* B 48 has also been evaluated in animal and *in vitro* models where it has been shown to suppress its synthesis, with the consequent reduction of CM rich in TGs. It has been considered that this inhibition may be due to a decrease in the expression of *apo* B 48 and/or increased post-translational degradation (Levy et al., 2006).

Apo C III is believed to contribute to the progression of atherosclerosis and CVD through a number of mechanisms, including the activation of proinflammatory pathways (Sacks et al., 2011). This apolipoprotein is a key factor in hypertriglyceridemia, mainly due to its inhibitory actions on LPL. It also inhibits receptor-mediated uptake of VLDL, CM and their remnants, slows down their purification rate and promotes the formation of small, dense, highly atherogenic LDL particles.

According to experimental findings, ω–3 PUFAs are associated with reductions in plasma TGs levels of approximately 25% to 34% (Harris, 1997), but these findings vary at baseline. Studies in individuals with severe hypertriglyceridemia (TGs 5.65 mmol /L [500 mg /dL]) reported reductions in TGs from 40% to 79% with EPA and DHA intake of 3 g/day (Harris, 1997; Phillipson et al., 1985). When baseline TGs levels are ≥2.0 mmol/L (≥177 mg/dL), the reduction rate is approximately 34%, while for subjects where baseline TGs levels are below 2.0 mmol/L, the reduction is up to 25%. With administration of ω–3 PUFAs in patients with CVD and TGs levels ≥1.70 mmol/L (150 mg/dL), reduction of up to 30% of TGs has been observed. Two of the three studies in patients with dyslipidemia reported a 20% to 33% reduction in TGs. With the administration of EPA and DHA in patients with diabetes, reduction of TGs between 25% and 45% has been observed, resulting in a dose-dependent effect (Nettleton and Katz, 2005). In middle-aged and elderly subjects, a diet rich in marine ω–3 PUFAs has been observed to decrease concentrations of lipoproteins rich in TGs (VLDL) by direct catabolism of the *apo* B 100, and it is reasonable to expect an increase in LDL levels (Ooi et al., 2012). In an unfavorable but to some extent expected degree, ω–3 PUFAs produce elevations in cLDL levels between 5% and 11% (Harris, 1997). However, this increase in cLDL levels from ω–3 PUFAs supplementation appears to be due to an increase in LDL particle size rather than to the number of LDL molecules. These nutritional supplements modify the composition of LDL cholesterol by increasing apolipoprotein B and decreasing lipoprotein levels, resulting in a less atherogenic molecule (von Schacky, 2000). Small LDL particles are considered as an important cardiovascular risk factor (Hulthe et al., 2000) and correlated with subclinical atherosclerosis as measured by the thickening of the middle intima (Lahdenpera et al., 1996).

Other findings in the field of dyslipidemias have shown that ω–3 PUFAs significantly reduce liver fat by 18% in women with hepatic steatosis (Cussons et al., 2009), and in obese dyslipidemic men with insulin resistance, combined ω–3 PUFAs and statins therapy provided the optimal change in lipid profile and HDL cholesterol (Chan et al., 2002). However, the impact of ω–3 PUFAs on HDL is variable. Some studies report discrete elevations in HDL cholesterol of approximately 1% to 3% (Harris, 1997), while other authors report a decrease in HDL of approximately 8%–14% (Weintraub, 2014).

Considerations for the use of PUFAs in the treatment of dyslipidemias include: (1) Long-chain omega-3 fatty acids may be a well-tolerated and effective alternative to fibrates and niacin, yet further large-scale clinical studies are required to evaluate their effects on cardiovascular outcomes and CVD risk reduction in patients with hypertriglyceridemia (Davidson et al., 2014). (2) Since they have different mechanisms of action and the efficacy profile against the different lipid components varies between ω–3 PUFAs and statins, combined therapy is expected to provide complementary benefits over lipid profile when given together. However, further long-term studies with clinical end points are needed to confirm the synergistic benefits of statins and omega-3 supplements on cardiovascular incidence and mortality (Minihane, 2013). (3) As for accepted therapeutic uses, the 2016 ESC/EAS Guidelines for the Management of Dyslipidemias, suggest that if TGs are not controlled by statins or fibrates, addition of ω–3 PUFAs may be considered, and these combinations are safe and well tolerated (Catapano et al., 2016). (4) Despite findings that ω–3 PUFAs supplements (fish oil) are highly effective in the treatment of hypertriglyceridemia, omega-3 dietary supplements are not approved by the Food and Drug Administration (FDA) for this purpose (Jellinger et al., 2017).

1.3 EFFECTS ON ATHEROSCLEROSIS

Atherosclerosis is an inflammatory disease of the vascular system and is a major cause of cardiovascular and cerebrovascular diseases. When considering the pathophysiological process of atheroma plaque development, it can be inferred that ω-3 PUFAs have beneficial effects, given their ability to modify both the lipid profile and the inflammatory process. However, variations in methodological designs of both experimental and clinical studies show controversial results in relation to the ability of ω-3 PUFAs in reversing or preventing atherosclerosis.

It is well known that the principal constituents of an atheroma plaque are LDLc, and some studies have demonstrated that, in addition to reducing TGs and VLDL, ω-3 PUFAs increase the concentration of floating and large LDL particles while reducing the concentration of dense atherogenic particles of LDL. In relation to this property, in a study conducted in low-density lipoprotein receptors knockout mice (LDLR-/-) subjected to a diet rich in cholesterol, fish oils supplements inhibited the development of atherosclerosis after 20 weeks of treatment (Zampolli et al., 2006). Using the model of apolipoprotein E knockout mice (*apo* E -/-), administration of 1% fish oil for 14 weeks did not modify atherogenesis (Xu et al., 2007). In contrast, another study using both models, where LDLR -/- and apo E -/- mice were administered a cholesterol-rich diet, treatment for 12 weeks with high concentrations of EPA decreased the plaque size (Matsumoto et al., 2008).

Although LDLs play a key role in the development of atheroma plaque, studies carried out in other animal models have shown that the development of atherosclerosis may be significantly inhibited by diet rich in ω-3 PUFAs, without this effect being associated with a reduction in lipids (Davis et al., 1987; Weiner et al., 1986). These findings lead us to consider other aspects related to the pathogenesis of plaque formation, which may be involved in the cardioprotective effects described for EPA and DHA. This forces us to reconsider the inflammatory nature of this vascular disease in which the excess of LDL deposited in the vascular intima layer induces excessive recruitment of monocytes, which are different from macrophages and allows elimination of the excess of sub-endothelial cholesterol. This process gives rise to the generation of foam cells capable of secreting proinflammatory mediators that characterize the pathogenesis of atherosclerosis and, precisely, one of the biological effects of ω-3 PUFAs is to alter the course of the inflammatory process.

The anti-inflammatory effects of ω-3 PUFAs are partly related to their attenuating effects on inflammatory prostanoids and leukotrienes. For example, in intestinal microvascular endothelial cells, DHA has been shown to reduce cyclooxygenase-2 (COX-2) expression and limits the production of PGE_2 and leukotriene B_4 (LTB_4) (Ibrahim et al., 2011). Similar effects have been observed in rats with colitis, where administration of fish oil rich in EPA and DHA inhibited the production of PGE_2 and LTB_4.

Monocyte adhesion to endothelial cells is mediated by adhesion molecules, such as intracellular adhesion molecule 1 (ICAM-1) and vascular adhesion molecule 1 (VCAM-1). In the models described above by Ibrahim et al. (2011), inhibition of eicosanoids was accompanied by reduction in VCAM-1 expression. In humans, fish oil (EPA and DHA) supplements have been found to lower circulating VCAM-1 levels in elderly subjects (Miles et al., 2001). In patients with metabolic syndrome (Met-S), EPA administration decreased plasma concentrations of soluble ICAM-1 and VCAM-1 (Yamada et al., 2008). However, more recently, in order to assess the effects of ω-3 PUFAs supplements on plasma concentrations of soluble adhesion molecules, a meta-analysis of randomized controlled trials was carried out. This study revealed that ω-3 PUFAs supplements did not affect plasma concentrations of VCAM-1 but instead, in both healthy and dyslipidemic patients, they did significantly reduce ICAM-1 (Yang et al., 2012).

In relation to other components of the atherosclerotic process, high concentrations of EPA and DHA are known to modulate the expression of critical genes associated with inflammation, including interleukin-1-alpha (IL-1α), interleukin-1-beta (IL-1β) and tumor necrosis factor alpha (TNF-α) (Calder, 2001; Honda et al., 2015; Zainal et al., 2009). TNF-α is an inflammatory cytokine that plays an important role in the development of atherosclerotic lesions. TNF-α is associated with upregulation of monocyte chemotactic protein-1 (MCP-1) and VCAM-1 expression, in addition to inducing expression of early growth response protein 1 (Egr-1), which is capable of inducing ICAM-1 production (Maltzman et al., 1996). Additionally, the IKK/NF-κB pathway (IkB kinases/nuclear factor κB) regulates the genes involved in inflammation, oxidative stress and endothelial dysfunction through TNF-α activation (Kumar et al., 2004; Rimbach et al., 2000).

ω-3 PUFAs regulate TNF-α activity by activating the G-coupled protein receptor 120 (GPR120). This receptor intervenes in the activation cascade of NF-κB mediated by the TNF-α receptor (TNFR). Post-activation signaling pathways of the GPR120 receptor in adipocytes inhibit the proinflammatory activity of TNF-α (Gregor and Hotamisligil, 2011).

Additionally, in macrophages inflammasome activation is inhibited by DHA via GPR120 (free fatty acid 4; FFA4), an effect that also appears to be mediated by inhibition of NF-κB. Thus, blocking the TNF-α pathway mediated by ω-3 PUFAs-related receptors is crucial in explaining the regulation of the formation and stability of atheroma plaque.

Emerging evidence shows that uncontrolled inflammation is a prominent characteristic of many cardiovascular diseases and that atherosclerosis may be seen as a state of failed resolution of inflammation (Serhan et al., 2007). Thus, it is known that from essential fatty acids endogenous chemical mediators are biosynthesized in different phases of inflammatory response, where resolution is one of them. In this sense, the production of a new class of lipid mediators called resolvins (products of interaction in the resolution phase) have also been associated with the anti-inflammatory effects of ω-3 PUFAs (Figure 1.3).

Resolvins of series E (RvE) and series D (RvD) are obtained from EPA and DHA, respectively (Serhan et al., 2002). RvE and RvD-resolvins are believed to have a similar function, which is primarily to block the migration and infiltration of neutrophils and monocytes, protecting tissues from damage by immune system cells (Serhan et al., 2002). Both types of resolvins, like their precursors, can inhibit the NF-κB by a receptor-dependent mechanism activated by PPARγ in addition to the involvement of membrane receptors (Liao et al., 2012).

It has been shown that RvE1 can act as an agonist of the ChemR23 receptor, also known as Chemokine-like receptor 1 (Kaur et al., 2010) (Figure 1.4). Rv-E1-mediated signaling on ChemR23 plays a role in mononuclear cellular migration to inflammed tissue as well as in resolving inflammation (Serhan et al., 2008b). Actions mediated by Rv-E1 through ChemR23 block signaling of NF-κB induced by TNF-α (Arita et al., 2005) and increase phagocytosis of apoptotic neutrophils (Ohira et al., 2010). In addition to ChemR23, it has been described that RvE1 interacts with leukotriene

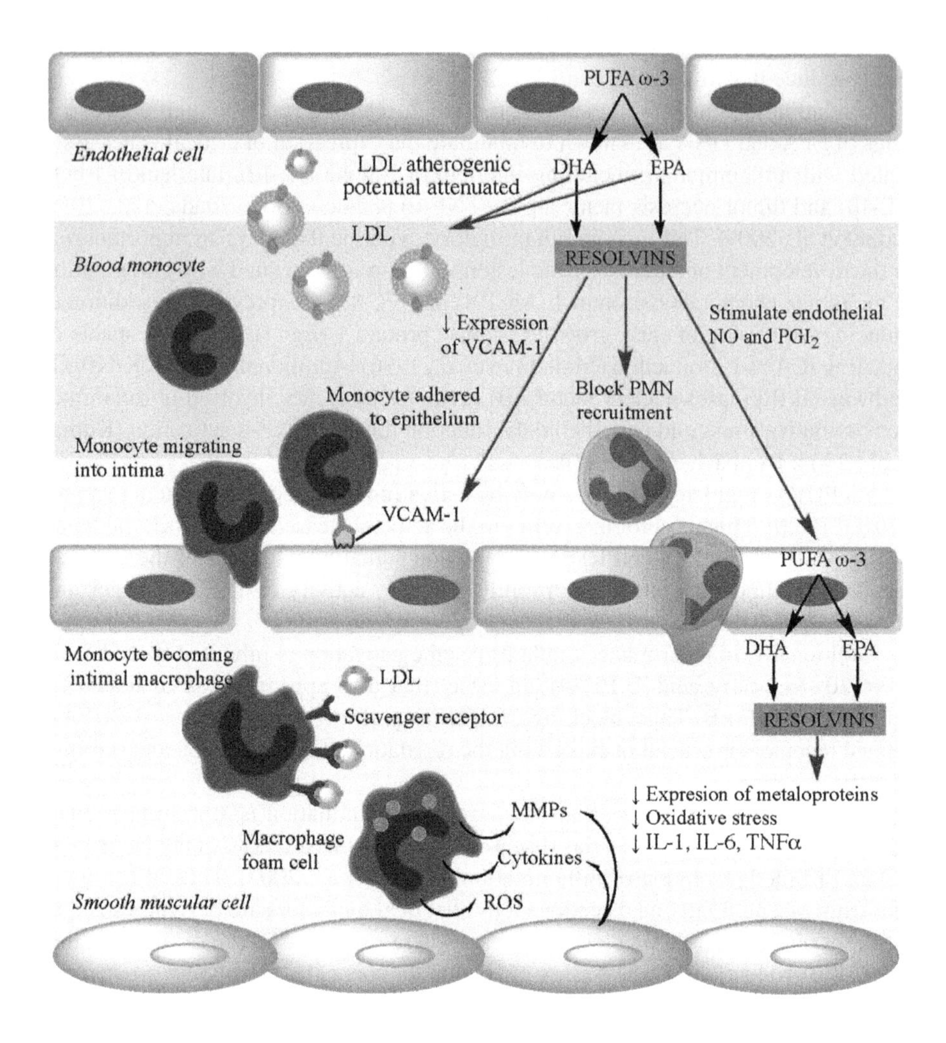

FIGURE 1.3 Atherosclerotic plaque development and the effects of ω-3 PUFAs and Resolvins. The innate and adaptive immunity co-operate in an inflammatory process that leads to vessel occlusion (1). Extravasation of LDL trigger inflammation that promotes endothelial cells (ECs) dysfunction (2). ECs expose adhesion molecules (VCAM) and increase permeability (3). Leukocyte adhesion and activation (4). Leukocytes infiltrate the intima (5). Activated smooth muscle cells produce cytokines and MMPs, which further change the microenvironment (6). The infiltrating monocytes transform into macrophages (7) producing MMPs, further cytokines and ROS (8). Macrophages and SMCs phagocytize lipid particles and become foam cells (9). ω-3 PUFAs and Resolvins reduce PMN recruitment and chemotaxis, induce a reduction of PMN activation and adhesion molecule expression, increase phagocytosis and clearance. DHA, docosahexanoic acid; EPA, eicosapentaenoic acid; IL-1, interleukin 1; IL-6, interleukin 6; LDL, low-density lipoprotein; MMPs, metalloproteinases; PGI_2, prostaglandin I2; PMN, polymorphonuclear neutrophil; ROS, reactive oxygen species; TNF-α, tumor necrosis factor α; VCAM-1, vascular adhesion molecule 1.

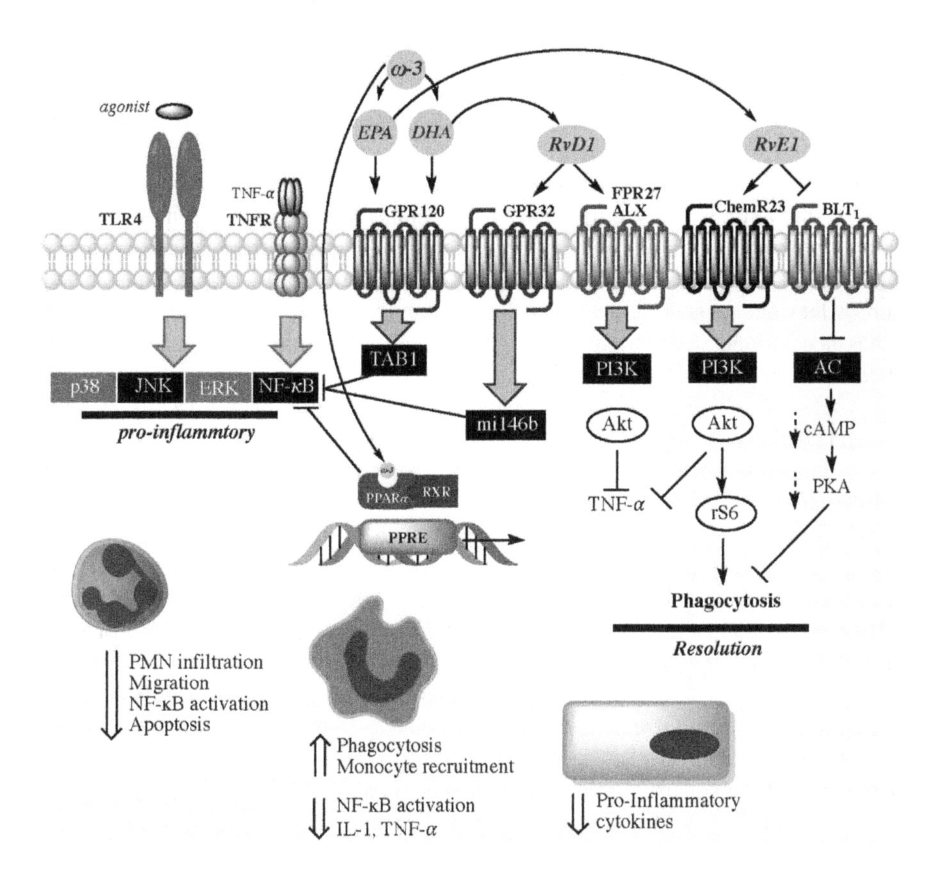

FIGURE 1.4 Illustration of the anti-inflammatory and pro-resolving effects of ω-3 PUFAs and Resolvins act on multiple G protein-coupled receptors to limit activation and recruitment of neutrophil, and stimulate nonphlogistic macrophage phagocytosis. Akt, Protein Kinase B; ALX/FPR2, lipoxin A4 receptor; BLT-1, leukotriene B4 receptor type 1; ChemR23, chemokine-like receptor 1; cAMP, Cyclic adenosine monophosphate; GPR32, G protein-coupled receptor 32; GPR120, G protein-coupled receptor 120; IL, interleukin; JNK, c-Jun N-terminal kinases; NFκB, nuclear factor kappa B; PI3K, phosphatidylinositol 3-kinase; PKA; protein kinase A; PMN, polymorphonuclear neutrophil; p38, P38 mitogen-activated protein kinases; PPAR, peroxisome proliferator-activated receptors; PPRE, Peroxisome proliferator hormone response element; RXR, retinoid X receptor; TLR4, Toll-like receptor 4; TNF-α, tumor necrosis factor.

B4 receptor type 1 (BLT_1), a receptor of LTB_4, and serves as a local buffer of BLT_1 signals in leukocytes (Arita et al., 2007).

Meanwhile, the actions of RvD1 are mediated by two G protein-coupled receptors (GPCR), ALX/FPR2 and GPR32, which also regulate specific microRNAs (miRNAs) and their target genes in new resolution circuits. Interaction with ALX/FPR2 signals could control infiltration of polymorphonuclear neutrophils (PMN) and stimulate macrophagocytosis of apoptotic PMNs (Chiang et al., 2006), while it is suggested that the GPR32 also plays a key role in mediating the effects of RvD1 on human macrophages (Schmid et al., 2016). Despite numerous findings that document the molecular actions associated with the anti-inflammatory effects of ω-3 PUFAs, crucial aspects of signaling mechanisms and downstream interactions remain to be elucidated.

The anti-inflammatory effects of ω-3 PUFAs seem to play an important role in both the prevention and stabilization of the plaque. This has been demonstrated in patients with symptomatic carotid atherosclerotic disease, where ω-3 PUFAs supplements are associated with less macrophage infiltration into the plaque and a firmer fibrous lamina, suggesting a greater plaque stability (Thies et al., 2003). Similarly, it has been found that patients with atherosclerotic disease, waiting for a carotid endarterectomy, receiving ω-3 PUFAs, specifically EPA, which is incorporated into atherosclerotic plaques and is associated with a reduction of foam cells, T cells, showed decreased inflammation and increased plaque stability.

It is known that plaque stability is determined by the production of matrix metalloproteinases (MMPs), proteases which are capable of degrading extracellular matrix proteins. The presence of MMPs generated from foam cells plays a definite role in the stability of atheroma plaque, as they are associated with degradation of the fibrous layer and rupture of the plaque. ω-3 PUFAs, evaluated in different cellular models, have shown that they are capable of decreasing expression of these proinflammatory markers (MMP-7, MMP-9, MMP-12) (Cawood et al., 2010), and these findings may represent an important mechanism by which ω-3 PUFAs reduce ischemic cardiovascular events by inducing plaque stabilization. However, marine ω-3 PUFAs in human studies show no effects on plasma MMP-9 levels, determined in patients with myocardial infarction (Aarsetoy et al., 2006) or in subjects at risk of coronary heart disease (Furenes et al., 2008), which again highlights inconsistencies between experimental observations and clinical findings.

Endothelial dysfunction, through increased inflammatory and thrombogenic molecular changes, has also been associated with atherosclerosis. Endothelial dysfunction is characterized by the alteration of the bioavailability of main endothelial derived relaxation factor, which affects endothelial-dependent vasodilation. Decreased bioavailability of nitric oxide (NO) has been associated with decreased expression of endothelial NO synthase (eNOS), reductions in eNOS activity and further degradation of NO by reactive oxygen species (Rossier et al., 2017; Harrison, 1997; Wilcox et al., 1997).

On the other hand, in certain disease states, endothelial dysfunction can also produce higher levels of eicosanoids and free radicals, which on the one hand affects the bioavailability of NO and on the other promotes abnormal contraction of blood vessels (Abeywardena, 2003; Higashi et al., 2009).

ω-3 PUFAs when incorporated into phospholipids, modulate both the composition and fluidity of the cell membrane. The endothelial cell membrane harbors lipid caveolae and lipid rafts where various receptors that regulate cell function are concentrated. ω-3 PUFAs modulate the composition of the caveolae, causing an increase in NO production and a reduction in proinflammatory mediators. In particular, DHA has been shown to improve endothelial-dependent vasorelaxation of aortic rings by increasing NO liberation (Lawson et al., 1991) and increasing the production of IL-1β-induced NO in vascular smooth muscle cells (Hirafuji et al., 2002). In addition to increasing NO production, ω-3 PUFAs reduce oxidative stress as a result of a direct modulatory effect on the sources of ROS. In hypercholesterolemic conditions and diabetes, ω-3 fatty acids improve endothelial function and reduce the vasoconstrictor response of vascular smooth muscle.

Although experimental findings appear to be consistent with the cardioprotective properties of ω-3 PUFAs, clinical data are not always convincing. In some studies, consumption of ω-3 PUFAs-enriched diet improved various functional biomarkers of endothelial activity, such as increased expression of the eNOS gene expression. Similarly, in studies carried out in the elderly, the circulatory biomarkers of endothelial dysfunction were also reduced. However, the general evidence of ω-3 PUFAs' effects on endothelial function is inconsistent, showing wide variations in the doses administered, characteristics of the mixtures (DHA and EPA) used as supplements, treated populations and evaluated parameters. Thus, adhesion molecules were affected in a variable manner, as discussed above. In healthy young adults, neither EPA nor DHA administered for a short period of time altered endothelin-1 (ET-1) or soluble E-selectin, VCAM-1 or ICAM-1 concentrations. Interestingly, a combination of EPA and DHA reduced both soluble E-selectin and VCAM-1.

Although there are contradictions between the studies published to date, most authors conclude that ω-3 PUFAs are a complement in preventing both the development as well as the progression of atherosclerosis.

1.4 EFFECTS ON METABOLIC SYNDROME, DIABETES, AND OBESITY

Metabolic syndrome (Met-S) is a group of coexisting and interrelated risk factors that include abdominal obesity, impaired glucose tolerance, hypertriglyceridemia, decreased serum HDL cholesterol and/or hypertension. The first observation related to this syndrome was insulin resistance, but given the complexity of the syndrome, the focus has been on its use as an epidemiological tool related to cardiovascular risk (Tune et al., 2017).

Although the definition of Met-S as a disease generates debate, it is considered to be a determining pathological stage in the development of diseases such as type 2 DM and atherosclerotic disease (Tune et al., 2017).

On the other hand, DM is a group of diseases characterized by elevated blood glucose levels due to insufficient production or action of insulin. This elevation of plasma glycemic level generates characteristic micro- and macro-vascular complications that include retinopathy, neuropathy, nephropathy, acute myocardial infarction (AMI) and stroke (Asmat et al., 2016).

Both syndromes are associated with obesity, which besides being a triggering factor, is characterized by excessive accumulation of adipose tissue particularly in the abdomen (Gonzalez-Muniesa et al., 2017). The relationship between these diseases is based on epidemiological studies, which have established a parallelism in terms of prevalence and increase in last decades (Grundy, 2008).

It is recognized that in all these pathologies, there are genetic factors involved that determine their predisposition. However, the environmental factors have great influence on the pathologies' development, especially a sedentary lifestyle and excess calories intake, where a positive energy balance leads to an increase in fat deposits in the adipose tissue with the consequent development of obesity and alterations in adipocyte functions (Rask-Madsen and Kahn, 2012).

Obese individuals present a chronic low-level inflammation, which has been associated with a chronic stage of oxidative stress, resulting from an imbalance between mechanisms of production and elimination of ROS, which are by-products of the respiratory chain within mitochondria, organelles responsible for energy production, carbohydrates, amino acids and lipids metabolism, and apoptosis. Other enzyme systems such as NADPH oxidase and peroxisomes can also generate ROS (Carrier, 2017).

In addition to this, the increase in the bioavailability of FFAs increases their cellular uptake inducing mitochondrial β-oxidation, which in turn generates a blockage at the substrate binding site, accumulation of toxic lipid intermediates and cellular dysfunction. This predominance of lipid metabolism at the expense of glucose utilization reduces the uptake of the last substrate and the synthesis of glycogen in the skeletal muscle, resulting in a chronic state of hyperglycemia that decreases the sensitivity to insulin (Verma and Hussain, 2017).

This rise in glycaemia, together with compensatory hyperinsulinemia, has been associated with insulin resistance, while the intolerance to glucose conduces to the glycation of circulating proteins and the formation of so-called advanced glycation end products (AGEs). The progression of these events causes a defect in the secretory function of beta pancreatic cells and ultimately leads to to apoptosis (Carrier, 2017; Verma and Hussain, 2017).

Ravussin and Smith (2002) showed that an increase in dietary fats favors the lipid storage within the liver, skeletal muscle and pancreas, unobserved under normal conditions (Ravussin and Smith, 2002).

All of the above suggests the importance of the inflammatory process as a basis for Met-S, diabetes and obesity in which the production of different proinflammatory cells from the immune system such as monocytes, macrophages, natural killer (NK) cells and lymphocytes induce cytokine secretion that perpetuates systemic inflammation. A large number of cytokines are related to the inflammatory process, but IL-1β, TNF-α, IL-17 and IL-6 are more important in this process (Pirola and Ferraz, 2017). Given the relevant role of the accumulation of adipose tissue in these conditions and its function as a regulating organ on energetic homeostasis, it is described that adipokines are released from the adipose tissue, some of which have proinflammatory activity (TNF-α y IL-6), while others generate anti-inflammatory actions (adiponectin).

Given the close pathological relationship between the described diseases and the elevated cardiovascular risk to patients, it is established that some lifestyle modifications such as a healthy diet, weight reduction and increased physical activity significantly reduce the development and progression of these diseases (Rochlani et al., 2017).

In relation to diets, ω-3 PUFAs have demonstrated their beneficial effects in metabolic diseases. Therefore, the utilization of ω-3 PUFAs as a supplement constitutes a strategy for treating these pathologies (Carpentier et al., 2006; Tortosa-Caparrós et al., 2017).

As diabetes, Met-S and obesity have common etiological factors (increased FFA, hyperglycemia, increased adipose tissue, increased free radicals and inflammation), the mechanisms by which ω-3 PUFAs exert their positive effects can be explained by three main factors: (a) lipid metabolism, (b) inflammation and (c) energetic metabolism.

a. *Lipid metabolism*: These effects have been described above in the section about the plasma lipid effect, where emphasis was placed on molecular actions leading to the reduction of TGs, a widely recognized effect of ω-3 PUFAs. These actions are summarized in Figure 1.2.
b. *Inflammation*: As previously stated, another relevant aspect in metabolic diseases is their interrelation with the inflammatory process, the latter being an essential component in the organism's response to infections or aggressions (Henson, 2005). The beneficial effects of ω-3 PUFAs on obesity and metabolic syndrome are due to their direct anti-inflammatory actions confirmed by experimental and clinical studies.

   It is recognized that saturated fatty acids (SFAs) directly stimulate the expression of inflammatory mediators such as TNF-α and IL-6 through binding to Toll-like receptor 4 (TLR-4) (Lee et al., 2001) and Toll-like receptor 2 (TLR-2) (Lee et al., 2004), mainly in adipocytes. The stimulation of these receptors has been associated with the activation of the NF-kB pathway, which is involved in diverse inflammatory pathologies (Lira et al., 2010). In this sense, the administration of ω-3 PUFAs has been shown to exert potent anti-inflammatory actions through the following mechanisms (Box 1.1).
c. *Energy metabolism*: As metabolic diseases such as diabetes and obesity have an imbalance between food intake, energy expenditure and fatty tissue accumulation, it is understandable that this factor should be improved. In this sense, the interest has been focused on adipose tissue, and the GRP120 receptor is postulated as a fundamental target in the regulation of metabolism in this tissue due to its wide expression. In a diet-induced obesity study (high-fat diet; HFD) in mice, activation of the GRP120 receptor by ω-3 PUFAs or the GW9508 agonist was shown to be associated with insulin sensitization and improvement in glucose intolerance. The same authors, using cell cultures of 3T3-L1 adipocytes and primary adipose tissue, reported a significant increase in activation of PI3K/Akt receptor pathway, which triggered the translocation of GLUT4 and an increase in the cellular uptake of glucose (Oh et al., 2010). These findings are reinforced by the publication of Liu et al. (2012), who report a reduction in the expression of GLUT4 and insulin receptor substrate (IRS) in 3T3-L1 cells with low expression of GRP120 (Liu et al., 2012). On the other hand, Ichimura et al. (2012), in GPR120 double knockout mice and fed with HFD, reported the development of hyperglycemia, glucose intolerance and insulin resistance as compared with the control group (Ichimura et al., 2012).

   An important observation in relation to the supply of HFD is the phenomenon of upregulation in the expression of the GPR120 and PPARγ receptors, which is suggested to provide a regulatory pathway for energy metabolism (Song et al., 2017). This pathway of GPR120-PPARγ also appears to intervene in the upregulation of vascular endothelial growth

BOX 1.1 MECHANISM OF ANTI-INFLAMMATORY ACTIONS OF ω-3 PUFAS

Facilitating the production of mediators known as resolvins

- Resolvins, metabolic products of PUFAs, generate potent anti-inflammatory actions at the sites of inflammation (Serhan et al., 2002).
- Anti-inflammatory effects include inhibition of NF-κB, blocking neutrophil migration, downregulating of adhesion molecules, reduction of respiratory burst, and promoting neutrophil apoptosis and clearance of apoptotic bodies by macrophages (Fritsche, 2015; Liao et al., 2012; Capo et al., 2018).

Affecting lipid rafts of the plasma membrane of immune system cells.

- ω-3 PUFAs induce changes in the size as well as the composition of the cell membrane (cholesterol and sphingomyelin) and, therefore, modify its biological properties, with intracellular signaling being significantly altered.
- ω-3 PUFAs suppress T-cell activation in murine models, and have an anti-inflammatory effect (Stillwell et al., 2005; Turk and Chapkin, 2013). Transduction signals by T cells mediated proinflammatory mechanism, such as epidermal growth factor (EGF), Ras activation (Isoform H) and PLCγ.

Stimulation of free fatty acid receptor 120 (FFA4 or GPR120)

- GPR120 belongs to the family of GPCR, which is expressed in various mammalian tissues, and under physiological conditions it is predominantly activated by intermediate and long-chain fatty acids; however, it has greater affinity for DHA and EPA (Miyamoto et al., 2016).
- GPR120 probably couples mainly with Gq/11 proteins, whose signaling mechanism is associated with an increase in intracellular Ca^{2+} in both human and mice cells (Moniri, 2016; Katsuma et al., 2005).
- Activation of GPR120 receptor is associated with the reduction of COX-2 gene expression and production of PGE_2, a mechanism involved in inflammatory response (Li et al., 2013).
- Studies with murine macrophage cell cultures showed that DHA generates a significant reduction in inflammatory response. In these cells, receptor activation leads to β-arrestine-2 recruitment forming a complex (DHA-GPR120-β-arrestine-2) that binds to the TAK-1 binding protein (TAB-1), which interferes with the phosphorylation and activation of TAK-1 (kinase 1 activated by TGF-β) (Oh et al., 2010; Im, 2016; Song et al., 2017).

Affecting hormones and mediators in adipose tissue

- In relation to the effect on leptin, a study in insulin-resistant animal model, found that the ω-3 PUFA-treated group showed an increase of 75% in plasma leptin levels compared to the control group. ω-3 PUFA treatment increases expression of GLUT4 transporters in adipose tissue (Peyron-Caso et al., 2002).

- Findings related to animal studies were described by Kondo et al. (2010), who developed a study in a small group of non-obese patients, determining that administration of PUFAs increases adiponectin levels; this effect being more pronounced in women than in men (Kondo et al., 2010).
- Incubation of human adipocytes with EPA and DHA increased adiponectin levels, an anti-inflammatory cytokine, without modifying leptin secretion. Only DHA showed a decrease in TNF-α levels (Romacho et al., 2015).

factor-A (VEGF-A) in adipocytes 3T3-L1, an effect that is related to the decrease in glucose intolerance and insulin sensitization in mice fed with HFD (Sun et al., 2012).

The demonstrated increase in GPR120 expression in several studies is crucial because of its significant effects on adipogenesis. However, it has been reported that obese patients have a reduction of both the receptor and its mRNA in visceral fatty tissue as compared with lean subjects, results that contradict the rest of the studies (Rodriguez-Pacheco et al., 2014). In addition, animal models demonstrated that receptor functionality was essential for proper adipogenesis. Therefore, obese people may be considered to lack functional receptors, and HFD consumption worsens this dysfunction and insulin resistance (Ichimura et al., 2014; Rodriguez-Pacheco et al., 2014). Unfortunately, at present, there are no tools available to measure the functionality of the receptor in patients, and therefore this assumption should be corroborated.

As mentioned above, the activation of the GRP120 receptor facilitates adipogenesis and lipogenesis within the adipose tissue, assuring metabolic homeostasis. In relation to the expression of this receptor in other tissues of metabolic relevance, there are studies that indicate its existence in murine and human pancreatic islets, in addition to β-cell lines (Fritsche, 2015; Taneera et al., 2012). The activation of this receptor by DHA in β-cells increases insulin secretion, although the subtype of GPR40 receptor is considered to be mainly responsible for this action. Taking into account its signaling mechanism, the GPR120 receptor implies an increase of intracellular calcium ($[Ca^{2+}i]$) levels; an insulinotropic effect would be plausible (Ozdener et al., 2014).

This aspect was considered by Zhang et al. (2017), who in both animal models and cell cultures unequivocally demonstrated for the first time the expression of these receptors in β-pancreatic cells where their function is to modulate the release of insulin through a PLC/Ca^{2+}-dependent signaling pathway. In this study, the authors demonstrated that oral administration of GPR120 receptor agonists in mice with DM-2 improves postprandial hyperglycemia and increases insulin secretion in the oral glucose tolerance test (OGT) (Zhang et al., 2017).

These findings can be extrapolated to other GPR120 agonist such as ω-3 PUFAs. In fact, while the administration of DHA increased the insulinotropic effect in obese non-diabetic mice (Raimondi et al., 2005), it reduced it in those with DM-2. In view of these findings, the authors propose that hyperlipidemia in OND and hyperglycemia in DM-2 could affect PPARγ expression in β-cells through different pathways, which are responsible for modifications in the expression of GPR120 (Liu et al., 2012).

The regulatory role in β-cell survival is another effect described for this receptor. Activation of the receptor in different cells such as enteroendocrines, adipocytes, macrophages and hypothalamic neurons induce Akt/ERK phosphorylation (Fritsche, 2015; Katsuma et al., 2005); this signaling in β-cells is closely related to their functionality and survival (Wijesekara et al., 2010). In this sense, Zhang et al. (2017) observed in rodent islets, that stimulation of GPR120 protects the cells against glucotoxicity and was dependent on the Akt/ERK pathway, which was increased in OND islets but reduced in diabetic islets. These results suggest the importance of cell survival in pathological conditions and the role that ω-3 PUFAs can provide during its utilization (Zhang et al., 2017).

Despite the functions already described for this receptor, it is expected that new roles will be elucidated which may be associated with the expression of the receptor in particular tissues as well as the signaling mechanisms underlying its activation.

The evidence for the beneficial effects of ω-3 PUFAs in improving obesity, Met-S and diabetes has been provided from studies in experimental animals. This is a limiting factor because there are differences among species in accumulating these fatty acids in the cell membranes, which depend directly on the amount ingested. Because their positive actions in these pathologies are related to the agonist effect on GRP120, very high doses will be required, which are impractical. For this reason long-term daily consumption of ω-3 PUFAs is highly recommended (Huang et al., 2016).

1.5 EFFECTS ON PLATELET FUNCTION AND THROMBOSIS

Both endothelial function and platelet activation determine acute thrombotic events. Decades of inflammatory vascular damage progressively increase the likelihood of triggering events, platelet adhesion and aggregation, and thrombus formation (Strong et al., 1999).

Agonists such as adenosine diphosphate (ADP), serotonin, thrombin, epinephrine and AA participate in the platelet activation process (Li et al., 2010). Arachidonic acid, the most important of the ω–6 PUFAs physiologically, is required as a constituent of membrane phospholipids and is the precursor of "Series 2" eicosanoids (PGD_2, PGE_2, PGF_2, PGI_2, or A_2 thromboxane (TXA_2)). The prostanoid production pathway begins with the mobilization of AA from membrane phospholipids by cytosolic A_2 phospholipase ($cPLA_2$). The next step is the production of PGH_2 prostaglandin endoperoxide from AA, conversion mediated by endoperoxide prostaglandin H synthase-1 or -2 (PGHS-1 or -2), also known as cyclooxygenase-1 or -2 (COX-1 or -2) (Simmons et al., 2004; Smith, 2005). The process of synthesis of eicosanoids culminates in the isomerization of PGH_2 into AA-derived "Series 2" products. The "3-series" prostanoids, including PGD_3, PGE_3, PGF_3, PGI_3 and TxA_3, are derived from EPA and DHA. Both ω–6 PUFAs and ω–3 PUFAs use the same metabolic pathways, which include specific enzymes such as lipoxygenase (LOX) (Brash, 1999; Kuhn et al., 2005) and cytochrome P450 (CYP450) (Capdevila and Falck, 2002) in addition to the COX pathway described above (Figure 1.5). Eicosanoids obtained by the metabolic pathway of PUFAs participate in different ways in this platelet aggregation process. Thus, while TXA induces amplification of the platelet response, prostaglandin I (PGI) acts by inhibiting platelet activation (FitzGerald, 1991).

The influence of the ratio ω–6/ω–3 on arterial thrombosis risk and the progression of atherosclerosis has been investigated using animal models (Yamashita et al., 2005). These studies demonstrated that, in addition to reducing TGs and LDL levels, the lower proportion of ω–6/ω–3 tested was more effective in suppressing thrombotic and atherosclerotic parameters.

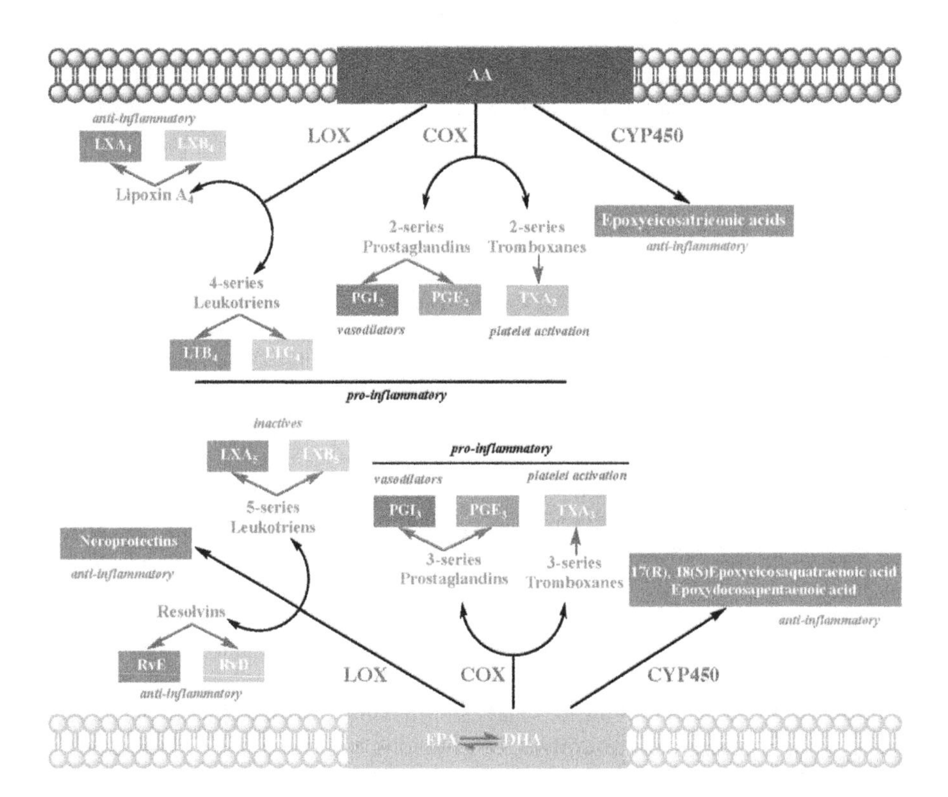

FIGURE 1.5 Eicosanoids and their biosynthetic origins. The metabolic pathway of eicosanoids produces active metabolic agents with vasoactive properties and anti- and proinflammatory effects, from arachidonic acid (AA), eicosapentaenoic acid (EPA) and docosahexaenoic acid (DHA). COX, cyclooxygenase; LTB_4, leukotriene B_4; LTC_4, leukotriene C_4; LOX, lipoxygenase; LXA, lipoxine A; LXB, lipoxine B; PG, prostaglandins; RvD, resolvin D; RvE, resolvin E; TXA, thromboxane A.

The antithrombotic properties observed with ω–3 PUFAs have been attributed to the incorporation of EPA and DHA in platelet phospholipids, displacing ω-6 PUFAs such as AA (Smith, 2005). EPA has been shown to induce profound changes in the biosynthesis of prostanoids and leukotrienes by competing with AA for COX and LOX binding sites. The efficiencies of prostanoid production can be significantly lower with EPA compared to AA, considering that COX-1 and COX-2 produce EPA oxygenation at only 10% and 30% of the rates obtained with AA (Wada et al., 2007).

In relation to the biological properties of prostanoids, it was initially proposed that TXA_2, originating from the AA pathway, was a potent pro-aggregator; whereas TXA_3, originating from the EPA-derived Series 3, was inactive. However, more recent studies have shown that both TX series have similar binding and stimulation capabilities of the thromboxane receptor (TP) (Wada et al., 2007). This does not allow us to sustain the theory of competition between EPA and AA for the production of pro-aggregant TXA. Nor does prostaglandins appear to tip the scale in favor of one or the other route, since both AA-derived (PGI_2) and Series 3 (PGI_3) prostacyclins generated from EPA have similar ability to inhibit platelet activation. The preferable route in favor of EPA's protective effects is the production of leukotrienes, while AA is metabolized by 5-LOX to the "4-series" of leukotriene (LTB_4), which induces inflammation and is a powerful chemoattractant agent of neutrophils. The same pathway produces LTB_5 from EPA, a metabolite of the "5-series," which is at least 30 times less potent than LTB_4 (Terano et al., 1984). This competition between AA and EPA for the production of leukotrienes provides an explanation for the anti-inflammatory effect of dietary ω-3 PUFAs, which, in turn, could also be considered in pro-thrombotic protection. In addition to competing with COX and LOX enzymes, ω-3 PUFA derivatives also compete with AA for other enzymatic systems involved in the synthesis of lipid derivatives with proinflammatory activity. EPA and DHA are also the parental fatty acids of new classes of lipid mediators, called resolvins, which have very potent anti-inflammatory properties and can play an essential role in protecting against various inflammatory diseases (Serhan et al., 2008a). Atherosclerosis, which in turn is associated with pro-thrombotic and atherogenic processes, would also be included in this group of inflammatory diseases.

On the other hand, CYP450 enzymes catalyze hydroxylation and epoxidation of AA in what was recently established as the "third branch" of the eicosanoid cascade (Capdevila and Falck, 2002). These enzymes catalyze the conversion of AA into epoxyeicosatrienoic acids (EETs). The CYP450 isoforms that metabolize AA also bind efficiently to EPA and DHA to produce epoxyeicosatetraenoic acids (EEQs) and epoxydocosapentaenoic acids (EDPs), respectively. These interactions can have important physiological implications and can provide a novel insight into the mechanisms of the vascular and cardiovascular protective effects of ω-3 PUFAs. Thus, it has been observed that the EET produced by AA epoxidation are inhibitors of thrombocyte adhesion to the vascular wall. However, in a ristocetin-induced thrombocyte aggregation model, inhibitory potency was found to be five times higher for EPA and DHA derivatives than for AA-derived EETs. These results provide new data on the very specific role of CYP450-dependent eicosanoids in preventing thromboembolic events. In addition, they suggest that the formation of EEQs and EDPs may contribute to the antithrombotic effects of ω-3 PUFAs (Jung et al., 2012).

Over the past two decades, several studies have reported the effects of ω-3 PUFAs on the composition of platelet lipids and platelet function (Kristensen et al., 1989). Dietary ω-3 fatty acids have been shown to inhibit both platelet aggregation and release of platelet TXA in response to collagen and ADP (Brox et al., 1983; Hirai et al., 1980). Other studies have shown an inhibition of thrombin-induced platelet aggregation (Ahmed and Holub, 1984) and adrenaline-induced platelet aggregation (Kristensen et al., 1987). In the presence of ω-3 PUFAs, the platelet response has also been inhibited by the addition of platelet-activating factor (PAF) (Codde et al., 1987). However, ω-3 PUFAs do not appear to affect the platelet aggregation induced by AA (Knapp et al., 1986).

These findings are complemented by other observations aimed at evaluating the activity of ω-3 PUFAs on endothelial activity, as the role of these cells in the production of vasoactive substances is recognized. Prostacyclins and NO, mediators produced by vascular endothelium, are factors that promote the relaxation of vascular smooth cells and the inhibition of the platelet activation pathways involved in the maintenance of homeostasis. In this way, endothelial function plays a protective or limiting role in the development of thrombosis, particularly in arterial thrombus formation.

In this sense, it is known that in human coronary endothelial cells, DHA regulates the function of eNOS and promotes NO synthesis by activating Akt (Stebbins et al., 2008). Therefore, it could be hypothesized that DHA-induced increases in bioavailability of NO improve endothelial function and blood flow, in addition to counteracting the pro-thrombotic effects of eicosanoids derived from active platelets.

These findings suggest that a reduction in platelet-vascular wall interaction may result from a diet rich in ω-3 PUFAs. However, it is clear that the mechanisms are complex and encompass more than a simple imbalance between eicosanoid derivatives with pro-aggregant or anti-aggregant properties. In any case, experimental data correlate with the effect of a diet rich in EPA and DHA on platelet phospholipid fatty acid composition, platelet aggregation and bleeding time.

In accordance with experimental data (one study conducted on healthy subjects), the EPA diet prolonged bleeding time (42%) and decreased platelet aggregation (Thorngren and Gustafson, 1981). This study also showed a decrease in sensitivity to ADP that persisted for several weeks after stopping the ω-3 PUFAs-rich diet. Similarly, in humans with high cardiovascular risk, it was observed that consumption of ω-3 PUFAs at an average dose of 3.65 g/d, is capable of reducing platelet aggregation induced by different agonists such as collagen, PAF and TXB_2 (Mori et al., 1997). There are other studies that support the antiplatelet activity of ω-3 PUFAs. For example, a reduction in platelet reactivity after DHA supplements for two weeks was reported in healthy male volunteers (Guillot et al., 2009). In addition, other experimental studies conclude that low ω-3 PUFA supplements lead to a state of platelet overactivity, increased blood viscosity and a tendency to thrombus development (Wan et al., 2010). Similarly, another study was conducted on healthy volunteers who received 3.4 g/d of EPA+DHA where clinical hematological parameters were measured along with collagen-stimulated platelet activation and protein phosphorylation. Treatment with ω-3 PUFAs produced a significant attenuation of collagen-mediated platelet signaling events and α-granule secretion (Larson et al., 2011).

These results based on platelet function suggest that ω-3 PUFAs incorporated into the diet generate a broad spectrum of favorable effects on factors associated with thrombosis risk. However, other data reported are not consistent with the antithrombotic properties described previously.

Unsatisfactory data have been observed in human trials regarding the activity of ω-3 PUFAs on platelet activity and coagulation (Balk et al., 2004; Wang et al., 2004). On the one hand, the reports include data that reveal that the consumption of n-3 ω-3 PUFAs has no effect on platelet aggregation or coagulation factors . This is in accordance with the report by Burgin-Maunder et al. (2015), who concluded that the administration of ω-3 PUFAs does not reduce the plasma concentration of von Wilbrand factor (vWF), nor its binding to collagen. These same authors in a study on human endothelial cells, had described that the ω-3 PUFAs modulate the release of von Wilbrand factor (vWF) (Burgin-Maunder et al., 2013). Thus, the evaluation of PUFAs on specific coagulation cascade factors, in particular for vWF, does not appear to be conclusive and new studies should be proposed in this field.

Based on the biological properties of ω-3 PUFAs described above, the debate on their safety in patients susceptible to bleeding is still open. This may be particularly relevant in view of the specific recommendations for people with a medical history of cardiovascular events or patients in preparation for surgery who frequently use antithrombotic drugs. In this sense, there has been no excess risk of clinical bleeding with fish or fish oil consumption, including among patients undergoing surgery or percutaneous intervention (Eritsland et al., 1995). On the other hand, in a study using 4 g/day doses of fish oil, change in bleeding time or number of bleeding episodes was not observed in patients on aspirin or warfarin therapy after coronary artery bypass (Eritsland et al., 1995). In a more recent analysis that included eight clinical studies conducted with enteral nutritional products containing fish oil as a source of ω-3 PUFAs, adverse events related to bleeding and effects on key clotting parameters were assessed. The authors found that there was no evidence of increased risk of bleeding with the use of fish oil-enriched medical nutrition in patients with moderate to severe disease (Jeansen et al., 2018). Despite these findings, it is suggested that due to the possibility of prolonged bleeding time, patients receiving warfarin therapy should be monitored and the dosage of anticoagulant adjusted if necessary.

1.6 EFFECTS ON BLOOD PRESSURE AND ENDOTHELIAL FUNCTION

High blood pressure (HBP), defined as the sustained rise in blood pressure higher than 140/90 mmHg, is the leading modifiable cause of death associated with cerebrovascular disease and is responsible for severe disability worldwide (WHO, 2013). Based on its etiology, HBP is usually classified as primary or secondary, the latter being of a specific origin such as primary hyperaldosteronism, pheochromocytoma, use of drugs, and so on, while the primary or essential hypertension has multifactorial causes that include genetic and environmental factors and alterations in the blood-pressure regulating systems (Rossier et al., 2017; Shrout et al., 2017).

Blood pressure (BP), determined by cardiac output (CO) and peripheral vascular resistance (PVR), is a physiological variable that depends on the heart, blood vessels and extracellular fluid volume, whose functions are regulated by the activity of the central nervous system (CNS), autonomic nervous system, kidneys and circulating hormonal factors such as the renin-angiotensin-aldosterone system (RAAS) (Lawton and Chatterjee, 2013).

Due to the many factors responsible for the regulation of BP and considering that the alteration of these factors is involved in the development of HBP, the therapeutic approach includes a large number of drugs with different mechanisms of action. These drugs aim to modify the underlying alterations and control the increase in BP while at the same time reducing the harmful effects of the pathology on target organs such as the heart, blood vessels, kidney and brain.

This pharmacotherapy has shown over time a significant reduction in cardiovascular morbidity and mortality (Chobanian, 2009; Ettehad et al., 2016). Therefore, it is imperative that hypertensive patients receive medication in addition to making lifestyle changes, which have also been shown to decrease BP (James et al., 2014).

Despite the existence of drugs with clinical efficacy and demonstrated safety, control of HBP is inadequate, as several authors point out that a large percentage of patients do not reach the established target (<140/90 mm Hg) and thus maintain an elevated cardiovascular risk (Ferdinand and Nasser, 2017; Laurent, 2017).

As indicated above, there is renewed interest in the search for adjunct therapies that guarantee a better antihypertensive response in patients. One of these possibilities is the use of nutritional supplements, which have been shown to demonstrate positive effects of diet on health. Among this group of agents are ω-3 PUFAs, which have pleiotropic properties in CVD (Turner and Spatz, 2016).

In this regard, animal, randomized and observational studies indicate that intake of ω-3 PUFAs modifies BP. However, the reduction of BP observed with these fatty acids is mild, but in other cases the results are inconsistent (Cabo et al., 2012).

Since BP is the result of CO and PVR, these fatty acids would be expected to exert their antihypertensive effect by modifying both parameters and the regulating variables. For this reason, studies have been conducted to evaluate the effects of ω-3 PUFAs on the following parameters that determine the BP.

1.6.1 Effect of ω-3 PUFAs on Cardiac Output

In relation to this parameter, a meta-analysis found a significant reduction in heart rate by 1.6 beats per minute (bpm) associated with the consumption of ω-3 PUFAs. The effect on CO appears to be independent of amount ingested; however, the treatment period is determinant, and the most pronounced effect has been observed in studies with periods of over 12 weeks of treatment (Mozaffarian et al., 2005). Despite this observed effect, its mechanism is not yet defined and may, therefore, be related to the ability of these compounds to modulate cardiomyocyte activity, associated with an increase in parasympathetic activity in the myocardium (Mozaffarian and Wu, 2011). This reduction in HR could favor the diastolic filling function, causing an increase in systolic volume, effects that are evident from the beginning of supplementation (Cabo et al., 2012). Considering that the effect on CO is minimal, it can be stated that the antihypertensive effect associated with ω-3 PUFAs is more related to the modification of PVR than it is with its actions on CO.

1.6.2 Effect of ω-3 PUFAs on Peripheral Vascular Resistance

This parameter is determined by the caliber of the resistance blood vessels, where the smooth muscular layer and vascular endothelium are most important

1. *Endothelial function:* ω-3 PUFAs increase the production and disposition of NO from endothelial cells and the effect on this vasodilatory mediator is due to the following actions:
   a. Alterations in the composition of phospholipids in the endothelial cell membrane, as a result of their incorporation. Within these cells there are the caveolae that contain eNOS bound to caveolin-1, a protein complex capable of inhibiting enzymatic function.

      In cell cultures incubated with EPA, eNOS activity is increased due to the dissociation of the enzyme from caveolin-1 (Zanetti et al., 2015). Another complementary study shows that EPA stimulates the phosphorylation of the eNOS- dependent pathway of adenosine monophosphate-activated protein kinase (AMPK), which increases NO. This surge in AMPK activity was related to the higher expression of the mitochondrial proton transporter (UCP-2), which has a regulatory impact in the production of NO (Enre and Nübel, 2010; Wu et al., 2012).

      In the case of DHA, some authors indicate similar effect on NO production, but its mechanism is by promoting the interaction between eNOS and Heat Shock Protein-90 (HSP-90), which activates the PKB/AKt pathway generating consecutive phosphorylation of eNOS (Stebbins et al., 2008).
   b. Reduction of asymmetric dimethylarginine (ADMA). A study carried out by Raimondi et al. (2005) in spontaneously hypertensive older rats reported that EPA-DHA supplementation reduced the concentration of ADMA, an endogenous inhibitor of eNOS intimately associated with HBP and other pathologies. This effect on ADMA is related to an increase in NO availability (Raimondi et al., 2005).
   c. eNOS gene expression. Various studies in laboratory animals have shown that administration of ω-3 PUFAs induces an increase in the eNOS gene expression which increases NO availability.
2. *Antioxidant effect:* This effect has been demonstrated by the ability of these fatty acids in decreasing the formation of reactive oxygen species such as peroxynitrite by modulating the activity of NADPH oxidase and inducible nitric oxide synthase (iNOS).
3. *Anti-inflammatory effect:* ω-3 PUFAs reduce the activity of various inflammatory mediators, such as NFκB, IL-6, IL-1 and C-reactive protein (CRP), associated with endothelial dysfunction, proliferation of vascular smooth muscle cells and production of free radicals (Wang et al., 2011b). The neutralization of these mediators would increase NO levels and confer the ω-3 PUFAs with a vasodilatory and antihypertensive effects.
4. *Vascular smooth muscle cells:* The capability of DHA to produce hypotensive effects in mice after intravenous administration has been described and is

associated with its ability to directly activate calcium-dependent and voltage-activated K+ (BK) channels (Hoshi et al., 2013b). These play an important role in vascular tone since their activation induces hyperpolarization, thus closing calcium-dependent-voltage channels, which results in vasodilation. This action on BK was confirmed by *in vitro* studies in coronary smooth muscle cells, arteries and cell culture, where the application of DHA increases this current, inducing a vasodilatory effect (Hoshi et al., 2013a; Lai et al., 2009).

In the case of DHA and its antihypertensive effect, it has been proposed that this effect may depend upon its conversion to active metabolites, EDPs, which have a greater hypotensive effect and better stability than eicosatrienoic acids (ETrAs) which are derivatives of AA (Wang et al., 2011a; Fer et al., 2008). Based on the above, Ulu et al. (2014) investigated the effect of an epoxy DHA, 19,20 epoxy-docosapentaenoic acid (19,20-EDP), on an angiotensin II-induced hypertension in mice, and demonstrated that administration of 19,20-EDP was involved in the antihypertensive effect of DHA (Ulu et al., 2014).

Despite the considerable evidence from experimental results, the effect of ω-3 PUFAs on BP in humans is not conclusive, since the interpretation of epidemiological studies is difficult, especially because of the accurate estimation of fish intake in the population (Mori, 2006). In this respect, the INTERMAP study that evaluated 4,680 individuals from Asia and Western countries demonstrated an important direct association between the consumption of these fatty acids and the reduction of BP (Ueshima et al., 2007).

From these observations, several studies have evaluated the effect of ω-3 PUFAs *vs* placebo in order to establish their antihypertensive effect. Thus, in early studies, intake of 3 g of ω-3 PUFAs in untreated hypertensive patients report reductions of −5.5 mmHg in systolic blood pressure (SBP) and −3.5 mmHg in diastolic blood pressure (DBP) (Turner and Spatz, 2016).

A recent meta-analysis of 70 randomized controlled trials (RCTs) published by Miller et al. (2014), which considered all possible sources of EPA and DHA administered in an average dose of 3.8 g/d, reported a significant reduction of BP in both hypertensive and normotensive subjects. In this meta-analysis, the effect was more pronounced in hypertensive patients, as shown in earlier studies, but no dose-dependent relationship was established (Miller et al., 2014).

Given the effect shown in clinical studies, other prospective studies have evaluated the possible impact of diet supplemented with these fatty acids on the development of HBP in normal subjects, indicating a significant reduction in the risk of developing HBP compared with subjects with low ω-3 PUFAs intake (Colussi et al., 2017).

With these results, it has been demonstrated that the effect of ω-3PUFAs on BP is minimal and is manifested mainly in hypertensive patients. However, ω-3PUFAs' ability to decrease the progression of this disease could be relevant for the population. In addition, there is a need for further studies to evaluate more broadly the antihypertensive efficacy of these agents using more strict criteria related to populations, sources and quantity of ω-3 PUFAs, duration and treatment modalities.

1.7 EFFECTS ON CARDIAC ARRHYTHMIAS

Cardiac arrhythmias are defined as an alteration in the onset or form of electrical impulses transmitted in the myocardium. They constitute a relevant health problem since, apart from being related to other cardiovascular pathologies such as heart failure (HF), myocardial ischemia and acute myocardial infarction, they can be responsible for sudden cardiac death (SCD) (Gaztanaga et al., 2012).

There are different types and ways of classification of cardiac arrhythmias. However, the most commonly used form is the conventional, which is based on heart rate or pacemaker disturbance. In relation to the latter criteria, arrhythmias are divided into: a) supraventricular arrhythmias, those originating in the atrium or above the atrioventricular node (AVN), and b) ventricular arrhythmias that occur in the ventricular tissue. Within this division, there are sub-classifications, some of which do not require treatment. Others, however, due to implicit risk of fatality, require a strict pharmacological treatment that guarantees the suppression of the altered rhythm (Li, 2015).

Two mechanisms involved in the genesis of arrhythmias are described: (1) increased normal and abnormal automaticity, (2) decreased driving performance.

Both mechanisms, in turn, are determined by cardiac electrophysiology that depends on the different ionic currents, a determining factor of the cardiac action potential. Therefore, most antiarrhythmic agents act by modifying these ionic currents or, in some particular cases, membrane receptors.

Within the conventional classification, three types are of particular clinical interest: atrial fibrillation (AF), ventricular tachycardia (VT) and ventricular fibrillation (VF), in which cases combined therapy is required, ranging from anticoagulant drugs to utilization of electrical cardioversion.

It is recognized that AF is the most common arrhythmia in the world population, with an estimated prevalence of 3% in adults. It is associated with factors such as age, valvular heart disease, coronary artery disease, diabetes mellitus (DM) and chronic kidney disease (Gomez-Outes et al., 2017; Kanaporis and Blatter, 2017).

Meanwhile, TV and VF are potentially lethal arrhythmias that can develop during AMI or as clinical sequelae of advanced cardiomyopathy, and are often associated with SCD (Jong-Ming Pang and Green, 2017). Because these types of arrhythmias generate important morbidity and mortality, and despite the existence of available treatments in recent decades, there is a need for additional effective strategies that provide better results for patients. In this sense, epidemiological data from studies of more than 40 years have demonstrated an inverse association between diets with a high content of fish fat and cardiac mortality (Daviglus et al., 1997).

Additionally, the main components of this diet, ω-3 PUFAs, had shown in clinical trials a significant reduction in mortality in post-infarction patients (Marchioli et al., 2002) or patients with high cholesterol levels receiving conventional treatment. These results indicated ω-3 PUFAs' safety in combination therapy and their antiarrhythmic potential (Yokoyama et al., 2007).

Based on the above, a large number of studies have been conducted in animals including rats, pigs and monkeys, and in cell cultures, to determine the precise mechanism of action by which ω-3 PUFAs produce their antiarrhythmic effect

(McLennan, 2014). These studies indicate that ω-3 PUFAs possess modulating properties of ionic currents in the cardiomyocyte, which are responsible for maintaining the electrophysiology of the heart and are involved in the origin of arrhythmias (Roy and Le Guennec, 2017; Endo and Arita, 2016). It has been proposed that the ability to modify the sodium, potassium and calcium channels can be a direct or indirect effect, the latter being due to the ability of ω-3 PUFAs to alter the constitution and function of phospholipids in the cellular membrane (Reiffel and McDonald, 2006). Other studies have reported that the administration of these agents does not cause changes in electrocardiographic parameters, making it difficult to explain, in a congruent manner, the observed molecular effects on ion channels and their relationship with changes in the electrocardiogram (Roy and Le Guennec, 2017).

Despite these discrepancies, research has continued, giving more emphasis on modulation of sodium, potassium and calcium channels.

1.7.1 Effects on Sodium Currents

Cellular electrophysiological studies on the effects of acute administration of ω-3 PUFAs in animals indicate an inhibitory effect of the ventricular sodium current, denoting a reduction in potential amplitude, reduction of the channel inactivation voltage and rapid increase of its inactivation and an increase in the activation threshold (Reiffel and McDonald, 2006; Roy and Le Guennec, 2017). These effects are evident in adult and neonatal cardiomyocytes.

In non-heart cells transfected with mutant sodium channels, Xiao et al. (2006) reported that ω-3 PUFAs induced their potent blockage. The effects of ω-3 PUFAs on $I_{\mathrm{Na\ late}}$ was greater than that on $I_{\mathrm{Na\ peak}}$. It had been demonstrated that an increase in persistent I_{Na} with a consequent increase on $[Ca^{2+}]i$ level, can cause arrhythmias and irreversible cell damage. Thus, the inhibition of $I_{\mathrm{Na\ late}}$ by ω-3 PUFAs might have a potential therapeutic value in certain patients with ischemia-induced arrhythmias. This action was limited to ω-3 PUFAs and was not produced by monounsaturated or saturated fatty acids (Xiao et al., 2006).

On the other hand, the longer exposition to these compounds in ventricular myocytes isolated from pigs and rats feeding with a diet rich in ω-3 PUFAs has not shown significant effects on sodium current. However, the antiarrhythmic effect is maintained, which may indicate that the effect on sodium channels is not the main mechanism for this therapeutic effect (Moreno et al., 2012). Additionally, Li et al. (2009) using human atrial myocytes, demonstrated similar inhibitory effects on this current, this effect being more pronounced with EPA than with DHA (Li et al., 2009).

1.7.2 Effects on Potassium Currents

In cardiomyocytes, there are different types of potassium currents which are responsible for the repolarization process during the action potential. Therefore, the modulation of these currents generates significant effects on the electrical activity of the heart. The principal potassium currents in the cardiac action potential are: transient outward current, ultra-fast, fast and slow repolarizing (I_{Kur}, I_{Kr} and I_{Ks}) and rectifying inward (I_{Kir}) current. These currents have a varied distribution according to the type of cardiac tissue. For example, I_{to} and I_{Kur} are located mainly in atria (Li et al., 2009), while I_{Ks}, I_{Kr} and I_{Kir} predominate in the ventricle (Song et al., 2013).

Despite the large number of potassium currents, few studies have been conducted to determine the effect of ω-3 PUFAs on these currents. The most information has been obtained from cellular studies with acute administration of these compounds without evaluating the effect of dietary supplements and their incorporation into the membranes (McLennan, 2014). Thus, DHA demonstrated an inhibitory effect on I_{to} current in transfected Chinese hamster ovary cells (Singleton et al., 1999). This effect was also observed in isolated human atrial cardiomyocytes, where treatment with DHA and EPA significantly inhibited this current (Li et al., 2009).

Another study in rat ventricular myocytes reported that DHA significantly inhibited repolarizing currents with the consequent prolongation of the duration of action potential (APD) and effective refractory period (ERP), effects that determine their ability to eliminate reentry inputs and cardiac arrhythmias (Song et al., 2013). The same study showed that DHA had no effect on I_{Kir}, indicating that DHA does not modify the resting potential in these cells.

Taking into consideration the inhibitory effect on different potassium currents, it can be inferred that this effect may explain, in part, the low incidence of supraventricular and ventricular arrhythmias associated with the use of ω-3 PUFAs (Mozaffarian et al., 2004; Leaf et al., 2005).

1.7.3 Effects on Calcium Channels

Two calcium currents are present in the heart: the transitory calcium current (I_{CaT}) present exclusively in the nodes, and the slow calcium current (I_{CaL}) expressed in all cells. The latter has the characteristics of generating a depolarization phase in nodal cells and sustaining the plateau during cardiac action potential in atrial and ventricular cells, allowing these cells to raise the $[Ca^{2+}]i$ levels necessary for mechanical contraction function. Moreover, the elevation in the level of this ion has been related to different arrhythmias such as those induced by digitalis, adrenergic stress and anoxia (Landstrom et al., 2017). In this sense, acute and long-term treatment with ω-3 PUFAs in rats prevented transient elevation and calcium overload in cardiomyocytes in an associated effect with blockage of these ionic currents (Rinaldi et al., 2002; Jahangiri et al., 2006). Several authors using mammalian cardiomyocytes have shown that the effect on calcium levels is due to the blocking of I_{CaT} and I_{CaL} currents, which reduces transitory elevations of this ion, shortens the plateau of action potential and decreases the occurrence of arrhythmias (McLennan, 2014).

Another important aspect of $[Ca^{2+}]i$ is the function of the sodium-calcium exchanger (NCX) responsible for a current (I_{NCX}) that can lead to the accumulation of calcium inside the cell, which can generate arrhythmias. Xiao et al., in HEK293t cell lines, demonstrated that ω-3 PUFAs suppress this activity. This finding is relevant because during ischemia, the accumulation of hydrogen ions stimulates the activity of the sodium/hydrogen exchanger, which leads to the accumulation of intracellular sodium and consequently increases the activity of NCX, causing the increase of calcium (Xiao et al., 2004).

Given the intimate relationship between calcium overload and arrhythmias, it is clear that blocking calcium currents is an important antiarrhythmic mechanism of ω-3 PUFAs.

Other mechanisms: it is recognized that these fatty acids accumulate in the cell membrane and modify its constitutive and functional properties. From this point of view, the functionality of membrane-attached proteins (ionic channels) can be altered without having a direct interaction between ω-3 PUFAs and these channels, thus affecting the ionic currents involved in cardiac action potentials (Mozaffarian and Wu, 2011, 2012).

With respect to the alteration of membrane composition, some authors suggest that the incorporation of ω-3 PUFAs in the cardiomyocyte membrane reduces the release of inositol triphosphate (1,4,5-triphosphate, IP3) in response to stimulation of alpha-adrenergic receptors in porcine cardiomyocytes (Nair et al., 2000) and rat hearts in response to ischemia (Anderson et al., 1996). This mediator, IP3, is a second messenger in signals resulting from the activation of PLC whose function is to increase the release of intracellular calcium; consequently, its reduction could have an implication in the antiarrhythmic effect.

All the mechanisms described above have been established mainly using isolated cardiomyocytes and cell cultures, but in some similar *in vitro* and *in vivo* models these results are inconsistent. Therefore, a definite mechanism has not been established and cannot be extrapolated to humans. These differences may be due to different cell types and the composition of the membrane, which could alter the effect of ω-3 PUFAs. Another factor to consider is the administration of these compounds, which may be acute (*in situ*) intravenously or on a long-term basis (supplied in the diet), where direct and indirect action (incorporation into cell membranes) may play a relevant role within their actions (McLennan, 2014; Roy and Le Guennec, 2017; Trimarco, 2012).

In relation to the possible antiarrhythmic effect of ω-3 PUFAs in humans, several clinical studies show evidence of their capacity to reduce the development of supraventricular, ventricular arrhythmias and SCD in different populations (patients with ischemia, with implanted defibrillators or AMI convalescent). However, there are also contradictory studies in this respect, and for this reason the antiarrhythmic potential should be carefully considered (Borghi and Pareo, 2012).

These conflicting results can be explained in part by the findings of Hu et al. (2016), who showed that there was no difference in the prevalence of AMI compared the Inuit population in Canada versus the United States and Canadian general populations. Authors associate this observation with levels of methylmercury (MeHg) present in the fish diet of the Inuit Canadian population. Thus, the intake and accumulation of mercury can favor lipid peroxidation counteracting the beneficial effect of ω-3 PUFAs (Hu et al., 2016).

A recent publication by Roy et al. (2015), indicates that the antiarrhythmic effect of DHA is related to the non-enzymatic oxidation carried out by ROS, where neuroprostanes are generated. These mediators are recognized biomarkers of oxidative stress and therefore would be expected to have a harmful effect. However, these researchers demonstrated experimentally that neuroprostanes, primarily 4RS-4F4t-neuroprostane and 10(S)-10-F4t-neuroprostane, can regulate the function of the ryanodine receptor (RyR2), decreasing calcium efflux from the sarcoplasmic reticulum leading to a reduction in calcium sparks and the risk of arrhythmias (Roy et al., 2015). Under this premise, it would be expected that in chronic conditions of oxidative stress, which are common in cardiovascular diseases (ischemia, atherosclerosis, AMI, cardiac post-surgery, etc.) (Luscher, 2015; Islam et al., 2016; Yalta and Yalta, 2018), ω-3 PUFAs generate the production of neuroprostanes which participate in the antiarrhythmic effects induced by these fatty acids.

Because of many factors and variables involved that can modify the results of preclinical and clinical studies with the use of these compounds, it is imperative that new studies consider all these factors so that ω-3 PUFAs' mechanism of action and their application as antiarrhythmic agents can be more precisely supported.

1.8 CONCLUSIONS

Recent developments confirm and extend the concept that ω-3 PUFAs are beneficial in the prevention of cardiovascular disease and sudden cardiac death. In experimental studies and animal models, ω-3 PUFAs modulate a variety of relevant biologic pathways, with several lines of evidence suggesting at least some differential benefits.

During regular consumption of ω-3 PUFAs, a potential decrease of endothelial dysfunction and prevention of CVD has been described through effects on endothelial metabolism, inflammation, thrombosis and arrhythmia. On the other hand, the effects shown on vascular function and inflammation could explain the anti-atherogenic properties of these fatty acids. Such is the consideration of this effect that if in the treatment of dyslipidemia, the LDL-c target cannot be achieved by lifestyle changes or treatment with statins, consideration should be given to adding supplementation with ω-3 PUFAs in order to help reduce cardiovascular risk. Associated with effects on smooth muscle and endothelial cells, the administration of ω-3 PUFAs has been shown to lower blood pressure in both animal models and hypertensive patients.

There is no question that research on ω-3 PUFAs has made significant progress in a number of areas. However, there is a lack of conclusive data from clinical and mechanistic studies on the potential benefits of ω-3 fatty acids for primary and secondary prevention in CVD.

On the basis of the available literature, ω-3 PUFAs should be considered as important components of a healthy diet and as a potential therapeutic modality in patients with coronary artery disease, particularly in populations at heightened risk of cardiovascular disease. These patients should be advised to eat a healthy dietary pattern that includes fatty fish. For individuals with low consumption of fish rich in ω-3 PUFAs, fish oil capsules as a source of EPA and DHA can be considered, given their long history of safety and the relationship between benefit and risk.

In conclusion, research on the biology of ω-3 PUFAs and its role in cardiovascular health is developing rapidly. However, harmonization of clinical criteria and parameters will be necessary in order to achieve consensus on beneficial cardiovascular effects of ω-3 PUFAs.

REFERENCES

Aarsetoy, H., T. Brugger-Andersen, Ø. Hetland, H. Grundt, and D. W. Nilsen. 2006. "Long term influence of regular intake of high dose n-3 fatty acids on CD40-ligand, pregnancy-associated plasma protein A and matrix metalloproteinase-9 following acute myocardial infarction." *Thromb. Haemost.* 95 (2):329–36. doi:10.1160/TH05-07-0497.

Abeywardena, M. Y. 2003. "Dietary fats, carbohydrates and vascular disease: Sri Lankan perspectives." *Atherosclerosis* 171 (2):157–61.

Ahmed, A. A., and B. J. Holub. 1984. "Alteration and recovery of bleeding times, platelet aggregation and fatty acid composition of individual phospholipids in platelets of human subjects receiving a supplement of cod-liver oil." *Lipids* 19 (8):617–24.

Anderson, K. E., X. J. Du, A. J. Sinclair, E. A. Woodcock, and A. M. Dart. 1996. "Dietary fish oil prevents reperfusion Ins(1,4,5)P3 release in rat heart: Possible antiarrhythmic mechanism." *Am. J. Physiol.* 271 (4 Pt 2):H1483–90.

Arita, M., F. Bianchini, J. Aliberti, A. Sher, N. Chiang, S. Hong, R. Yang, N. A. Petasis, and C. N. Serhan. 2005. "Stereochemical assignment, antiinflammatory properties, and receptor for the omega-3 lipid mediator resolvin E1." *J. Exp. Med.* 201 (5):713–22. doi:10.1084/jem.20042031.

Arita, M., T. Ohira, Y. P. Sun, S. Elangovan, N. Chiang, and C. N. Serhan. 2007. "Resolvin E1 selectively interacts with leukotriene B4 receptor BLT1 and ChemR23 to regulate inflammation." *J. Immunol.* 178 (6):3912–7.

Asmat, U., K. Abad, and K. Ismail. 2016. "Diabetes mellitus and oxidative stress-A concise review." *Saudi Pharm. J.* 24 (5). doi:10.1016/j.jsps.2015.03.013.

Backes, J., D. Anzalone, D. Hilleman, and J. Catini. 2016. "The clinical relevance of omega-3 fatty acids in the management of hypertriglyceridemia." *Lipids Health Dis.* 15 (1):118. doi:10.1186/s12944-016-0286-4.

Baker, E. J., E. A. Miles, G. C. Burdge, P. Yaqoob, and P. C. Calder. 2016. "Metabolism and functional effects of plant-derived omega-3 fatty acids in humans." *Prog. Lipid Res.* 64:30–56. doi:10.1016/j.plipres.2016.07.002.

Balk, E., M. Chung, A. Lichtenstein, P. Chew, B. Kupelnick, A. Lawrence, D. DeVine, and J. Lau. 2004. "Effects of omega-3 fatty acids on cardiovascular risk factors and intermediate markers of cardiovascular disease." *Evid Rep Technol Assess (Summ)* 93 (93):1–6.

Bernert, J. T., Jr., and H. Sprecher. 1975. "Studies to determine the role rates of chain elongation and desaturation play in regulating the unsaturated fatty acid composition of rat liver lipids." *Biochim. Biophys. Acta* 398 (3):354–63.

Borghi, C., and I. Pareo. 2012. "Omega-3 in antiarrhythmic therapy: Cons position." *High Blood Press. Cardiovasc. Prev.* 19 (4):207–11. doi:10.1007/BF03297632.

Brash, A. R. 1999. "Lipoxygenases: Occurrence, functions, catalysis, and acquisition of substrate." *J. Biol. Chem.* 274 (34):23679–82.

Brox, J. H., J. E. Killie, B. Osterud, S. Holme, and A. Nordoy. 1983. "Effects of cod liver oil on platelets and coagulation in familial hypercholesterolemia (type IIa)." *Acta Med. Scand.* 213 (2):137–44.

Burdge, G. C., A. E. Jones, and S. A. Wootton. 2002. "Eicosapentaenoic and docosapentaenoic acids are the principal products of alpha-linolenic acid metabolism in young men*." *Br. J. Nutr.* 88 (4):355–63. doi:10.1079/BJN2002662.

Burgin-Maunder, C. S., P. R. Brooks, and F. D. Russell. 2013. "Omega-3 fatty acids modulate Weibel-Palade body degranulation and actin cytoskeleton rearrangement in PMA-stimulated human umbilical vein endothelial cells." *Mar. Drugs* 11 (11):4435–50. doi:10.3390/md11114435.

Burgin-Maunder, C. S., P. R. Brooks, D. Hitchen-Holmes, and F. D. Russell. 2015. "Moderate dietary supplementation with Omega-3 fatty acids does not impact plasma von Willebrand factor profile in mildly hypertensive subjects." *BioMed Res. Int.* 2015:394871. doi:10.1155/2015/394871.

Cabo, J., R. Alonso, and P. Mata. 2012. "Omega-3 fatty acids and blood pressure." *Br. J. Nutr.* 107 (Suppl 2):S195–200. doi:10.1017/S0007114512001584.

Calder, P. C. 2001. "Polyunsaturated fatty acids, inflammation, and immunity." *Lipids* 36 (9):1007–24.

Capdevila, J. H., and J. R. Falck. 2002. "Biochemical and molecular properties of the cytochrome P450 arachidonic acid monooxygenases." *Prostaglandins Other Lipid Mediat.* 68–69:325–44.

Capo, X., M. Martorell, C. Busquets-Cortes, S. Tejada, J. A. Tur, A. Pons, and A. Sureda. 2018. "Resolvins as proresolving inflammatory mediators in cardiovascular disease." *Eur. J. Med. Chem.* doi:10.1016/j.ejmech.2017.07.018.

Carpentier, Y. A., L. Portois, and W. J. Malaisse. 2006. "n-3 fatty acids and the metabolic syndrome." *Am. J. Clin. Nutr.* 83 (6 Suppl):1499S–504S.

Carrier, A. 2017. "Metabolic syndrome and oxidative stress: A complex relationship." *Antioxid. Redox Signal.* 26 (9):429–31. doi:10.1089/ars.2016.6929.

Catapano, A. L., I. Graham, G. De Backer, O. Wiklund, M. J. Chapman, H. Drexel, A. W. Hoes, C. S. Jennings, U. Landmesser, T. R. Pedersen, Z. Reiner, G. Riccardi, M. R. Taskinen, L. Tokgozoglu, W. M. Verschuren, C. Vlachopoulos, D. A. Wood, and J. L. Zamorano. 2016. "2016 ESC/EAS Guidelines for the Management of Dyslipidaemias: The Task Force for the Management of Dyslipidaemias of the European Society of Cardiology (ESC) and European Atherosclerosis Society (EAS) Developed with the special contribution of the European Association for Cardiovascular Prevention & Rehabilitation (EACPR)." *Atherosclerosis* 253:281–344. doi:10.1016/j.atherosclerosis.2016.08.018.

Cawood, A. L., R. Ding, F. L. Napper, R. H. Young, J. A. Williams, M. J. Ward, O. Gudmundsen, et al. 2010. "Eicosapentaenoic acid (EPA) from highly concentrated n-3 fatty acid ethyl esters is incorporated into advanced atherosclerotic plaques and higher plaque EPA is associated with decreased plaque inflammation and increased stability." *Atherosclerosis* 212 (1):252–9. doi:10.1016/j.atherosclerosis.2010.05.022.

Chan, D. C., G. F. Watts, T. A. Mori, P. H. Barrett, L. J. Beilin, and T. G. Redgrave. 2002. "Factorial study of the effects of atorvastatin and fish oil on dyslipidaemia in visceral obesity." *Eur. J. Clin. Invest.* 32 (6):429–36.

Chiang, N., C. N. Serhan, S. E. Dahlen, J. M. Drazen, D. W. Hay, G. E. Rovati, T. Shimizu, T. Yokomizo, and C. Brink. 2006. "The lipoxin receptor ALX: Potent ligand-specific and stereoselective actions in vivo." *Pharmacol. Rev.* 58 (3):463–87. doi:10.1124/pr.58.3.4.

Chobanian, A. V. 2009. "Shattuck lecture. The hypertension paradox--More uncontrolled disease despite improved therapy." *N. Engl. J. Med.* 361 (9):878–87. doi:10.1056/NEJMsa0903829.

Clarke, S. D. 2001. "Polyunsaturated fatty acid regulation of gene transcription: A molecular mechanism to improve the metabolic syndrome." *J. Nutr.* 131 (4):1129–32.

Codde, J. P., R. Vandongen, T. A. Mori, L. J. Beilin, and K. J. Hill. 1987. "Can the synthesis of platelet-activating factor, a potent vasodilator and pro-aggregatory agent, be altered by dietary marine oils?" *Clin. Exp. Pharmacol. Physiol.* 14 (3):197–202.

Colussi, G., C. Catena, M. Novello, N. Bertin, and L. A. Sechi. 2017. "Impact of omega-3 polyunsaturated fatty acids on vascular function and blood pressure: Relevance for cardiovascular outcomes." *Nutr. Metab. Cardiovasc. Dis.* 27 (3):191–200. doi:10.1016/j.numecd.2016.07.011.

Cussons, A. J., G. F. Watts, T. A. Mori, and B. G. Stuckey. 2009. "Omega-3 fatty acid supplementation decreases liver fat content in polycystic ovary syndrome: A randomized controlled trial employing proton magnetic resonance spectroscopy." *J. Clin. Endocrinol. Metab.* 94 (10):3842–8. doi:10.1210/jc.2009-0870.

Das, U. N. 2006. "Essential Fatty acids - A review." *Curr. Pharm. Biotechnol.* 7 (6):467–82.

Davidson, M. H. 2006. "Mechanisms for the hypotriglyceridemic effect of marine omega-3 fatty acids." *Am. J. Cardiol.* 98 (4A):27i–33i. doi:10.1016/j.amjcard.2005.12.024.

Davidson, M. H., A. K. Phillips, D. Kling, and K. C. Maki. 2014. "Addition of omega-3 carboxylic acids to statin therapy in patients with persistent hypertriglyceridemia." *Expert Rev. Cardiovasc. Ther.* 12 (9):1045–54. doi:10.1586/14779072.2014.942640.

Daviglus, M. L., J. Stamler, A. J. Orencia, A. R. Dyer, K. Liu, P. Greenland, M. K. Walsh, D. Morris, and R. B. Shekelle. 1997. "Fish consumption and the 30-year risk of fatal myocardial infarction." *N. Engl. J. Med.* 336 (15):1046–53. doi:10.1056/NEJM199704103361502.

Davis, H. R., R. T. Bridenstine, D. Vesselinovitch, and R. W. Wissler. 1987. "Fish oil inhibits development of atherosclerosis in rhesus monkeys." *Arteriosclerosis* 7 (5):441–9.

De Caterina, R., R. Madonna, R. Zucchi, and M. T. La Rovere. 2003. "Antiarrhythmic effects of omega-3 fatty acids: from epidemiology to bedside." *Am. Heart J.* 146 (3):420–30. doi:10.1016/S0002-8703(03)00327-2.

de Lorgeril, M., S. Renaud, N. Mamelle, P. Salen, J. L. Martin, I. Monjaud, J. Guidollet, P. Touboul, and J. Delaye. 1994. "Mediterranean alpha-linolenic acid-rich diet in secondary prevention of coronary heart disease." *Lancet* 343 (8911):1454–9.

de Urquiza, A. M., S. Liu, M. Sjoberg, R. H. Zetterstrom, W. Griffiths, J. Sjovall, and T. Perlmann. 2000. "Docosahexaenoic acid, a ligand for the retinoid X receptor in mouse brain." *Science* 290 (5499):2140–4.

DeFilippis, A. P., and L. S. Sperling. 2006. "Understanding omega-3's." *Am. Heart J.* 151 (3):564–70. doi:10.1016/j.ahj.2005.03.051.

Desvergne, B., L. Michalik, and W. Wahli. 2006. "Transcriptional regulation of metabolism." *Physiol. Rev.* 86 (2):465–514. doi:10.1152/physrev.00025.2005.

Dhe-Paganon, S., K. Duda, M. Iwamoto, Y. I. Chi, and S. E. Shoelson. 2002. "Crystal structure of the HNF4 alpha ligand binding domain in complex with endogenous fatty acid ligand." *J. Biol. Chem.* 277 (41):37973–6. doi:10.1074/jbc.C200420200.

Dimitrow, P. P., and M. Jawien. 2009. "Pleiotropic, cardioprotective effects of omega-3 polyunsaturated fatty acids." *Mini Rev. Med. Chem.* 9 (9):1030–9.

Endo, J., and M. Arita. 2016. "Cardioprotective mechanism of omega-3 polyunsaturated fatty acids." *J. Cardiol.* 67 (1):22–7. doi:10.1016/j.jjcc.2015.08.002.

Enre, Y., and T. Nübel. 2010. "Uncoupling protein UCP2: When mitochondrial activity meets immunity." *FEBS Lett.* 584 (8):1437–42. doi:10.1016/j.febslet.2010.03.014.

Eritsland, J., H. Arnesen, I. Seljeflot, and P. Kierulf. 1995. "Long-term effects of n-3 polyunsaturated fatty acids on haemostatic variables and bleeding episodes in patients with coronary artery disease." *Blood Coagul. Fibrinolysis* 6 (1):17–22.

Ettehad, D., C. A. Emdin, A. Kiran, S. G. Anderson, T. Callender, J. Emberson, J. Chalmers, A. Rodgers, and K. Rahimi. 2016. "Blood pressure lowering for prevention of cardiovascular disease and death: A systematic review and meta-analysis." *Lancet* 387 (10022):957–67. doi:10.1016/S0140-6736(15)01225-8.

Fer, M., Y. Dreano, D. Lucas, L. Corcos, J. P. Salaun, F. Berthou, and Y. Amet. 2008. "Metabolism of eicosapentaenoic and docosahexaenoic acids by recombinant human cytochromes P450." *Arch. Biochem. Biophys.* 471 (2):116–25. doi:10.1016/j.abb.2008.01.002.

Ferdinand, K. C., and S. A. Nasser. 2017. "Management of essential hypertension." *Cardiol. Clin.* 35 (2):231–46. doi:10.1016/j.ccl.2016.12.005.

FitzGerald, G. A. 1991. "Mechanisms of platelet activation: Thromboxane A2 as an amplifying signal for other agonists." *Am. J. Cardiol.* 68 (7):11B–5B.

Fritsche, K. L. 2015. "The science of fatty acids and inflammation." *Adv. Nutr.* 6 (3):293S–301S. doi:10.3945/an.114.006940.

Furenes, E. B., I. Seljeflot, S. Solheim, E. M. Hjerkinn, and H. Arnesen. 2008. "Long-term influence of diet and/or omega-3 fatty acids on matrix metalloproteinase-9 and pregnancy-associated plasma protein-A in men at high risk of coronary heart disease." *Scand. J. Clin. Lab. Invest.* 68 (3):177–84. doi:10.1080/00365510701663350.

Gaztanaga, L., F. E. Marchlinski, and B. P. Betensky. 2012. "Mechanisms of cardiac arrhythmias." *Rev Esp Cardiol (Engl Ed)* 65 (2):174–85. doi:10.1016/j.recesp.2011.09.018.

Goldberg, I. J., R. H. Eckel, and R. McPherson. 2011. "Triglycerides and heart disease: Still a hypothesis?" *Arterioscler. Thromb. Vasc. Biol.* 31 (8):1716–25. doi:10.1161/ATVBAHA.111.226100.

Gomez-Outes, A., M. L. Suarez-Gea, and J. M. Garcia-Pinilla. 2017. "Causes of death in atrial fibrillation: Challenges and opportunities." *Trends Cardiovasc. Med.* 27 (7):494–503. doi:10.1016/j.tcm.2017.05.002.

Gonzalez-Muniesa, P., M. A. Martinez-Gonzalez, F. B. Hu, J. P. Despres, Y. Matsuzawa, R. J. F. Loos, L. A. Moreno, G. A. Bray, and J. A. Martinez. 2017. "Obesity." *Nat. Rev. Dis. Prim.* 3:17034. doi:10.1038/nrdp.2017.34.

Gregor, M. F., and G. S. Hotamisligil. 2011. "Inflammatory mechanisms in obesity." *Annu. Rev. Immunol.* 29:415–45. doi:10.1146/annurev-immunol-031210-101322.

Grundy, S. M. 2008. "Metabolic syndrome pandemic." *Arterioscler. Thromb. Vasc. Biol.* 28 (4):629–36. doi:10.1161/ATVBAHA.107.151092.

Guillot, N., E. Caillet, M. Laville, C. Calzada, M. Lagarde, and E. Vericel. 2009. "Increasing intakes of the long-chain omega-3 docosahexaenoic acid: Effects on platelet functions and redox status in healthy men." *FASEB J.* 23 (9):2909–16. doi:10.1096/fj.09-133421.

Harris, W. S. 1997. "n-3 fatty acids and serum lipoproteins: Human studies." *Am. J. Clin. Nutr.* 65 (5 Suppl):1645S–54S.

Harris, W. S. 2013. "Are n-3 fatty acids still cardioprotective?" *Curr. Opin. Clin. Nutr. Metab. Care* 16 (2):141–9. doi:10.1097/MCO.0b013e32835bf380.

Harrison, D. G. 1997. "Cellular and molecular mechanisms of endothelial cell dysfunction." *J. Clin. Invest.* 100 (9):2153–7. doi:10.1172/JCI119751.

Henson, P. M. 2005. "Dampening inflammation." *Nat. Immunol.* 6 (12):1179–81. doi:10.1038/ni1205-1179.

Higashi, Y., K. Noma, M. Yoshizumi, and Y. Kihara. 2009. "Endothelial function and oxidative stress in cardiovascular diseases." *Circ. J.* 73 (3):411–8.

Hirafuji, M., T. Machida, M. Tsunoda, A. Miyamoto, and M. Minami. 2002. "Docosahexaenoic acid potentiates interleukin-1beta induction of nitric oxide synthase through mechanism involving p44/42 MAPK activation in rat vascular smooth muscle cells." *Br. J. Pharmacol.* 136 (4):613–9. doi:10.1038/sj.bjp.0704768.

Hirai, A., T. Hamazaki, T. Terano, T. Nishikawa, Y. Tamura, A. Kamugai, and J. Jajiki. 1980. "Eicosapentaenoic acid and platelet function in Japanese." *Lancet* 2 (8204):1132–3.

Honda, K. L., S. Lamon-Fava, N. R. Matthan, D. Wu, and A. H. Lichtenstein. 2015. "Docosahexaenoic acid differentially affects TNFalpha and IL-6 expression in LPS-stimulated RAW 264.7 murine macrophages." *Prostaglandins Leukot. Essent. Fatty Acids* 97:27–34. doi:10.1016/j.plefa.2015.03.002.

Horton, J. D., J. L. Goldstein, and M. S. Brown. 2002. "SREBPs: Activators of the complete program of cholesterol and fatty acid synthesis in the liver." *J. Clin. Invest.* 109 (9):1125–31. doi:10.1172/JCI15593.

Hoshi, T., Y. Tian, R. Xu, S. H. Heinemann, and S. Hou. 2013a. "Mechanism of the modulation of BK potassium channel complexes with different auxiliary subunit compositions by the omega-3 fatty acid DHA." *Proc. Natl. Acad. Sci. U. S. A.* 110 (12):4822–7. doi:10.1073/pnas.1222003110.

Hoshi, T., B. Wissuwa, Y. Tian, N. Tajima, R. Xu, M. Bauer, S. H. Heinemann, and S. Hou. 2013b. "Omega-3 fatty acids lower blood pressure by directly activating large-conductance Ca(2)(+)-dependent K(+) channels." *Proc. Natl. Acad. Sci. U. S. A.* 110 (12):4816–21. doi:10.1073/pnas.1221997110.

Hu, X. F., B. D. Laird, and H. M. Chan. 2016. "Mercury diminishes the cardiovascular protective effect of omega-3 polyunsaturated fatty acids in the modern diet of Inuit in Canada." *Environ. Res.* 152:470–7. doi:10.1016/j.envres.2016.06.001.

Huang, C. W., Y. S. Chien, Y. J. Chen, K. M. Ajuwon, H. M. Mersmann, and S. T. Ding. 2016. "Role of n-3 polyunsaturated fatty acids in Ameliorating the Obesity-Induced Metabolic Syndrome in Animal Models and Humans." *Int. J. Mol. Sci.* 17 (10). doi:10.3390/ijms17101689.

Hulthe, J., L. Bokemark, J. Wikstrand, and B. Fagerberg. 2000. "The metabolic syndrome, LDL particle size, and atherosclerosis: The Atherosclerosis and insulin Resistance (AIR) study." *Arterioscler. Thromb. Vasc. Biol.* 20 (9):2140–7.

Ibrahim, A., K. Mbodji, A. Hassan, M. Aziz, N. Boukhettala, M. Coeffier, G. Savoye, P. Dechelotte, and R. Marion-Letellier. 2011. "Anti-inflammatory and anti-angiogenic effect of long chain n-3 polyunsaturated fatty acids in intestinal microvascular endothelium." *Clin. Nutr.* 30 (5):678–87. doi:10.1016/j.clnu.2011.05.002.

Ichimura, A., A. Hirasawa, O. Poulain-Godefroy, A. Bonnefond, T. Hara, L. Yengo, I. Kimura, et al. 2012. "Dysfunction of lipid sensor GPR120 leads to obesity in both mouse and human." *Nature* 483 (7389):350–4. doi:10.1038/nature10798.

Ichimura, A., T. Hara, and A. Hirasawa. 2014. "Regulation of energy homeostasis via GPR120." *Front. Endocrinol. (Lausanne)* 5 (111). doi:10.3389/fendo.2014.00111.

Im, D. S. 2016. "Functions of omega-3 fatty acids and FFA4 (GPR120) in macrophages." *Eur. J. Pharmacol.* 785:36–43. doi:10.1016/j.ejphar.2015.03.094.

Islam, M. A., F. Alam, M. Solayman, M. I. Khalil, M. A. Kamal, and S. H. Gan. 2016. "Dietary phytochemicals: Natural swords combating inflammation and oxidation-mediated degenerative diseases." *Oxid. Med. Cell. Longev.* 2016:5137431. doi:10.1155/2016/5137431.

Jahangiri, A., W. R. Leifert, K. L. Kind, and E. J. McMurchie. 2006. "Dietary fish oil alters cardiomyocyte Ca2+ dynamics and antioxidant status." *Free Radic. Biol. Med.* 40 (9):1592–602. doi:10.1016/j.freeradbiomed.2005.12.026.

James, P. A., S. Oparil, B. L. Carter, W. C. Cushman, C. Dennison-Himmelfarb, J. Handler, D. T. Lackland, et al. 2014. "2014 Evidence-based guideline for the management of high blood pressure in adults: Report from the panel members appointed to the Eighth Joint National Committee (JNC 8)." *JAMA* 311 (5):507–20. doi:10.1001/jama.2013.284427.

Jeansen, S., R. F. Witkamp, J. A. Garthoff, A. van Helvoort, and P. C. Calder. 2018. "Fish oil LC-PUFAs do not affect blood coagulation parameters and bleeding manifestations: Analysis of 8 clinical studies with selected patient groups on omega-3-enriched medical nutrition." *Clin. Nutr.* doi:10.1016/j.clnu.2017.03.027.

Jellinger, P. S., Y. Handelsman, P. D. Rosenblit, Z. T. Bloomgarden, V. A. Fonseca, A. J. Garber, G. Grunberger, et al. 2017. "American Association of Clinical Endocrinologists and American College of endocrinology guidelines for management of dyslipidemia and prevention of cardiovascular disease." *Endocr. Pract.* 23 (Suppl 2):1–87. doi:10.4158/EP171764.APPGL.

Jong-Ming Pang, B., and M. S. Green. 2017. "Epidemiology of ventricular tachyarrhythmia: Any changes in the past decades?" *Herzschrittmacherther. Elektrophysiol.* 28 (2):143–8. doi:10.1007/s00399-017-0503-5.

Jump, D. B., and S. D. Clarke. 1999. "Regulation of gene expression by dietary fat." *Annu. Rev. Nutr.* 19:63–90. doi:10.1146/annurev.nutr.19.1.63.

Jung, F., C. Schulz, F. Blaschke, D. N. Muller, C. Mrowietz, R. P. Franke, A. Lendlein, and W. H. Schunck. 2012. "Effect of cytochrome P450-dependent epoxyeicosanoids on Ristocetin-induced thrombocyte aggregation." *Clin. Hemorheol. Microcirc.* 52 (2–4): 403–16. doi:10.3233/CH-2012-1614.

Kanaporis, G., and L. A. Blatter. 2017. "Alternans in atria: Mechanisms and clinical relevance." *Medicina (Kaunas)* 53 (3):139–49. doi:10.1016/j.medici.2017.04.004.

Katsuma, S., N. Hatae, T. Yano, Y. Ruike, M. Kimura, A. Hirasawa, and G. Tsujimoto. 2005. "Free fatty acids inhibit serum deprivation-induced apoptosis through GPR120 in a murine enteroendocrine cell line STC-1." *J. Biol. Chem.* 280 (20):19507–15. doi:10.1074/jbc.M412385200.

Kaur, J., R. Adya, B. K. Tan, J. Chen, and H. S. Randeva. 2010. "Identification of chemerin receptor (ChemR23) in human endothelial cells: Chemerin-induced endothelial angiogenesis." *Biochem. Biophys. Res. Commun.* 391 (4):1762–8. doi:10.1016/j.bbrc.2009.12.150.

Knapp, H. R., I. A. Reilly, P. Alessandrini, and G. A. FitzGerald. 1986. "In vivo indexes of platelet and vascular function during fish-oil administration in patients with atherosclerosis." *N. Engl. J. Med.* 314 (15):937–42. doi:10.1056/NEJM198604103141501.

Kondo, K., K. Morino, Y. Nishio, M. Kondo, T. Fuke, S. Ugi, H. Iwakawa, A. Kashiwagi, and H. Maegawa. 2010. "Effects of a fish-based diet on the serum adiponectin concentration in young, non-obese, healthy Japanese subjects." *J. Atheroscler. Thromb.* 17 (6):628–37.

Kris-Etherton, P. M., W. S. Harris, L. J. Appel, and Nutrition Committee. 2003. "Fish consumption, fish oil, omega-3 fatty acids, and cardiovascular disease." *Arterioscler. Thromb. Vasc. Biol.* 23 (2):e20–30.

Kristensen, S. D., E. B. Schmidt, H. R. Andersen, and J. Dyerberg. 1987. "Fish oil in angina pectoris." *Atherosclerosis* 64 (1):13–9.

Kristensen, S. D., E. B. Schmidt, and J. Dyerberg. 1989. "Dietary supplementation with n-3 polyunsaturated fatty acids and human platelet function: A review with particular emphasis on implications for cardiovascular disease." *J. Intern. Med. Suppl.* 731:141–50.

Kuhn, H., J. Saam, S. Eibach, H. G. Holzhutter, I. Ivanov, and M. Walther. 2005. "Structural biology of mammalian lipoxygenases: Enzymatic consequences of targeted alterations of the protein structure." *Biochem. Biophys. Res. Commun.* 338 (1):93–101. doi:10.1016/j.bbrc.2005.08.238.

Kumar, A., Y. Takada, A. M. Boriek, and B. B. Aggarwal. 2004. "Nuclear factor-kappaB: Its role in health and disease." *J. Mol. Med. (Berl.)* 82 (7):434–48. doi:10.1007/s00109-004-0555-y.

Labreuche, J., P. J. Touboul, and P. Amarenco. 2009. "Plasma triglyceride levels and risk of stroke and carotid atherosclerosis: A systematic review of the epidemiological studies." *Atherosclerosis* 203 (2):331–45. doi:10.1016/j.atherosclerosis.2008.08.040.

Lahdenpera, S., M. Syvanne, J. Kahri, and M. R. Taskinen. 1996. "Regulation of low-density lipoprotein particle size distribution in NIDDM and coronary disease: Importance of serum triglycerides." *Diabetologia* 39 (4):453–61.

Lai, L. H., R. X. Wang, W. P. Jiang, X. J. Yang, J. P. Song, X. R. Li, and G. Tao. 2009. "Effects of docosahexaenoic acid on large-conductance Ca2+-activated K+ channels and voltage-dependent K+ channels in rat coronary artery smooth muscle cells." *Acta Pharmacol. Sin.* 30 (3):314–20. doi:10.1038/aps.2009.7.

Landstrom, A. P., D. Dobrev, and X. H. T. Wehrens. 2017. "Calcium Signaling and Cardiac Arrhythmias." *Circ. Res.* 120 (12):1969–93. doi:10.1161/CIRCRESAHA.117.310083.

Larson, M. K., G. C. Shearer, J. H. Ashmore, J. M. Anderson-Daniels, E. L. Graslie, J. T. Tholen, J. L. Vogelaar, et al. 2011. "Omega-3 fatty acids modulate collagen signaling in human platelets." *Prostaglandins Leukot. Essent. Fatty Acids* 84 (3–4):93–8. doi:10.1016/j.plefa.2010.11.004.

Laurent, S. 2017. "Antihypertensive drugs." *Pharmacol. Res.* 124:116–25. doi:10.1016/j.phrs.2017.07.026.

Lawson, D. L., J. L. Mehta, K. Saldeen, P. Mehta, and T. G. Saldeen. 1991. "Omega-3 polyunsaturated fatty acids augment endothelium-dependent vasorelaxation by enhanced release of EDRF and vasodilator prostaglandins." *Eicosanoids* 4 (4):217–23.

Lawton, W., and K. Chatterjee. 2013. "Antihhypertensive drugs." In *Cardiac drugs: an evidence-based approach*, 478. New Delhi 110 002, India: Jaypee Brothers Medical Publishers (P) Ltd.

Leaf, A., C. M. Albert, M. Josephson, D. Steinhaus, J. Kluger, J. X. Kang, B. Cox, et al. 2005. "Prevention of fatal arrhythmias in high-risk subjects by fish oil n-3 fatty acid intake." *Circulation* 112 (18):2762–8. doi:10.1161/CIRCULATIONAHA.105.549527.

Lee, J. Y., K. H. Sohn, S. H. Rhee, and D. Hwang. 2001. "Saturated fatty acids, but not unsaturated fatty acids, induce the expression of cyclooxygenase-2 mediated through Toll-like receptor 4." *J. Biol. Chem.* 276 (20):16683–9. doi:10.1074/jbc.M011695200.

Lee, J. Y., L. Zhao, H. S. Youn, A. R. Weatherill, R. Tapping, L. Feng, W. H. Lee, K. A. Fitzgerald, and D. H. Hwang. 2004. "Saturated fatty acid activates but polyunsaturated fatty acid inhibits Toll-like receptor 2 dimerized with Toll-like receptor 6 or 1." *J. Biol. Chem.* 279 (17):16971–9. doi:10.1074/jbc.M312990200.

Levy, E., S. Spahis, E. Ziv, A. Marette, M. Elchebly, M. Lambert, and E. Delvin. 2006. "Overproduction of intestinal lipoprotein containing apolipoprotein B-48 in Psammomys obesus: Impact of dietary n-3 fatty acids." *Diabetologia* 49 (8):1937–45. doi:10.1007/s00125-006-0315-3.

Li, R. Y. 2015. "Overview of cardiac arrhythmias and drug therapy." In *Cardiovascular diseases: from molecular pharmacology to evidence-based therapeutics*, 504, Hoboken, New Jersey: John Wiley & Sons, Inc.

Li, G. R., H. Y. Sun, X. H. Zhang, L. C. Cheng, S. W. Chiu, H. F. Tse, and C. P. Lau. 2009. "Omega-3 polyunsaturated fatty acids inhibit transient outward and ultra-rapid delayed rectifier K+currents and Na+current in human atrial myocytes." *Cardiovasc. Res.* 81 (2):286–93. doi:10.1093/cvr/cvn322.

Li, Z., M. K. Delaney, K. A. O'Brien, and X. Du. 2010. "Signaling during platelet adhesion and activation." *Arterioscler. Thromb. Vasc. Biol.* 30 (12):2341–9. doi:10.1161/ATVBAHA.110.207522.

Li, X., Y. Yu, and C. D. Funk. 2013. "Cyclooxygenase-2 induction in macrophages is modulated by docosahexaenoic acid via interactions with free fatty acid receptor 4 (FFA4)." *FASEB J.* 27 (12):4987–97. doi:10.1096/fj.13-235333.

Liao, Z., J. Dong, W. Wu, T. Yang, T. Wang, L. Guo, L. Chen, D. Xu, and F. Wen. 2012. "Resolvin D1 attenuates inflammation in lipopolysaccharide-induced acute lung injury through a process involving the PPARgamma/NF-kappaB pathway." *Respir. Res.* 13:110. doi:10.1186/1465-9921-13-110.

Lira, F. S., J. C. Rosa, G. D. Pimentel, V. A. Tarini, R. M. Arida, F. Faloppa, E. S. Alves, et al. 2010. "Inflammation and adipose tissue: Effects of progressive load training in rats." *Lipids Health Dis.* 9:109. doi:10.1186/1476-511X-9-109.

Liu, D., L. Wang, Q. Meng, H. Kuang, and X. Liu. 2012. "G-protein coupled receptor 120 is involved in glucose metabolism in fat cells." *Cell. Mol. Biol. (Noisy-le-grand)*:OL1757.

Luscher, T. F. 2015. "Ageing, inflammation, and oxidative stress: Final common pathways of cardiovascular disease." *Eur. Heart J.* 36 (48):3381–3. doi:10.1093/eurheartj/ehv679.

Maltzman, J. S., J. A. Carmen, and J. G. Monroe. 1996. "Transcriptional regulation of the Icam-1 gene in antigen receptor- and phorbol ester-stimulated B lymphocytes: Role for transcription factor EGR1." *J. Exp. Med.* 183 (4):1747–59.

Marchioli, R., F. Barzi, E. Bomba, C. Chieffo, D. Di Gregorio, R. Di Mascio, M. G. Franzosi, et al. 2002. "ISSI-Prevenzione Investigators." *Circulation* 105 (16):1897–903.

Matsumoto, M., M. Sata, D. Fukuda, K. Tanaka, M. Soma, Y. Hirata, and R. Nagai. 2008. "Orally administered eicosapentaenoic acid reduces and stabilizes atherosclerotic lesions in ApoE-deficient mice." *Atherosclerosis* 197 (2):524–33. doi:10.1016/j.atherosclerosis.2007.07.023.

McLennan, P. L. 2014. "Cardiac physiology and clinical efficacy of dietary fish oil clarified through cellular mechanisms of omega-3 polyunsaturated fatty acids." *Eur. J. Appl. Physiol.* 114 (7):1333–56. doi:10.1007/s00421-014-2876-z.

Michaud, S. E., and G. Renier. 2001. "Direct regulatory effect of fatty acids on macrophage lipoprotein lipase: Potential role of PPARs." *Diabetes* 50 (3):660–6.

Miles, E. A., F. Thies, F. A. Wallace, J. R. Powell, T. L. Hurst, E. A. Newsholme, and P. C. Calder. 2001. "Influence of age and dietary fish oil on plasma soluble adhesion molecule concentrations." *Clin. Sci. (Lond.)* 100 (1):91–100.

Miller, P. E., M. Van Elswyk, and D. D. Alexander. 2014. "Long-chain omega-3 fatty acids eicosapentaenoic acid and docosahexaenoic acid and blood pressure: A meta-analysis of randomized controlled trials." *Am. J. Hypertens.* 27 (7):885–96. doi:10.1093/ajh/hpu024.

Minihane, A. M. 2013. "Fish oil omega-3 fatty acids and cardio-metabolic health, alone or with statins." *Eur. J. Clin. Nutr.* 67 (5):536–40. doi:10.1038/ejcn.2013.19.

Miyamoto, J., S. Hasegawa, M. Kasubuchi, A. Ichimura, A. Nakajima, and I. Kimura. 2016. "Nutritional signaling via free fatty acid receptors." *Int. J. Mol. Sci.* 17 (4):450. doi:10.3390/ijms17040450.

Mohrhauer, H., K. Christiansen, M. V. Gan, M. Deubig, and R. T. Holman. 1967. "Chain elongation of linoleic acid and its inhibition by other fatty acids in vitro." *J. Biol. Chem.* 242 (19):4507–14.

Moniri, N. H. 2016. "Free-fatty acid receptor-4 (GPR120): Cellular and molecular function and its role in metabolic disorders." *Biochem. Pharmacol.* 110–111 (111):1–15. doi:10.1016/j.bcp.2016.01.021.

Moreno, C., A. Macías, A. Prieto, A. de la Cruz, T. González, and C. Valenzuela. 2012. "Effects of n-3 polyunsaturated fatty acids on Cardiac Ion Channels." *Front. Physiol.* 3 (245):245. doi:10.3389/fphys.2012.00245.

Mori, T. A. 2006. "Omega-3 fatty acids and hypertension in humans." *Clin. Exp. Pharmacol. Physiol.* 33 (9):842–6. doi:10.1111/j.1440-1681.2006.04451.x.

Mori, T. A. 2014. "Dietary n-3 PUFA and CVD: A review of the evidence." *Proc. Nutr. Soc.* 73 (1):57–64. doi:10.1017/S0029665113003583.

Mori, T. A., L. J. Beilin, V. Burke, J. Morris, and J. Ritchie. 1997. "Interactions between dietary fat, fish, and fish oils and their effects on platelet function in men at risk of cardiovascular disease." *Arterioscler. Thromb. Vasc. Biol.* 17 (2):279–86.

Morton, A. M., J. D. Furtado, J. Lee, W. Amerine, M. H. Davidson, and F. M. Sacks. 2016. "The effect of omega-3 carboxylic acids on apolipoprotein CIII-containing lipoproteins in severe hypertriglyceridemia." *J. Clin. Lipidol.* 10 (6):1442–1451.e4. doi:10.1016/j.jacl.2016.09.005.

Mozaffarian, D., and J. H. Wu. 2011. "Omega-3 fatty acids and cardiovascular disease: Effects on risk factors, molecular pathways, and clinical events." *J. Am. Coll. Cardiol.* 58 (20):2047–67. doi:10.1016/j.jacc.2011.06.063.

Mozaffarian, D., and J. H. Wu. 2012. "(n-3) fatty acids and cardiovascular health: Are effects of EPA and DHA shared or complementary?" *J. Nutr.* 142 (3):614S–25S. doi:10.3945/jn.111.149633.

Mozaffarian, D., B. M. Psaty, E. B. Rimm, R. N. Lemaitre, G. L. Burke, M. F. Lyles, D. Lefkowitz, and D. S. Siscovick. 2004. "Fish intake and risk of incident atrial fibrillation." *Circulation* 110 (4):368–73. doi:10.1161/01.CIR.0000138154.00779.A5.

Mozaffarian, D., A. Geelen, I. A. Brouwer, J. M. Geleijnse, P. L. Zock, and M. B. Katan. 2005. "Effect of fish oil on heart rate in humans: A meta-analysis of randomized controlled trials." *Circulation* 112 (13):1945–52. doi:10.1161/CIRCULATIONAHA.105.556886.

Nair, S. S., J. Leitch, and M. L. Garg. 2000. "Suppression of inositol phosphate release by cardiac myocytes isolated from fish oil-fed pigs." *Mol. Cell. Biochem.* 215 (1–2):57–64.

Nettleton, J. A., and R. Katz. 2005. "n-3 long-chain polyunsaturated fatty acids in type 2 diabetes: A review." *J. Am Diet Assoc.* 105 (3):428–40. doi:10.1016/j.jada.2004.11.029.

Oh, D. Y., S. Talukdar, E. J. Bae, T. Imamura, H. Morinaga, W. Fan, P. Li, et al. 2010. "GPR120 is an omega-3 fatty acid receptor mediating potent anti-inflammatory and insulin-sensitizing effects." *Cell* 142 (5):687–98. doi:10.1016/j.cell.2010.07.041.

Ohira, T., M. Arita, K. Omori, A. Recchiuti, T. E. Van Dyke, and C. N. Serhan. 2010. "Resolvin E1 receptor activation signals phosphorylation and phagocytosis." *J. Biol. Chem.* 285 (5):3451–61. doi:10.1074/jbc.M109.044131.

Ooi, E. M., T. W. Ng, G. F. Watts, D. C. Chan, and P. H. Barrett. 2012. "Effect of fenofibrate and atorvastatin on VLDL apoE metabolism in men with the metabolic syndrome." *J. Lipid Res.* 53 (11):2443–9. doi:10.1194/jlr.P029223.

Ou, J., H. Tu, B. Shan, A. Luk, R. A. DeBose-Boyd, Y. Bashmakov, J. L. Goldstein, and M. S. Brown. 2001. "Unsaturated fatty acids inhibit transcription of the sterol regulatory element-binding protein-1c (SREBP-1c) gene by antagonizing ligand-dependent activation of the LXR." *Proc. Natl. Acad. Sci. U. S. A.* 98 (11):6027–32. doi:10.1073/pnas.111138698.

Ozdener, M. H., S. Subramaniam, S. Sundaresan, O. Sery, T. Hashimoto, Y. Asakawa, P. Besnard, N. A. Abumrad, and N. A. Khan. 2014. "CD36- and GPR120-mediated Ca(2)(+) signaling in human taste bud cells mediates differential responses to fatty acids and is altered in obese mice." *Gastroenterology* 146 (4):995–1005. doi:10.1053/j.gastro.2014.01.006.

Pegorier, J. P., C. Le May, and J. Girard. 2004. "Control of gene expression by fatty acids." *J. Nutr.* 134 (9):2444S–9S.

Peyron-Caso, E., M. Taverna, M. Guerre-Millo, A. Veronese, N. Pacher, G. Slama, and S. W. Rizkalla. 2002. "Dietary (n-3) polyunsaturated fatty acids up-regulate plasma leptin in insulin-resistant rats." *J. Nutr.* 132 (8):2235–40.

Phillipson, B. E., D. W. Rothrock, W. E. Connor, W. S. Harris, and D. R. Illingworth. 1985. "Reduction of plasma lipids, lipoproteins, and apoproteins by dietary fish oils in patients with hypertriglyceridemia." *N. Engl. J. Med.* 312 (19):1210–6. doi:10.1056/NEJM198505093121902.

Pirola, L., and J. C. Ferraz. 2017. "Role of pro- and anti-inflammatory phenomena in the physiopathology of type 2 diabetes and obesity." *World J. Biol. Chem.* 8 (2):120–8. doi:10.4331/wjbc.v8.i2.120.

Postic, C., R. Dentin, P. D. Denechaud, and J. Girard. 2007. "ChREBP, a transcriptional regulator of glucose and lipid metabolism." *Annu. Rev. Nutr.* 27:179–92. doi:10.1146/annurev.nutr.27.061406.093618.

Prieur, X., H. Coste, and J. C. Rodriguez. 2003. "The human apolipoprotein AV gene is regulated by peroxisome proliferator-activated receptor-alpha and contains a novel farnesoid X-activated receptor response element." *J. Biol. Chem.* 278 (28):25468–80. doi:10.1074/jbc.M301302200.

Raimondi, L., M. Lodovici, F. Visioli, L. Sartiani, L. Cioni, C. Alfarano, G. Banchelli, et al. 2005. "n-3 polyunsaturated fatty acids supplementation decreases asymmetric dimethyl arginine and arachidonate accumulation in aging spontaneously hypertensive rats." *Eur. J. Nutr.* 44 (6):327–33. doi:10.1007/s00394-004-0528-5.

Rask-Madsen, C., and C. R. Kahn. 2012. "Tissue-specific insulin signaling, metabolic syndrome, and cardiovascular disease." *Arterioscler. Thromb. Vasc. Biol.* 32 (9):2052–9. doi:10.1161/ATVBAHA.111.241919.

Ravussin, E., and S. R. Smith. 2002. "Increased fat intake, impaired fat oxidation, and failure of fat cell proliferation result in ectopic fat storage, insulin resistance, and type 2 diabetes mellitus." *Ann. N Y Acad. Sci.* 967:363–78.

Reiffel, J. A., and A. McDonald. 2006. "Antiarrhythmic effects of omega-3 fatty acids." *Am. J. Cardiol.* 98 (4A):50i–60i. doi:10.1016/j.amjcard.2005.12.027.

Rimbach, G., G. Valacchi, R. Canali, and F. Virgili. 2000. "Macrophages stimulated with IFN-gamma activate NF-kappa B and induce MCP-1 gene expression in primary human endothelial cells." *Mol. Cell Biol. Res. Commun.* 3 (4):238–42. doi:10.1006/mcbr.2000.0219.

Rinaldi, B., P. Di Pierro, M. R. Vitelli, M. D'Amico, L. Berrino, F. Rossi, and A. Filippelli. 2002. "Effects of docosahexaenoic acid on calcium pathway in adult rat cardiomyocytes." *Life Sci.* 71 (9):993–1004.

Robinson, J. G., and N. J. Stone. 2006. "Antiatherosclerotic and antithrombotic effects of omega-3 fatty acids." *Am. J. Cardiol.* 98 (4A):39i–49i. doi:10.1016/j.amjcard.2005.12.026.

Rochlani, Y., N. V. Pothineni, S. Kovelamudi, and J. L. Mehta. 2017. "Metabolic syndrome: Pathophysiology, management, and modulation by natural compounds." *Ther. Adv. Cardiovasc. Dis.* 11 (8):215–25. doi:10.1177/1753944717711379.

Rodriguez-Pacheco, F., S. Garcia-Serrano, E. Garcia-Escobar, C. Gutierrez-Repiso, J. Garcia-Arnes, S. Valdes, M. Gonzalo, et al. 2014. "Effects of obesity/fatty acids on the expression of GPR120." *Mol. Nutr. Food Res.* 58 (9):1852–60. doi:10.1002/mnfr.201300666.

Romacho, T., P. Glosse, I. Richter, M. Elsen, M. H. Schoemaker, E. A. van Tol, and J. Eckel. 2015. "Nutritional ingredients modulate adipokine secretion and inflammation in human primary adipocytes." *Nutrients* 7 (2):865–86. doi:10.3390/nu7020865.

Rossier, B. C., M. Bochud, and O. Devuyst. 2017. "The hypertension pandemic: An evolutionary perspective." *Physiology (Bethesda)* 32 (2):112–25. doi:10.1152/physiol.00026.2016.

Roy, J., and J. Y. Le Guennec. 2017. "Cardioprotective effects of omega 3 fatty acids: Origin of the variability." *J. Muscle Res. Cell Motil.* 38 (1):25–30. doi:10.1007/s10974-016-9459-z.

Roy, J., C. Oger, J. Thireau, J. Roussel, O. Mercier-Touzet, D. Faure, E. Pinot, et al. 2015. "Nonenzymatic lipid mediators, neuroprostanes, exert the antiarrhythmic properties of docosahexaenoic acid." *Free Radic. Biol. Med.* 86:269–78. doi:10.1016/j.freeradbiomed.2015.04.014.

Sacks, F. M., C. Zheng, and J. S. Cohn. 2011. "Complexities of plasma apolipoprotein C-III metabolism." *J. Lipid Res.* 52 (6):1067–70. doi:10.1194/jlr.E015701.

Schmid, M., C. Gemperle, N. Rimann, and M. Hersberger. 2016. "Resolvin D1 polarizes primary human macrophages toward a proresolution phenotype through GPR32." *J. Immunol.* 196 (8):3429–37. doi:10.4049/jimmunol.1501701.

Schwartz, E. A., and P. D. Reaven. 2012. "Lipolysis of triglyceride-rich lipoproteins, vascular inflammation, and atherosclerosis." *Biochim. Biophys. Acta* 1821 (5):858–66. doi:10.1016/j.bbalip.2011.09.021.

Serhan, C. N., S. Hong, K. Gronert, S. P. Colgan, P. R. Devchand, G. Mirick, and R. L. Moussignac. 2002. "Resolvins: A family of bioactive products of omega-3 fatty acid transformation circuits initiated by aspirin treatment that counter proinflammation signals." *J. Exp. Med.* 196 (8):1025–37.

Serhan, C. N., S. D. Brain, C. D. Buckley, D. W. Gilroy, C. Haslett, L. A. O'Neill, M. Perretti, A. G. Rossi, and J. L. Wallace. 2007. "Resolution of inflammation: State of the art, definitions and terms." *FASEB J.* 21 (2):325–32. doi:10.1096/fj.06-7227rev.

Serhan, C. N., N. Chiang, and T. E. Van Dyke. 2008a. "Resolving inflammation: Dual anti-inflammatory and pro-resolution lipid mediators." *Nat. Rev. Immunol.* 8 (5):349–61. doi:10.1038/nri2294.

Serhan, C. N., S. Yacoubian, and R. Yang. 2008b. "Anti-inflammatory and proresolving lipid mediators." *Annu. Rev. Pathol.* 3:279–312. doi:10.1146/annurev.pathmechdis.3.121806.151409.

Shrout, T., D. W. Rudy, and M. T. Piascik. 2017. "Hypertension update, JNC8 and beyond." *Curr. Opin. Pharmacol.* 33:41–6. doi:10.1016/j.coph.2017.03.004.

Simmons, D. L., R. M. Botting, and T. Hla. 2004. "Cyclooxygenase isozymes: The biology of prostaglandin synthesis and inhibition." *Pharmacol. Rev.* 56 (3):387–437. doi:10.1124/pr.56.3.3.

Simopoulos, A. P. 1991. "Omega-3 fatty acids in health and disease and in growth and development." *Am. J. Clin. Nutr.* 54 (3):438–63.

Simopoulos, A. P. 2006. "Evolutionary aspects of diet, the omega-6/omega-3 ratio and genetic variation: Nutritional implications for chronic diseases." *Biomed. Pharmacother.* 60 (9):502–7. doi:10.1016/j.biopha.2006.07.080.

Singleton, C. B., S. M. Valenzuela, B. D. Walker, H. Tie, K. R. Wyse, J. A. Bursill, M. R. Qiu, S. N. Breit, and T. J. Campbell. 1999. "Blockade by N-3 polyunsaturated fatty acid of the Kv4.3 current stably expressed in Chinese hamster ovary cells." *Br. J. Pharmacol.* 127 (4):941–8. doi:10.1038/sj.bjp.0702638.

Skulas-Ray, A. C., P. Alaupovic, P. M. Kris-Etherton, and S. G. West. 2015. "Dose-response effects of marine omega-3 fatty acids on apolipoproteins, apolipoprotein-defined lipoprotein subclasses, and Lp-PLA2 in individuals with moderate hypertriglyceridemia." *J. Clin. Lipidol.* 9 (3):360–7. doi:10.1016/j.jacl.2014.12.001.

Smith, W. L. 2005. "Cyclooxygenases, peroxide tone and the allure of fish oil." *Curr. Opin. Cell Biol.* 17 (2):174–82. doi:10.1016/j.ceb.2005.02.005.

Song, Y. J., P. S. Dong, H. L. Wang, J. H. Zhu, S. Y. Xing, Y. H. Han, R. X. Wang, and W. P. Jiang. 2013. "Regulatory functions of docosahexaenoic acid on ion channels in rat ventricular myocytes." *Eur. Rev. Med. Pharmacol. Sci.* 17 (19):2632–8.

Song, T., Y. Yang, Y. Zhou, H. Wei, and J. Peng. 2017. "GPR120: A critical role in adipogenesis, inflammation, and energy metabolism in adipose tissue." *Cell. Mol. Life Sci.* 74 (15):2723–33. doi:10.1007/s00018-017-2492-2.

Sprecher, H., Q. Chen, and F. Q. Yin. 1999. "Regulation of the biosynthesis of 22:5n-6 and 22:6n-3: A complex intracellular process." *Lipids* 34 (Suppl):S153–6.

Stebbins, C. L., J. P. Stice, C. M. Hart, F. N. Mbai, and A. A. Knowlton. 2008. "Effects of dietary decosahexaenoic acid (DHA) on eNOS in human coronary artery endothelial cells." *J. Cardiovasc. Pharmacol. Ther.* 13 (4):261–8. doi:10.1177/1074248408322470.

Stillwell, W., S. R. Shaikh, M. Zerouga, R. Siddiqui, and S. R. Wassall. 2005. "Docosahexaenoic acid affects cell signaling by altering lipid rafts." *Reprod. Nutr. Dev.* 45 (5):559–79. doi:10.1051/rnd:2005046.

Strong, J. P., G. T. Malcom, C. A. McMahan, R. E. Tracy, W. P. Newman, 3rd, E. E. Herderick, and J. F. Cornhill. 1999. "Prevalence and extent of atherosclerosis in adolescents and young adults: Implications for prevention from the pathobiological Determinants of Atherosclerosis in Youth Study." *JAMA* 281 (8):727–35.

Sun, K., I. Wernstedt Asterholm, C. M. Kusminski, A. C. Bueno, Z. V. Wang, J. W. Pollard, R. A. Brekken, and P. E. Scherer. 2012. "Dichotomous effects of VEGF-A on adipose tissue dysfunction." *Proc. Natl. Acad. Sci. U. S. A.* 109 (15):5874–9. doi:10.1073/pnas.1200447109.

Szostak-Wegierek, D., L. Klosiewicz-Latoszek, W. B. Szostak, and B. Cybulska. 2013. "The role of dietary fats for preventing cardiovascular disease. A review." *Rocz. Panstw. Zakl. Hig* 64 (4):263–9.

Taneera, J., S. Lang, A. Sharma, J. Fadista, Y. Zhou, E. Ahlqvist, A. Jonsson, et al. 2012. "A systems genetics approach identifies genes and pathways for type 2 diabetes in human islets." *Cell Metab.* 16 (1):122–34. doi:10.1016/j.cmet.2012.06.006.

Terano, T., J. A. Salmon, and S. Moncada. 1984. "Biosynthesis and biological activity of leukotriene B5." *Prostaglandins* 27 (2):217–32.

Thies, F., J. M. Garry, P. Yaqoob, K. Rerkasem, J. Williams, C. P. Shearman, P. J. Gallagher, P. C. Calder, and R. F. Grimble. 2003. "Association of n-3 polyunsaturated fatty acids with stability of atherosclerotic plaques: A randomised controlled trial." *Lancet* 361 (9356):477–85. doi:10.1016/S0140-6736(03)12468-3.

Thorngren, M., and A. Gustafson. 1981. "Effects of 11-week increases in dietary eicosapentaenoic acid on bleeding time, lipids, and platelet aggregation." *Lancet* 2 (8257):1190–3.

Tortosa-Caparrós, E., D. Navas-Carrillo, F. Marín, and E. Orenes-Piñero. 2017. "Anti-inflammatory effects of omega 3 and omega 6 polyunsaturated fatty acids in cardiovascular disease and metabolic syndrome." *Crit. Rev. Food Sci. Nutr.* 57 (16):3421–9. doi: 10.1080/10408398.2015.1126549.

Trimarco, B. 2012. "Omega-3 in antiarrhythmic therapy : Pros position." *High Blood Press. Cardiovasc. Prev.* 19 (4):201–5. doi:10.1007/BF03297631.

Tune, J. D., A. G. Goodwill, D. J. Sassoon, and K. J. Mather. 2017. "Cardiovascular consequences of metabolic syndrome." *Transl. Res.* 183:57–70. doi:10.1016/j.trsl.2017.01.001.

Turk, H. F., and R. S. Chapkin. 2013. "Membrane lipid raft organization is uniquely modified by n-3 polyunsaturated fatty acids." *Prostaglandins Leukot. Essent. Fatty Acids.* 88 (1):43–7. doi:10.1016/j.plefa.2012.03.008.

Turner, J. M., and E. S. Spatz. 2016. "Nutritional supplements for the treatment of hypertension: A practical guide for clinicians." *Curr. Cardiol. Rep.* 18 (12):126. doi:10.1007/s11886-016-0806-x.

Ueshima, H., J. Stamler, P. Elliott, Q. Chan, I. J. Brown, M. R. Carnethon, M. L. Daviglus, et al. 2007. "Food omega-3 fatty acid intake of individuals (total, linolenic acid, long-chain) and their blood pressure: INTERMAP study." *Hypertension* 50 (2):313–9. doi:10.1161/HYPERTENSIONAHA.107.090720.

Ulu, A., K. S. S. Lee, C. Miyabe, J. Yang, B. G. Hammock, H. Dong, and B. D. Hammock. 2014. "An omega-3 epoxide of docosahexaenoic acid lowers blood pressure in angiotensin-II-dependent hypertension." *J. Cardiovasc. Pharmacol.* 64 (1):87–99. doi:10.1097/FJC.0000000000000094.

Verma, S., and M. E. Hussain. 2017. "Obesity and diabetes: An update." *Diabetes Metab. Syndr.* 11 (1):73–9. doi:10.1016/j.dsx.2016.06.017.

von Schacky, C. 2000. "n-3 fatty acids and the prevention of coronary atherosclerosis." *Am. J. Clin. Nutr.* 71 (1 Suppl):224S–7S.

Vu-Dac, N., K. Schoonjans, B. Laine, J. C. Fruchart, J. Auwerx, and B. Staels. 1994. "Negative regulation of the human apolipoprotein A-I promoter by fibrates can be attenuated by the interaction of the peroxisome proliferator-activated receptor with its response element." *J. Biol. Chem.* 269 (49):31012–8.

Wada, M., C. J. DeLong, Y. H. Hong, C. J. Rieke, I. Song, R. S. Sidhu, C. Yuan, et al. 2007. "Enzymes and receptors of prostaglandin pathways with arachidonic acid-derived versus eicosapentaenoic acid-derived substrates and products." *J. Biol. Chem.* 282 (31):22254–66. doi:10.1074/jbc.M703169200.

Wallis, J. G., J. L. Watts, and J. Browse. 2002. "Polyunsaturated fatty acid synthesis: What will they think of next?" *Trends Biochem. Sci.* 27 (9):467.

Wan, J. B., L. L. Huang, R. Rong, R. Tan, J. Wang, and J. X. Kang. 2010. "Endogenously decreasing tissue n-6/n-3 fatty acid ratio reduces atherosclerotic lesions in apolipoprotein E-deficient mice by inhibiting systemic and vascular inflammation." *Arterioscler. Thromb. Vasc. Biol.* 30 (12):2487–94. doi:10.1161/ATVBAHA.110.210054.

Wang, C., M. Chung, A. Lichtenstein, E. Balk, B. Kupelnick, D. DeVine, A. Lawrence, and J. Lau. 2004. "Effects of omega-3 fatty acids on cardiovascular disease." *Evid. Rep. Technol. Assess. Summ.* 94:1–8.

Wang, C., W. S. Harris, M. Chung, A. H. Lichtenstein, E. M. Balk, B. Kupelnick, H. S. Jordan, and J. Lau. 2006. "n-3 Fatty acids from fish or fish-oil supplements, but not alpha-linolenic acid, benefit cardiovascular disease outcomes in primary- and secondary-prevention studies: A systematic review." *Am. J. Clin. Nutr.* 84 (1):5–17.

Wang, R. X., Q. Chai, T. Lu, and H. C. Lee. 2011a. "Activation of vascular BK channels by docosahexaenoic acid is dependent on cytochrome P450 epoxygenase activity." *Cardiovasc. Res.* 90 (2):344–52. doi:10.1093/cvr/cvq411.

Wang, T. M., C. J. Chen, T. S. Lee, H. Y. Chao, W. H. Wu, S. C. Hsieh, H. H. Sheu, and A. N. Chiang. 2011b. "Docosahexaenoic acid attenuates VCAM-1 expression and NF-κB activation in TNF-α-treated human aortic endothelial cells." *J. Nutr. Biochem.* 22 (2):187–94. doi:10.1016/j.jnutbio.2010.01.007.

Weiner, B. H., I. S. Ockene, P. H. Levine, H. F. Cuenoud, M. Fisher, B. F. Johnson, A. S. Daoud, et al. 1986. "Inhibition of atherosclerosis by cod-liver oil in a hyperlipidemic swine model." *N. Engl. J. Med.* 315 (14):841–6. doi:10.1056/NEJM198610023151401.

Weintraub, H. S. 2014. "Overview of prescription omega-3 fatty acid products for hypertriglyceridemia." *Postgrad. Med.* 126 (7):7–18. doi:10.3810/pgm.2014.11.2828.

WHO. 2013. *A global brief on hypertension.* Geneva 27, Switzerland: World Health Organization.

Wijesekara, N., M. Krishnamurthy, A. Bhattacharjee, A. Suhail, G. Sweeney, and M. B. Wheeler. 2010. "Adiponectin-induced ERK and Akt phosphorylation protects against pancreatic beta cell apoptosis and increases insulin gene expression and secretion." *J. Biol. Chem.* 285 (44):33623–31. doi:10.1074/jbc.M109.085084.

Wilcox, J. N., R. R. Subramanian, C. L. Sundell, W. R. Tracey, J. S. Pollock, D. G. Harrison, and P. A. Marsden. 1997. "Expression of multiple isoforms of nitric oxide synthase in normal and atherosclerotic vessels." *Arterioscler. Thromb. Vasc. Biol.* 17 (11):2479–88.

Wisely, G. B., A. B. Miller, R. G. Davis, A. D. Thornquest, Jr., R. Johnson, T. Spitzer, A. Sefler, et al. 2002. "Hepatocyte nuclear factor 4 is a transcription factor that constitutively binds fatty acids." *Structure* 10 (9):1225–34.

Wu, Y., C. Zhang, Y. Dong, S. Wang, P. Song, B. Viollet, and M. H. Zou. 2012. "Activation of the AMP-activated protein kinase by eicosapentaenoic acid (EPA, 20:5 n-3) improves endothelial function in vivo." *PLOS ONE* 7 (4). doi:10.1371/journal.pone.0035508.

Xiao, Y. F., Q. Ke, Y. Chen, J. P. Morgan, and A. Leaf. 2004. "Inhibitory effect of n-3 fish oil fatty acids on cardiac Na+/Ca2+ exchange currents in HEK293t cells." *Biochem. Biophys. Res. Commun.* 321 (1):116–23. doi:10.1016/j.bbrc.2004.06.114.

Xiao, Y. F., L. Ma, S. Y. Wang, M. E. Josephson, G. K. Wang, J. P. Morgan, and A. Leaf. 2006. "Potent block of inactivation-deficient Na+ channels by n-3 polyunsaturated fatty acids." *Am. J. Physiol. Cell Physiol.* 290 (2):C362–70. doi:10.1152/ajpcell.00296.2005.

Xu, H. E., M. H. Lambert, V. G. Montana, D. J. Parks, S. G. Blanchard, P. J. Brown, D. D. Sternbach, et al. 1999. "Molecular recognition of fatty acids by peroxisome proliferator-activated receptors." *Mol. Cell* 3 (3):397–403.

Xu, Z., N. Riediger, S. Innis, and M. H. Moghadasian. 2007. "Fish oil significantly alters fatty acid profiles in various lipid fractions but not atherogenesis in apo E-KO mice." *Eur. J. Nutr.* 46 (2):103–10. doi:10.1007/s00394-006-0638-3.

Yalta, T., and K. Yalta. 2018. "Systemic inflammation and arrhythmogenesis: A review of mechanistic and clinical perspectives." *Angiology*:3319717709380. doi:10.1177/0003319717709380.

Yamada, H., M. Yoshida, Y. Nakano, T. Suganami, N. Satoh, T. Mita, K. Azuma, et al. 2008. "In vivo and in vitro inhibition of monocyte adhesion to endothelial cells and endothelial adhesion molecules by eicosapentaenoic acid." *Arterioscler. Thromb. Vasc. Biol.* 28 (12):2173–9. doi:10.1161/ATVBAHA.108.171736.

Yamashita, T., E. Oda, T. Sano, T. Yamashita, Y. Ijiru, J. C. Giddings, and J. Yamamoto. 2005. "Varying the ratio of dietary n-6/n-3 polyunsaturated fatty acid alters the tendency to thrombosis and progress of atherosclerosis in apoE-/- LDLR-/- double knockout mouse." *Thromb. Res.* 116 (5):393–401. doi:10.1016/j.thromres.2005.01.011.

Yang, Y., N. Lu, D. Chen, L. Meng, Y. Zheng, and R. Hui. 2012. "Effects of n-3 PUFA supplementation on plasma soluble adhesion molecules: A meta-analysis of randomized controlled trials." *Am. J. Clin. Nutr.* 95 (4):972–80. doi:10.3945/ajcn.111.025924.

Yokoyama, M., H. Origasa, M. Matsuzaki, Y. Matsuzawa, Y. Saito, Y. Ishikawa, S. Oikawa, et al. 2007. "Effects of eicosapentaenoic acid on major coronary events in hypercholesterolaemic patients (JELIS): a randomised open-label, blinded endpoint analysis." *Lancet* 369 (9567):1090–8. doi:10.1016/S0140-6736(07)60527-3.

Yoshikawa, T., H. Shimano, N. Yahagi, T. Ide, M. Amemiya-Kudo, T. Matsuzaka, M. Nakakuki, et al. 2002. "Polyunsaturated fatty acids suppress sterol regulatory element-binding protein 1c promoter activity by inhibition of liver X receptor (LXR) binding to LXR response elements." *J. Biol. Chem.* 277 (3):1705–11. doi:10.1074/jbc.M105711200.

Yuan, X., T. C. Ta, M. Lin, J. R. Evans, Y. Dong, E. Bolotin, M. A. Sherman, B. M. Forman, and F. M. Sladek. 2009. "Identification of an endogenous ligand bound to a native orphan nuclear receptor." *PLOS ONE* 4 (5). doi:10.1371/journal.pone.0005609.

Zainal, Z., A. J. Longman, S. Hurst, K. Duggan, B. Caterson, C. E. Hughes, and J. L. Harwood. 2009. "Relative efficacies of omega-3 polyunsaturated fatty acids in reducing expression of key proteins in a model system for studying osteoarthritis." *Osteoarthritis Cartilage* 17 (7):896–905. doi:10.1016/j.joca.2008.12.009.

Zampolli, A., A. Bysted, T. Leth, A. Mortensen, R. De Caterina, and E. Falk. 2006. "Contrasting effect of fish oil supplementation on the development of atherosclerosis in murine models." *Atherosclerosis* 184 (1):78–85. doi:10.1016/j.atherosclerosis.2005.04.018.

Zanetti, M., A. Grillo, P. Losurdo, E. Panizon, F. Mearelli, L. Cattin, R. Barazzoni, and R. Carretta. 2015. "Omega-3 polyunsaturated fatty acids: Structural and functional effects on the vascular wall." *BioMed Res. Int.* 2015:791978. doi:10.1155/2015/791978.

Zhang, D., W. Y. So, Y. Wang, S. Y. Wu, Q. Cheng, and P. S. Leung. 2017. "Insulinotropic effects of GPR120 agonists are altered in obese diabetic and obese non-diabetic states." *Clin. Sci. (Lond.)* 131 (3):247–60. doi:10.1042/CS20160545.

2 Natural Polyketides to Prevent Cardiovascular Disease

Statins

George Hanna and Mark T. Hamann

CONTENTS

2.1 INTRODUCTION

The discovery and development of fungi-produced mevastatin, followed by lovastatin, and their subsequent development into pharmaceuticals represents the most successful utilization of natural polyketides for the treatment of disease. Specifically, for the treatment of lipid-related cardiovascular disease (CVD), the number-one cause of death in the modern world. The success of statins is not based only on economics—they are the most profitable pharmaceuticals to ever enter the market. They are also viewed as a success because they are effective medicines. The various statin analogues are directly involved in significantly lowering the risk of mortality as a result of CVD.

From the microbial ecology point of view, statins are employed to inhibit a nearby microbe's ability to synthesize cholesterol, a key component in properly functioning cellular membranes. This insight has led to the discovery and isolation of the compound mevastatin by Akira Endo et al. from the fungi *Penicillium citrium*, as described in his landmark publication in 1976 (Endo, 2010). Even though mevastatin never made it to market, its discovery is still heralded as a major milestone in the use of natural products as pharmaceuticals because of the resultant insights into the role statins naturally play in influencing cholesterol synthesis. In the following decade, statins were subject to intensive research, basic and clinical, which culminated in the discovery of additional analogues and the development of the first statin drug.

The notable mode of action for statins, to which their success in treating cardiovascular disease is attributed, is a two-front effect on lipid levels in the blood: reducing the liver's levels of lipid synthesis and a simultaneous increase in the rate of cellular absorption of circulating lipids. Thus, statins act by lowering the total amount of circulating lipids in the blood. This is achieved by the structural similarities shared between statins' active moiety and the substrate for the rate-limiting enzyme in cholesterol biosynthesis, the 3-hydroxy-3-methylglutaryl coenzyme A (HMG-CoA) reductase. The competitive inhibition of this integral enzyme results in the hindrance of the body's endogenous source of cholesterol, the mevalonate pathway. Most notably, low-density lipoproteins (LDL), which are established causative agents of atherosclerosis, a cumulative lifelong condition that can result in impaired cardiovascular health.

This chapter will not attempt to address the numerous controversies that surround the use of statins with relation to cardiovascular disease and their associated side effects. Rather, the focus of this chapter is on the nature of statins, their structure and how they interact with relevant biological processes. This should provide some perspective on the use of polyketides as pharmaceuticals. While statin intervention may be straightforward life-or-death in a microbial world, these molecules operate in the context of a more complex system, the human body, which has carefully regulated mechanisms to maintain function. It becomes more difficult to predict how the introduction of statins will affect the organism as a whole. Regardless, statins have proven to be the most consistent and effective lipid-lowering agent on the market since their introduction and are a valuable case study for the role of polyketides in disease prevention.

2.2 CARDIOVASCULAR DISEASE

Diseases affecting the heart or blood vessels, cardiovascular diseases (CVD), are responsible for more deaths globally than any other health condition. Over the past 25 years that number has grown from 12.3 million deaths per year in 1990 to nearly 17.9 million in 2015 (Naghavi, 2015; Wang et al., 2016). In the developed world, this can account for up to 30% of total deaths. There has been only one year in the past 100 years in which CVD has not been the leading global cause of death, 1918. That was the year in which the most devastating influenza pandemic in recorded history swept the globe, killing tens of millions of people. This comparison effectively illuminates cardiovascular disease for what it truly is, a preventable pandemic now spanning over a century. However, between 2003 and 2013 the United States saw a 29% decline in CVD-related deaths. This time line corresponds with a marked increase, from 18% to 26%, in the number of people prescribed a statin to control blood cholesterol (Gu et al., 2014).

A particularly pressing issue is the challenge the developing world faces with respect to combating the various factors that contribute to the development of CVD. These factors may be less avoidable in the developing world, and treatment there is not as widely available as it is in the developed world. Some of the factors that contribute to the development of CVD are smoking, diabetes, sedentary lifestyle, obesity, poor diet, excessive alcohol consumption and high blood cholesterol. Various treatments have been created to improve cardiovascular health. These treatments include both lifestyle adjustments and medicinal interventions. Statin drugs can attribute their widespread success to their ability to reduce the levels of blood cholesterol as a means of reducing the patient's risk of CVD-related mortality. This success is, in part, compounded by the fact that significant lifestyle changes, though recommended, are not required to achieve a lipid-lowering effect.

It is paramount that the pathological mechanisms that contribute to the onset and manifestation of cardiovascular diseases are thoroughly understood for future development of more precise treatments. The resulting knowledge combined with an understanding of the mechanism of action and associated side effects of medical and lifestyle interventions should be aimed at continuous development of better treatments. This is extremely relevant with regard to statin drugs that are already among the most successful pharmaceuticals of all time and continue to increase in use for both primary and secondary prevention of CVD. The successful development of more effective statins that have less severe adverse side effects stands to benefit tens of millions of people annually. This is particularly true considering some cohorts of people react negatively to one statin drug but are fine when prescribed another. Increasing the options from which physicians can choose may improve the overall quality of care for patients receiving statins. The utilization of novel sources of statins represents a valuable avenue for the manufacture of not only more effective analogues but also as a means of lowering the cost of production, which would improve availability in developing nations (Mendis et al., 2011).

2.3 HYPERLIPIDEMIA AND ATHEROSCLEROSIS

The progression from hypercholesterolemia and hyperlipidemia to atherosclerosis is among the leading causes of CVD-related deaths. Dyslipidemia is characterized by abnormal blood levels of lipids (cholesterol, triglycerides or both) and lipoproteins, the protein-lipid-complexes that allow for lipid transport through extracellular water and subsequent functionality through cellular interactions. The lipoproteins are often distinguished by their density, which is a function of the lipid-to-protein ratio where higher lipid content corresponds to a lower density. The characterization of lipoproteins ranges from high to low to very low, and are commonly referred to as HDLs, LDLs and VLDLs for high-density or low-density and very low-density lipoproteins, respectively. Numerous clinical studies have identified causative relationships between high levels of LDL and CVD. Statin intervention has shown consistent lowering of both blood plasma cholesterol and triglycerides (Stancu and Sima, 2001; Subramanian and Chait, 2012).

The causes of hyperlipidemia can result from both acquired and inherited conditions, with high-fat diet as the leading cause and endocrine disorders playing a role in the manifestation of the disease state. Hypothyroidism is considered a major influencer due to the associated hormone imbalance and the associated biochemical characteristics of the condition that results in the elevation of total cholesterol and circulating LDLs (Arikan et al., 2012). Additionally, hypercholesterolemia can result from cholestasis due to the effect the condition has on the secretion of bile salts from the liver, the body's predominant natural pathway for the removal of excess cholesterol (Wagner et al., 2009).

Directly related to high levels of lipids circulating in the blood, the inflammatory disease atherosclerosis and the resultant buildup of arterial lesions and plaque is a major contributor to cardiovascular disease and mortality via myocardial infarction (Ellis et al., 2005). The mechanisms that result in the disease state are not fully understood because of the complex suite of interactions between the body's lipid-regulating mechanisms and the immune response that results in lipid-induced inflammation. Atherosclerosis is thought to develop gradually throughout the course of a person's life and is aggravated by injury (connected to smoking, diabetes, high blood pressure, etc.) to the endothelial cell layer of the arteries. This allows the intrusion of lipoprotein, predominately LDL, into the arterial intima. As LDL accumulates in the extracellular spaces of the arterial intima, the lipids are subject to modifications that result in variations in how the body's immune systems responds to them. These include glycation and oxidation through enzymatic biochemical processes that are involved in cellular signaling and influence the systemic response. Circulating monocytes and subsequently macrophages respond to the modified lipid molecules, ultimately resulting in the development of foam cells. The ensuing immune response includes a cytokine-induced inflammation and the production of growth factors. These growth factors trigger the migration of smooth muscle cells that form a fibrous tissue layer surrounding the lipid-induced atherosclerotic lesion (Gu et al., 2012; Koga and Aikawa, 2012; Kzhyshkowska et al., 2012). Ultimately, the cumulative effects of the lipid intrusion, inflammation and buildup of fibrous muscle tissue over the lipid results in the narrowing of the affected artery and a decrease in its ability to deliver oxygenated blood to the heart and other organs. Not only are the systemic effects of this disease a concern for long-term health but they also lead to an increased likelihood of a myocardial infarction (Osterud and Bjorklid, 2003; Steinbrecher, 1999).

It is thought that atherosclerosis is a lifelong condition that increases in severity with age and can be negatively influenced by environmental stressors or poor diet. A major issue in the diagnosis and treatment of atherosclerosis is the subtle development of its associated symptoms, often developing asymptomatically for long periods of time. This characteristic makes it disproportionately more of a risk to lower-income demographics and in developing regions of the world where people do not have access to long-term routine medical examinations that are essential to the timely diagnosis and intervention of atherosclerosis. As such, atherosclerosis causes one in five deaths in a number of developing countries (Ellis et al., 2005; Talayero and Sacks, 2011).

It is important to note that much of what is known about the causes of atherosclerosis and the mechanisms of drug intervention is based on experimental animal models. However, there is evidence that many of the key contributors, cytokines, growth factors and bioactive molecules, discussed above, are expressed similarly in humans. What is certain is that there is a direct connection between lipid levels and the progression of atherosclerosis. Due to the ability of statin drugs to significantly lower circulating LDL levels, they are routinely prescribed to reduce symptomatic development and the potential for mortality (Borshch et al., 2012).

Atherosclerosis is considered a multi-front disease involving factors other than the lipoprotein load, progressing the perception of statins past their historical paradigm, simply as lipid-lowering agents is integral to the success of their continued prolific use. As such, the role of statins in the treatment of atherosclerosis includes how they influence two other mechanisms related to the state of the disease: systemic inflammatory response and plaque modulation. Due to the role of lipid levels in garnering an immune response and contributing to the formation of arterial plaques, it is likely that statins play a role in influencing how they progress atherosclerosis manifestation. However, the degree to which statins contribute to each of the intertwined mechanisms is still the subject of intense debate and research. Ultimately, parsing out the proportional role statins play in treating lipid levels, immune response (inflammation) and the modulation of arterial plaques is essential to the further development of this key influencer of global health. Not only will it aid in the development of future analogues that may possess increased specificity or potency but it will also provide valuable insight for physicians to apply to determining which patients should be considered for a statin prescription, something that will result in more precise treatment overall (Blaha and Martin, 2013).

2.4 DISCOVERY AND DEVELOPMENT OF STATINS

The discovery of ML-236B, now referred to as mevastatin, and the subsequent development of its analogues as a treatment for hyperlipidemia represents a significant milestone in the employment of polyketides as pharmaceuticals. This breakthrough is underscored by the extensive amount of time and significant research efforts between the key events leading up to the eventual FDA approval of the first statin analogue, lovastatin, for prescription and introduction to the market in 1986.

The role that arterial cholesterol plays in the manifestation of cardiovascular disease was first postulated by the German pathologist, Rudolf Virchow, in the mid-19th century. However, it would take nearly 100 years before the Framingham Heart Study would formally publish correlative evidence between high levels of cholesterol and cardiovascular disease (Tobert, 2003; Mahmood et al., 2014). As a result, throughout the 1950s and 1960s the development of cholesterol-lowering drugs received a considerable amount of basic and clinical research attention. Despite decades of research, the growing list of hopeful treatments was matched by a lengthy list of adverse side effects and marginal effectiveness. This fact is due, in part, to the targeting of the late steps in the cholesterol biosynthesis pathway which resulted in the accumulation of the lipophilic sterol by-products not easily degraded by endogenous metabolism. These results eventually led to the identification of the first and rate-limiting enzyme in the cholesterol pathway, HMG-CoA reductase (HMGR), as an ideal target for drug intervention.

In the early 1970s, Dr. Akira Endo developed a research unit focused on isolating antibiotics that inhibit HMGR, an enzyme key to the survival of many microbes. It is significant to note that this focus was fixated around the speculation that the inhibition of this pathway could serve as a defense mechanism, employed by fungi, against these microbes. The project took nearly two years and around 6,000 microbial strains. Among those, only a handful that showed promising HMGR inhibitory activity, particularly *Penicillium citrinum*, which was isolated from rice purchased at a vendor in Kyoto. Solvent extraction, silica gel chromatography and crystallization of the *P. citrinum* fermentation broth eventually yielded ML-236B, now referred to as mevastatin (Endo, 2004).

Shortly after the isolation of mevastatin, another potent HMGR inhibitor was discovered by Dr. Albert Chen. Once again supporting the value of polyketides as a source of novel compounds, lovastatin, was isolated from another fungi, *Aspergillus terreus* (Alberts et al., 1980). Over the course of the next several years, between 1980 and 1987, mevastatin would eventually be removed from clinical trials and lovastatin would take its place as the leading drug candidate, showing marked reductions of low-density lipoprotein (LDL) cholesterol and no significant adverse side effects in short-term animal trials. Despite initial concerns, in 1987 the US FDA approved lovastatin due to its impressive reduction of apolipoprotein-B-containing lipoproteins and plasma triglycerides paired with increases of high-density lipoproteins (HDL). Lovastatin was introduced as a means for addressing cholesterol levels other than by dietary modifications.

Statins, "a wonderful gift from nature" as Dr. Endo referred to them, would eventually, though not free from controversy, pave the way for the development of a number of statin analogues, including the mevastatin derivative metabolite, pravastatin, and a semi-synthetic derivative of lovastatin, simvastatin (Endo, 2004). The resulting statin drugs have had major socioeconomic impacts, culminating in Lipitor (the synthetic atorvastatin), which broke the record for the highest annual sales of any pharmaceutical in history, hitting $12.8 billion dollars in 2008. Statin drugs make up 93% of lipid-lowering drug prescriptions and are a mainstay in reducing the risk of cardiovascular related mortality, with over a quarter of adults in the United States over the age of 40 using a statin prescription (Brar et al., 2013; Kit et al., 2014).

2.5 MECHANISM OF ACTION

The cholesterol biosynthetic pathway (Figure 2.1) that takes place in the liver is responsible for the production of approximately two-thirds of the body's cholesterol (endogenous); the other third is obtained through the diet (exogenous). Starting with acetyl-CoA, the pathway involves more than 25 enzymes and results in the production of several products that influence a variety of systems in the body. The first step in the endogenous cholesterol pathway, converting 3-hydroxy-3-methylglutaryl-coenzyme A (HMG-CoA) into mevalonate through a reduction reaction is the rate-limiting step in the pathway and is the subject of inhibition by all statin molecules. The result is the hindrance of cholesterol biosynthesis and a buildup of HMG-CoA, which can be easily metabolized, because it does not contain the sterol ring associated with later products in the cholesterol pathway, making it a relatively innocuous point of inhibition (Steinberg et al., 1961).

Statin molecules contain a moiety that structurally resembles HMG-CoA, which results in their ability to bind to the HMG-CoA reductase (HMGR) active site and act as competitive inhibitors (Endo et al., 1976). The competitive inhibition prevents the reductive deacylation of HMG-CoA, which results in CoA and mevalonate. This disrupts the pathway that otherwise would culminate in the production of a variety

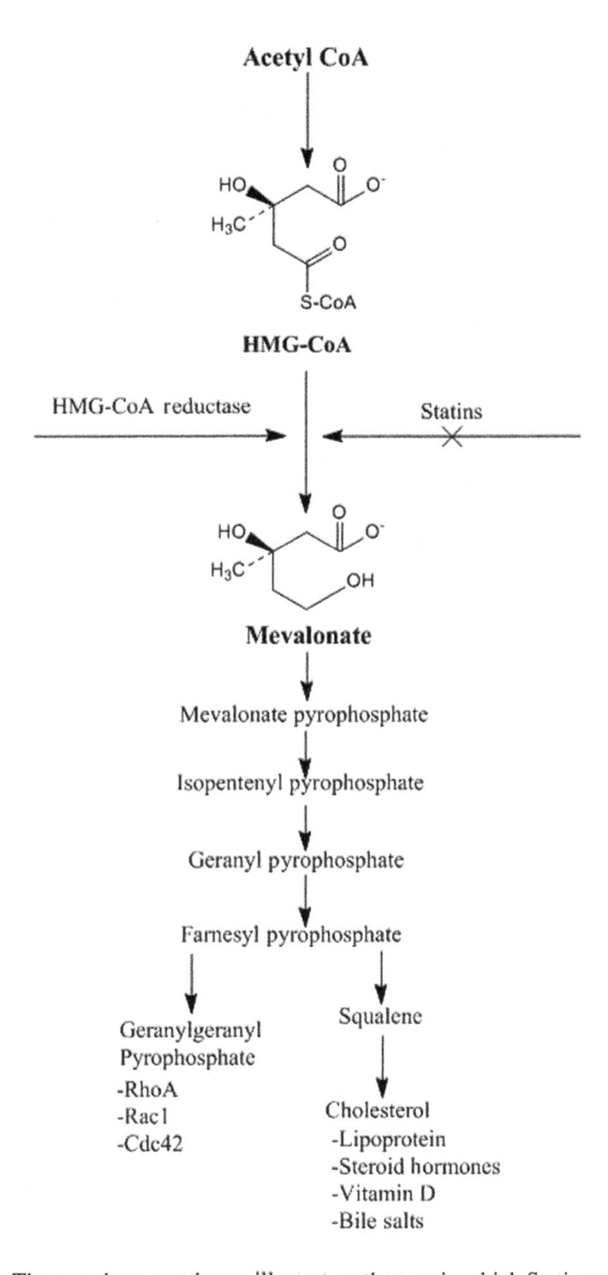

FIGURE 2.1 The mevalonate pathway, illustrating the step in which Statins and HMG-CoA reductase compete.

of products, notably cholesterol. The others will be addressed later in this section. In the normally functioning mevalonate pathway, HMG-CoA first binds to the reductase enzyme, followed by the binding of NADPH, facilitating the reduction reaction. It is the structural similarity between the lactone group of statins and HMG-CoA that imbues in them their inhibitory property. The inhibitory potency of statins lies in the variable hydroxy-hexahydro naphthalene ring central structure that varies between statins. It is important to note that this region varies drastically between natural and synthetic analogues and less so between just the naturally derived analogues. This variable region's activity is derived from its interaction with a hydrophobic pocket adjacent to HMGR's active site, resulting in extremely tight binding. Statins possess nano-molar affinity for HMGR, whereas the natural substrate, HMG-CoA, is in the micro-molar range (Stancu and Sima, 2001).

An additional key characteristic of the statins' mechanism of action is the systemic cellular response to the reduction of endogenously produced cholesterol circulating in the blood. The decrease in circulating cholesterol triggers the activation of a proteolytic mechanism that releases Sterol Regulatory Element-Binding Proteins (SREBPs). The release of these membrane-bound transcription factors allows their translocation into the nucleus. Once there, SREBPs bind to specific DNA sequences that initiate the transcription of genes that code for LDL receptors. Thus, in addition to the decrease in the production of cholesterol by inhibiting the mevalonate pathway, a transcriptional response results in the increase in uptake of circulating cholesterol and lowering plasma LDL levels even further (Sehayek et al., 1994).

Additionally, there are pleiotropic effects that the inhibition of the mevalonate pathway may incur as a result of this pathway's role in the production of a number of key isoprenoid intermediaries. The effect is due to the role that prenylation, the addition of hydrophobic molecules to a protein, plays in protein trafficking and cellular signaling by facilitating protein-membrane interactions. First, farnesyl pyrophosphate signals protein transport to the lumen of the endoplasmic reticulum for further modification and is implicated in interactions involving Ras family proteins. Second, geranylgeranyl pyrophosphate plays a role in the modulation of the RhoA, Rac and Cdc signaling pathways. These are integral GTPase proteins for cellular growth, proliferation and migration that are employed ubiquitously across normal cell types and are often the subject of mutation in tumorigenic cells (Liao and Laufs, 2005).

It is important to note the extent of the role statins play in the modulation of prenylation by inhibiting the mevalonate pathway is incompletely understood and is not clinically attributed to their efficacy as a pharmaceutical. Currently, the role that statin drugs play in lowering LDL levels in the blood is their most significant mode of action and reason for their utilization in mitigating CVD-related mortality (Zhou and Liao, 2010).

2.6 STRUCTURE

Statins can be divided into three categories: natural (Figure 2.2), semi-synthetic and synthetic (Figure 2.3). The natural statins and semi-synthetic statins possess a highly conserved polyketide portion, the hydroxy-hexahydro naphthalene ring central structure that varies based on the addition of side chains. The addition of the side chains occurs at the C6 and C8 position. The synthetic statins only share similarity with the natural statins with the inclusion of the HMG-CoA-like moiety, which is essential to their function, varying greatly in the rest of their structure. The variance influences the properties of the molecule that are attributed to its physical characteristics and functionality (Table 2.1). These include bio-availability, half-life, metabolism,

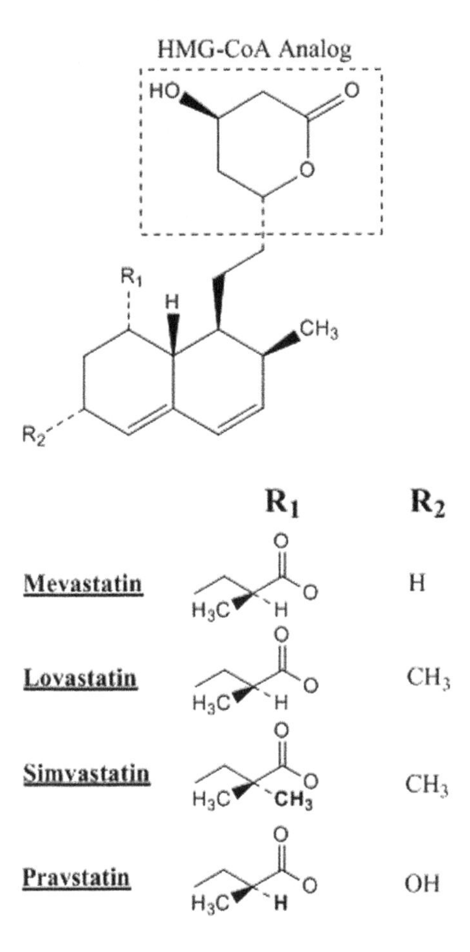

FIGURE 2.2 Structures of statin analogs isolated from nature.

protein binding, affinity, excretion routes and lipophilicity. Ultimately, the influential structural components need to be considered when determining the efficacy of a statin analogue and should be considered when addressing the potential relation to side effects (Nirogi et al., 2007)

2.7 MICROBIAL STATIN PRODUCTION

Of the seven common commercially available statins, two are produced by microbial fermentation, lovastatin and pravastatin, and a third is a semi-synthetic derivative of lovastatin, simvastatin. Combined, these three analogues make up around half of the statin prescriptions in the United States. Thus, there is still considerable incentive

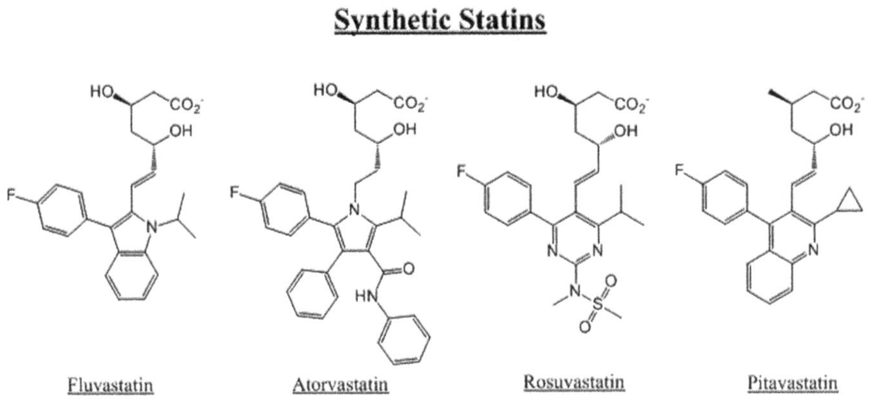

Synthetic Statins

Fluvastatin Atorvastatin Rosuvastatin Pitavastatin

FIGURE 2.3 Structures of synthetic statin analogs used as pharmaceuticals.

TABLE 2.1
Commonly Prescribed Statins, Their Production Source, Dosage and Solubility

Statin Analogue	Source	Dosage (mg)	Solubility
Lovastatin	Fermentation	10–80	Lipophilic
Simvastatin	Semi-synthetic	5–80	Lipophilic
Pravastatin	Fermentation	10–80	Hydrophilic
Fluvastatin	Synthetic	20–80	Lipophilic
Atorvastatin	Synthetic	10–80	Lipophilic
Rosuvastatin	Synthetic	5–40	Hydrophilic
Pitavastatin	Synthetic	2–4	Moderately lipophilic

for the continued pursuit of natural statin products (Findlay, 2007). The prospect of utilizing agro-industrial by-products for sustainable fermentation-based statin production and the utilization of novel microorganisms that produce either higher yields or new analogues of natural statins have received attention in recent years. A notable benefit for pushing the microbial production of statins forward is lowering the cost of development, which would in turn increase statin availability to financially limited populations that would otherwise opt out of their use (Bizukojc et al., 2007; Vilches-Ferrón et al., 2005; Seraman et al., 2010). The advent of patent expiration adds additional incentive for more independent investigation and development of production methods, novel statins and value-added chemical scaffolds for synthesis. Additionally, the realization of the high CVD risk in developing countries and those countries' contribution to long-term market growth are a key aspect to motivating the accommodation of this emerging market (Barrios-González and Miranda, 2010).

Several filamentous fungi have been identified as sources for the production of natural statins. These include *Monascus* spp., *Penicillium* spp., and *Aspergillus* spp.

However, there are notable variations in the production levels and metabolite profiles between species within each genus. Additionally, some strains are more favorable than others because of the additional metabolites they produce and their associated toxicities. The screening method employed to determine the production of statins is simple, an activity assay against the yeast *Candida albicans*, which cannot tolerate growth in their presence. This is based on measurements of the inhibition zone on plates containing the yeast after the introduction of statin impregnated paper (Auclair et al., 2001; Manzoni et al., 1998; Komagata et al., 1989; Manzoni and Rollini, 2002). However, high-throughput metabolomic and metagenomic techniques offer new methods for developing ways to analyze the microbial production of statins. These include the development of PCR primers to specifically detect statin-producing microbial strains (Seraman et al., 2010; Vilches-Ferrón et al., 2005; Kim et al., 2011). The progress of the last decade in high-throughput sequencing and genetic modification technologies has resulted in monumental leaps forward in the arsenal of molecular tools that can be utilized for natural product drug development and production.

Mutant strains of known statin producers have already successfully resulted in increased rates of production. This is highlighted by a comparison of the isolated *Aspergillus terreus* and a UV mutant counterpart increasing the production of lovastatin from 400 to 2,200 mg per liter of fermentation broth. In addition to mutant strains, the fine-tuning of culture media has proved useful for significantly improving the rate of production. The key components of any media are the carbon source, the nitrogen source and their defined ratios and levels of metabolic availability, which can influence the production of a given compound of interest. It has been found that nitrogen limitation increases the amount of carbon diverted to the lovastatin metabolic pathway (Li et al., 2011; Casas López et al., 2003; Kumar et al., 1998, Kumar et al., 2000, Kumar et al., 2011) and that the use of glucose decreases lovastatin synthesis (Miyake et al., 2006). It is apparent there is still more to learn about the nature of statins in the microbial world. Given the widespread use and important role that statins play in the modern developed world, any information furthering the effective utilization of this natural polyketide will have broad impact (Brar et al., 2013).

2.8 STATINS IN FOOD

It is apparent that several food products used in traditional medicine, mostly originating in China, are known to possess statins: red yeast rice, oyster mushrooms and pu-erh tea. However, in the United States because statins are recognized by the FDA as a drug, they are excluded from use as defined dietary supplements, making the sale of red yeast rice containing statins illegal and puts pu-erh tea in a sort of legal grey area (Mark, 2010). Despite the legality of manufacturing and selling statin-containing food products in the United States, worldwide these products have been used for centuries to treat a variety of health conditions, including hyperlipidemia. Research, particularly epidemiological studies, regarding dietary sources of statins may provide valuable insights into their long-term use and the associated side effects of diet-derived statins.

Unlike red yeast rice and pu-erh tea, which both contain statin-producing fermentation steps in their production process that subjects them to regulation, mushrooms belonging to the genus *Pleurotus* readily contain statins in the wild. During the production and isolation of statins by filamentous micromycetes, it is common for the majority of the statin isolate to be obtained not from the culture filtrate but from the vegetative tissue known as mycelium. Thus, given that edible fungi can produce statin compounds, their consumption may be a viable option for their incorporation into a normal diet. The production of statins by micromycetes has been well studied since the discovery of the first fungi-derived analogues in the 1970s. In comparison, research regarding macromycetes as producers of statins has lagged considerably (Alarcon et al., 2003). Recently, edible and medicinal mushrooms have been investigated using modern chemical content analyses regarding their beneficial properties for treating diseases such as hypertension, hypercholesterolemia, atherosclerosis and some cancers. This research is especially important for patients that cannot tolerate prescribed forms of statins, such as those receiving certain HIV retroviral therapies (Eckard and McComsey, 2015).

Representatives from over a dozen genera of mushrooms have been shown to contain metabolites known to be beneficial in treating cardiovascular disease, including GABA, ergothioneine and lovastatin. The metabolites were shown to be present in both the mycelium and fruiting bodies of most species, though at varying levels. The edible mushrooms in the genus *Pleurotus*, commonly referred to and marketed as "oyster mushrooms," are likely the most widely used in culinary applications and are among those that have been shown to produce notable levels of statins (Chen et al., 2012).

The lipid-lowering effects of the edible mushroom *Pleurotus ostreatus* have been shown in both animal studies and small-scale human trials (Schneider et al., 2011; Kressel et al. 2011). These findings warrant further investigation into the potential benefits of dietary supplements containing macromycete-derived nutraceuticals, taking into consideration how side effects vary between the natural statin analogues and related metabolites contained in nutraceutical supplements and long-term treatment with synthetically derived statins.

2.9 CONCLUSION AND FUTURE PERSPECTIVES

The increasingly ubiquitous use and unmatched effectiveness of the statins for the primary and secondary prevention of cardiovascular disease in the modern world, particularly developing nations, makes continued progress in research of these polyketides and their derivatives paramount for the well-being of billions of people. Considering the relative infancy of statin drugs and their rapid expansion into the pharmaceutical market, continued long-term clinical trials and epidemiological studies will reveal valuable trends to better inform physicians. Atherosclerosis, the global leading cause of death, is systemic and complex by nature, and increased precision in both detection and subsequent intervention will result in improved outcomes for the patients involved. The potency and effectiveness of statins in reducing lipid levels in the body by influencing both the endogenous and exogenous cholesterol pathways makes them a truly remarkable gift from nature. However, as is the case with atherosclerosis, the complexity and broad influence of lipid-associated biological processes warrant thorough investigation into a holistic view regarding the effect of statins and the cascade of processes they influence in the reduction of CVD-related mortality. Additionally, continued investigation into the pleiotropic effects of statins on normal cell function, inflammatory response and tumorigenesis may provide insights into strategies for developing additional statin therapeutics. The advent of high-throughput molecular technologies has led to the development of numerous tools that are ripe for adaptation to be applied to furthering the field of natural product drug development. These include novel techniques for identifying and isolating drug analogues and refining existing production methods to increase the diversity and availability of treatment options. The vast number and diversity of patients already receiving statin therapy and rising demand in the developing world for them underline the fact that statins represent ideal candidates for such research.

REFERENCES

Alarcon, Julio, Sergio Aguila, Patricia Arancibia-Avila, Oscar Fuentes, Enrique Zamorano-Ponce, and Margarita Hernández. "Production and purification of statins from Pleurotus ostreatus (Basidiomycetes) strains." *Zeitschrift Fur Naturforschung. C, Journal of Biosciences* 58, no. 1–2 (2003): 62–64.

Alberts, A.W., J. Chen, G. Kuron, V. Hunt, J. Huff, C. Hoffman, J. Rothrock, et al. "Mevinolin: A highly potent competitive inhibitor of hydroxymethylglutaryl-coenzyme A reductase and a cholesterol-lowering agent." *Proceedings of the National Academy of Sciences* 77, no. 7 (1980): 3957–3961.

Arikan, Senay, Mithat Bahceci, Alpaslan Tuzcu, Fatma Celik, and Deniz Gokalp. "Postprandial hyperlipidemia in overt and subclinical hypothyroidism." *European Journal of Internal Medicine* 23, no. 6 (2012): e141–ee145.

Auclair, Karine, Jonathan Kennedy, C. Richard Hutchinson, and John C. Vederas. "Conversion of cyclic nonaketides to lovastatin and compactin by a lovC deficient mutant of Aspergillus terreus." *Bioorganic & Medicinal Chemistry Letters* 11, no. 12 (2001): 1527–1531.

Barrios-González, Javier, and Roxana U. Miranda. "Biotechnological production and applications of statins." *Applied Microbiology and Biotechnology* 85, no. 4 (2010): 869–883.

Bizukojc, Marcin, Beata Pawlowska, and Stanislaw Ledakowicz. "Supplementation of the cultivation media with B-group vitamins enhances lovastatin biosynthesis by Aspergillus terreus." *Journal of Biotechnology* 127, no. 2 (2007): 258–268.

Blaha, Michael J., and Seth S. Martin. "How do statins work?: Changing paradigms with implications for statin allocation." *Journal of the American College of Cardiology* 62, no. 25 (2013): 2392–2394.

Borshch, V.N., E.R. Andreeva, S.G. Kuz'min, and I.N. Vozovikov. "New medicines and approaches to treatment of atherosclerosis." *Russian Journal of General Chemistry* 82, no. 3 (2012): 554–563.

Brar, Satinder Kaur, Gurpreet Singh Dhillon, and Carlos Ricardo Soccol, eds. *Bitransformation of waste biomass into high value biochemicals.* Springer Science & Business Media (2013).

Casas López, J.L., J.A. Sánchez Pérez, J.M. Fernández Sevilla, F.G. Acién Fernández, E. Molina Grima, and Y. Chisti. "Production of lovastatin by Aspergillus terreus: Effects of the C: N ratio and the principal nutrients on growth and metabolite production." *Enzyme and Microbial Technology* 33, no. 2–3 (2003): 270–277.

Chen, Shin-Yu, Kung-Jui Ho, Yun-Jung Hsieh, Li-Ting Wang, and Jeng-Leun Mau. "Contents of lovastatin, γ-aminobutyric acid and ergothioneine in mushroom fruiting bodies and mycelia." *LWT-Food Science and Technology* 47, no. 2 (2012): 274–278.

Eckard, Allison Ross, and Grace A. McComsey. "The role of statins in the setting of HIV infection." *Current HIV/AIDS Reports* 12, no. 3 (2015): 305–312.

Ellis, Jeffrey T., Deborah L. Kilpatrick, Paul Macke Consigny, Santosh Prabhu, and Syed F.A. Hossainy. "Therapy considerations in drug-eluting stents." *Critical Reviews in Therapeutic Drug Carrier Systems* 22, no. 1 (2005).

Endo, Akira. "The origin of the statins." In *International Congress Series* 1262: 3–8. Elsevier (2004).

Endo, Akira. "A historical perspective on the discovery of statins." *Proceedings of the Japan Academy*, Series B 86, no. 5 (2010): 484–493.

Endo, Akira, Masao Kuroda, and Kazuhiko Tanzawa. "Competitive inhibition of 3-hydroxy-3-methylglutaryl coenzyme A reductase by ML-236A and ML-236B fungal metabolites, having hypocholesterolemic activity." *FEBS Letters* 72, no. 2 (1976): 323–326.

Findlay, S. "The statin drugs. Prescription and price trends October 2005 to December 2006 and potential cost savings to Medicare from increased use of lower cost statins." *Consumer Reports* (2007).

Gu, Hong-feng, Chao-ke Tang, and Yong-zong Yang. "Psychological stress, immune response, and atherosclerosis." *Atherosclerosis* 223, no. 1 (2012): 69–77.

Gu, Qiuping, Ryne Paulose-Ram, Vicki L. Burt, and Brian K. Kit. "Prescription cholesterol-lowering medication use in adults aged 40 and over: United States, 2003–2012." NCHS data brief 177 (2014): 1–8.

Kim, J.S., Y.M. Youk, H.S. Ko, and D.H. Kang. "PCR primer for detecting lovastatin-producing strain and a detection method using the same," *Korean Patent Abstracts* (2011) 1020110044613.

Koga, Jun-Ichiro, and Masanori Aikawa. "Crosstalk between macrophages and smooth muscle cells in atherosclerotic vascular diseases." *Vascular Pharmacology* 57, no. 1 (2012): 24–28.

Komagata, Daisuke, Hideaki Shimada, Shigeo Murakawa, and Akira Endo. "Biosynthesis of monacolins: Conversion of monacolin L to monacolin J by a monooxygenase of Monascus ruber." *The Journal of Antibiotics* 42, no. 3 (1989): 407–412.

Kressel, Gaby, Annette Meyer, Ulrich Krings, Ralf G. Berger, and Andreas nigoHahn. "Lipid lowering effects of oyster mushroom (*Pleurotus ostreatus*) in humans." *Journal of Functional Foods* 3, no. 1 (2011): 17– 24.

Kumar, Yatendra, Rajesh Kumar Thaper, Satyananda Misra, S.M. Dileep Kumar, and Jag Mohan Khanna. "Process for manufacturing simvastatin from lovastatin or mevinolinic acid." U.S. Patent 5,763,646, issued June 9 (1998).

Kumar, M. Sitaram, Swapan K. Jana, V. Senthil, V. Shashanka, S. Vijay Kumar, and A. K. Sadhukhan. "Repeated fed-batch process for improving lovastatin production." *Process Biochemistry* 36, no. 4 (2000): 363–368.

Kumar, Sanjay, Nalini Srivastava, and James Gomes. "The effect of lovastatin on oxidative stress and antioxidant enzymes in hydrogen peroxide intoxicated rat." *Food and Chemical Toxicology: an International Journal Published for the British Industrial Biological Research Association* 49, no. 4 (2011): 898–902.

Kzhyshkowska, Julia, Claudine Neyen, and Siamon Gordon. "Role of macrophage scavenger receptors in atherosclerosis." *Immunobiology* 217, no. 5 (2012): 492–502.

Li, Mei, Shi-Weng Li, Hong-Ping Song, Jia-Li Feng, and Xi-Sheng Tai. "Induction of a high-yield lovastatin mutant of Aspergillus terreus by 12 C 6+ heavy-ion beam irradiation and the influence of culture conditions on lovastatin production under submerged fermentation." *Applied Biochemistry and Biotechnology* 165, no. 3–4 (2011): 913–925.

Liao, James K., and Ulrich Laufs. "Pleiotropic effects of statins." *Annual Review of Pharmacology and Toxicology* 45 (2005): 89–118.

Mahmood, Syed S., Daniel Levy, Ramachandran S. Vasan, and Thomas J. Wang. "The Framingham Heart Study and the epidemiology of cardiovascular disease: A historical perspective." *Lancet* 383, no. 9921 (2014): 999–1008.

Manzoni, M., and M. Rollini. Biosynthesis and biotechnological production of statins by filamentous fungi and application of these cholesterol-lowering drugs. *Applied Microbiology and Biotechnology* 58, no. 5(2002): 555–564.

Manzoni, Matilde, Manuela Rollini, Silvia Bergomi, and Valeria Cavazzoni. "Production and purification of statins from Aspergillus terreus strains." *Biotechnology Techniques* 12, no. 7 (1998): 529–532.

Mark, David A. "All red yeast rice products are not created equal—or legal." *The American Journal of Cardiology* 106, no. 3 (2010): 448.

Mendis, Shanthi, Pekka Puska, Bo Norrving, and World Health Organization (2011). Global atlas on cardiovascular disease prevention and control (PDF). World Health Organization in collaboration with the World Heart Federation and the World Stroke Organization. pp. 3–18. ISBN 978-92-4-156437-3. Archived (PDF) from the original on 2014-08-17.

Miyake, Tsuyoshi, Kumiko Uchitomi, Ming-Yong Zhang, Isato Kono, Nobuyuki Nozaki, Hiroyuki Sammoto, and Kenji Inagaki. "Effects of the principal nutrients on lovastatin production by Monascus pilosus." *Bioscience, Biotechnology, and Biochemistry* 70, no. 5 (2006): 1154–1159.

Naghavi, M. "Global, regional, and national age-sex specific all-cause and cause-specific mortality for 240 causes of death, 1990–2013: A systematic analysis for the Global Burden of Disease Study 2013." *Lancet* 385, no. 9963 (2015): 117–171.

Nirogi, Ramakrishna, Koteshwara Mudigonda, and Vishwottam Kandikere. "Chromatography–mass spectrometry methods for the quantitation of statins in biological samples." *Journal of Pharmaceutical and Biomedical Analysis* 44, no. 2 (2007): 379–387.

Osterud, Bjarne, and Eirik Bjorklid. "Role of monocytes in atherogenesis." *Physiological Reviews* 83, no. 4 (2003): 1069–1112.

Schneider, Inga, Gaby Kressel, Annette Meyer, Ulrich Krings, Ralf G. Berger, and Andreas Hahn. "Lipid lowering effects of oyster mushroom (*Pleurotus ostreatus*) in humans." *Journal of Functional Foods* 3, no. 1 (2011): 17–24.

Sehayek, E., E. Butbul, R. Avner, H. Levkovitz, and S. Eisenberg. "Enhanced cellular metabolism of very low density lipoprotein by simvastatin. A novel mechanism of action of HMG-CoA reductase inhibitors." *European Journal of Clinical Investigation* 24, no. 3 (1994): 173–178.

Seraman, Subhagar, Aravindan Rajendran, and Viruthagiri Thangavelu. "Statistical optimization of anticholesterolemic drug lovastatin production by the red mold Monascus purpureus." *Food and Bioproducts Processing* 88, no. 2 (2010): 266–276.

Stancu, Camelia, and Anca Sima. "Statins: Mechanism of action and effects." *Journal of Cellular and Molecular Medicine* 5, no. 4 (2001): 378–387.

Steinberg, Daniel, Joel Avigan, and Eugene B. Feigelson. "Effects of triparanol (MER-29) on cholesterol biosynthesis and on blood sterol levels in man." *The Journal of Clinical Investigation* 40, no. 5 (1961): 884–893.

Steinbrecher, U.P. "Receptors for oxidized low density lipoprotein." *Biochimica et Biophysica Acta* 1436, no. 3: 279–298 (1999).

Subramanian, Savitha, and Alan Chait. "Hypertriglyceridemia secondary to obesity and diabetes." *Biochimica et Biophysica Acta* 1821, no. 5 (2012): 819–825.

Talayero, Beatriz G., and Frank M. Sacks. "The role of triglycerides in atherosclerosis." *Current Cardiology Reports* 13, no. 6 (2011): 544–552.

Tobert, Jonathan A. "Lovastatin and beyond: The history of the HMG-CoA reductase inhibitors." *Nature Reviews. Drug Discovery* 2, no. 7 (2003): 517–526.

Vilches-Ferrón, M.A., J.L. Casas López, J.A. Sánchez Perez, J.M. Fernández Sevilla, and Y. Chisti. "Rapid screening of Aspergillus terreus mutants for overproduction of lovastatin." *World Journal of Microbiology and Biotechnology* 21, no. 2 (2005): 123–125.

Wagner, Martin, Gernot Zollner, and Michael Trauner. "New molecular insights into the mechanisms of cholestasis." *Journal of Hepatology* 51, no. 3 (2009): 565–580.

Wang et al. "GBD 2015 Mortality and Causes of Death, Collaborators." "Global, regional, and national life expectancy, all-cause mortality, and cause-specific mortality for 249 causes of death, 1980-2015: A systematic analysis for the Global Burden of Disease Study 2015," *Lancet* 388, no. 10053 (2016): 1459–1544. doi:10.1016/S0140-6736(16)31012-1. PMID 27733281.

Zhou, Qian, and James K. Liao. "Pleiotropic effects of statins. Basic research and clinical perspectives." *Circulation Journal: Official Journal of the Japanese Circulation Society* 74, no. 5 (2010): 818–826.

3 Impact of Dietary Polyphenols on Arterial Stiffness

Outline of Contributing Mechanisms

Tess De Bruyne, Lynn Roth, Harry Robberecht, Luc Pieters, Guido De Meyer and Nina Hermans

CONTENTS

3.1 INTRODUCTION: ARTERIAL STIFFNESS

During the last two decades, the role of arterial stiffness in the development of cardiovascular diseases has received much attention (Laurent et al., 2006). The term 'arterial stiffness' denotes the slope of the pressure-volume relationship in arteries ($\Delta P/\Delta V$) (Quinn et al., 2012). Increased arterial stiffness, as a marker of vascular ageing, indicates a reduced capability of arteries to expand and contract in response to pressure changes (Hamilton et al., 2007; Laurent et al., 2006; Della Corte et al., 2016). This degenerative process, under the effect of ageing and risk factors, affects mainly the extracellular matrix of elastic arteries (Palombo and Kozakova, 2016). Stiffening of the arterial wall leads to fundamental changes in central hemodynamics, with increased pulsatile strain on the microcirculation, leading to detrimental consequences for end-organ function (Lyle and Raaz, 2017; O'Rourke and Hashimoto, 2007).

3.1.1 Origins and Mechanisms

Normal elastic arteries have a smoothing function, assuring a steady blood flow in peripheral tissues (Quinn et al., 2012). Throughout adult life, the arterial tissue loses its elasticity, primarily due to degeneration of the extracellular matrix under the influence of ageing and other risk factors, thus causing arterial stiffness (Palombo and Kozakova, 2016). The relative contribution of predominant vascular wall proteins, collagen and elastin determines the stiffness of the vascular wall (Laurent et al., 2006). Progressive degeneration of the extracellular matrix in the media layer-causes loss of elasticity and thus loss of the biomechanical buffering capacity. This degeneration is characterized by elastin fatigue fracture and collagen deposition and cross-linking, fibrosis, inflammation, necrosis and calcification (Laurent et al., 2006; Palombo and Kozakova, 2016; Cecelja and Chowienczyk, 2016).

A decrease in the elastin/collagen ratio in the media layer progresses under the influence of altered lysyl oxidase (LOX) and matrix metalloproteinase (MMP) activity (Lyle and Raaz, 2017). Elastin becomes progressively fragmented and degraded over time, collagen increases and collagen-elastin cross-links increase under the influence of S-nitrosylation and advanced glycation end products. Increased angiotensin II (Ang II) signaling contributes to the collagen and advanced glycation endpoints (AGEs) accumulation and the elastin degradation. Glycation and AGE formation is thought to have a profound effect on enhancing age-related arterial stiffening. Additionally, upon activation of the receptor of AGE (RAGE), intracellular reactive oxygen species levels are increased through upregulation of NAD(P)H oxidase expression and further contribute to the oxidative stress associated with arterial stiffness. Moreover, fragmented elastin serves as nidus for microdisposition of Ca: vascular smooth muscle cells transform into osteoblast-type cells, inducing vascular calcification which further increases the stiffening process (Avolio, 2013; Lyle and Raaz, 2017; Sell and Monnier, 2012; Smulyan et al., 2016; Quinn et al., 2012; Cecelja and Chowienczyk, 2016; Vlassopoulos et al., 2014). While the calcification process in arterial stiffness affects the media layer and is associated with ageing, diabetes and renal disease, a similar calcification in the intima, on the other hand, results in atherosclerosis and the development of atherosclerotic plaques (Mozos and Luca, 2017).

Arterial stiffness and atherosclerosis thus share some common pathophysiological mechanisms and could be viewed as two synergic processes in the development of vascular changes underlying cardiovascular disease (Cecelja and Chowienczyk, 2016; Palombo and Kozakova, 2016).

Stiffened arteries contribute to increased systolic blood pressure (SBP), amplified by superposition of prematurely reflected pulse waves (Smulyan et al., 2016). Vascular stiffness is closely interrelated with elevated blood pressure: increased BP is a consequence of increased arterial stiffness but also a cause for the reduction of arterial elasticity. Therefore, the unraveling of contributing mechanisms is complicated considerably (Papaioannou et al., 2017). High SBP affects primarily microvasculature in vulnerable end-organs like brain and kidney (Smulyan et al., 2016; Lilamand et al., 2014). Therefore, arterial stiffness severity is also a sensitive predictor of cognitive impairment (Li et al., 2017b).

Furthermore, the coordinated functioning of the heart and arterial tree becomes impaired, leading to progressive hemodynamic alterations, including a rise in SBP and a fall in diastolic blood pressure (DBP), and hence, elevated pulse pressure (Lyle and Raaz, 2017; O'Rourke and Hashimoto, 2007; Willum-Hansen et al., 2006). Raised SBP requires increased left ventricular (LV) workload, with a need for increased coronary perfusion and oxygen. However, the diastolic pressure determining coronary flow is insufficient, causing LV dysfunction and hypertrophy, and subsequent development of heart failure (Smulyan et al., 2016; Lyle and Raaz, 2017). Figure 3.1 shows vascular and haemodynamic properties of stiffened arteries in comparison to normal ones.

Arterial stiffness (AS) is an independent risk factor for cardiovascular (CV) morbidity and mortality (Van Bortel, 2016; Lyle and Raaz, 2017; Cecelja and Chowienczyk, 2016), and is considered a surrogate end point in clinical trials (Cardoso and Salles, 2016; Della Corte et al., 2016). AS causes an increase in SBP due to premature reflection of waves, increasing central pulse pressure (Laurent et al., 2006). The consequent damage of microvasculature leads to end-organ dysfunction (Lilamand et al., 2014).

AS is thus also associated with LV hypertrophy, worsening of coronary ischaemia and increased fatigue of coronary wall tissues, all known risk factors for CV events (Laurent et al., 2006; Hamilton et al., 2007).

3.1.2 Factors Involved in the Development and Progression of Arterial Stiffness

Age and related processes are the main determinants of stiffness in elastic arteries, which are correlated with nutritional and lifestyle factors, and with subsequent age-associated disorders like metabolic syndrome, type 2 diabetes, hypertension, atherosclerosis and renal disease, implying thereby metabolic factors in its pathogenesis (Avolio, 2013; Li et al., 2017a; Sell and Monnier, 2012; Wu et al., 2015).

Degeneration and remodeling of elastic components of the arterial wall usually become important after the age of 55, concomitant to a decrease of intracellular magnesium (Wu et al., 2015). Furthermore, calcification, apoptosis, inflammation

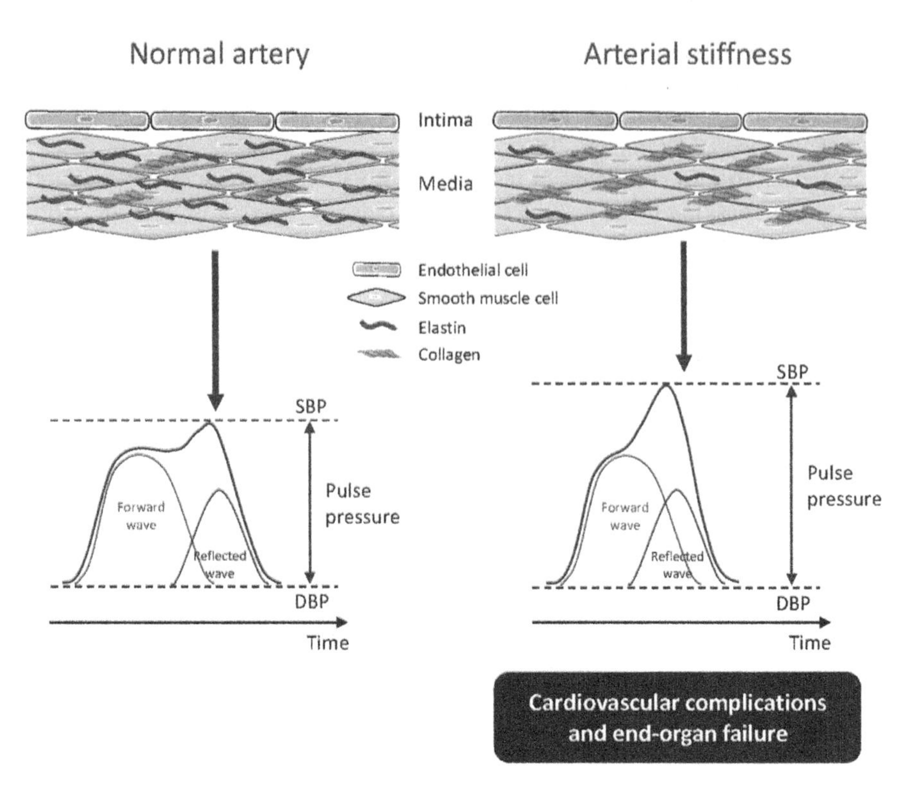

FIGURE 3.1 Arterial stiffness: vascular and haemodynamic properties in comparison with normal arteries. (SBD: systolic blood pressure, DBP: diastolic blood pressure.)

and oxidative and nitrosative stress, genetic influences, as well as reduced autophagy, add to the age-related stiffening (Lyle and Raaz, 2017; Li et al., 2017a; Sasaki et al., 2017; Sell and Monnier, 2012; Wu et al., 2015).

Oxidative stress is defined as an imbalance between reactive oxygen species (ROS) formation and elimination in favor of pro-oxidant processes. During vascular oxidative stress, it has been demonstrated that, in the vascular wall, many enzymatic systems produce reactive oxygen species (ROS), including nicotinamide adenine dinucleotide phosphate (NADPH) oxidase, mitochondrial enzymes, dysfunctional endothelial nitric oxide synthase (eNOS) and xanthine oxidase (XO). Vascular cells also have several antioxidant systems to counteract ROS generation: superoxide dismutase (SOD), catalase (CAT), glutathione peroxidase (GPx), paraoxonase (PONs), thioredoxin (TRX) peroxidase and heme oxygenase (Hmox) (Santilli et al., 2015).

Oxidative and nitrosative stress contribute to the etiology due to oxidative damage to lipids, proteins and DNA in endothelial cells and uncoupling of nitric oxide (NO) synthase, leading to endothelial dysfunction. Increased stress and angiotensin II induce increased ROS production in vascular smooth muscle cells (VSMCs). Elevated levels of superoxide radical anion react with NO to produce peroxynitrite, which is, together with other highly reactive species, responsible for vascular degeneration and dysfunction. Moreover, altered blood flow also increases ROS production, and mitochondrial oxidative stress and SOD 2 deficiency induce aortic stiffening (Lyle and Raaz, 2017). Stiff arteries potentially induce a positive feedback mechanism downregulating eNOS and upregulating endothelin-1, thus further increasing wall stiffness (Avolio, 2013).

Several markers of oxidative stress have been associated with increased arterial stiffness, although a causative link has sometimes been questioned, due to the experimental complexity of antioxidant clinical trials. Nevertheless, changes in malondialdehyde, superoxide dismutase, vascular adhesion protein-1, oxidized LDL, isoprostanes and thiobarbituric acid-reactive substances (TBARS) have been reported in different test and patient groups with vascular stiffness (Mozos and Luca, 2017).

Inflammation is involved in arterial stiffness development by multiple mechanisms. Low-grade inflammation impairs endothelial function by reducing NO bioavailability and increasing endothelin-1, thus contributing to progressive arterial stiffening, which, in turn, further impairs endothelial function (Avolio, 2013). The contribution of an inflammatory status is furthermore reflected in the role of MMPs in elastin degradation, the overexpression of lectin-like oxidized low-density lipoprotein receptor 1 (LOX-1) (by an NF-κB dependent mechanism), increasing uptake of oxidized low-density lipoprotein (oxLDL), the transdifferentiation of VSMCs into an osteoblastic phenotype under inflammatory conditions, influence of cytokines, increased AGEs synthesis, C-reactive protein (CRP: inhibits endothelial NO synthase, increases cytokine expression, increases generation of ROS), influence of adhesion molecules, microRNAs and so on. (Mozos et al., 2017a). Chronic low-grade inflammation interacts synergistically with oxidative stress, but the order and relationship between these events are uncertain (LaRocca et al., 2017).

Autophagy is a complex cellular process. It starts with the sequestration of cytoplasmic constituents into double-membraned vesicles, termed autophagosomes. Fusion of autophagosomes with lysosomes results in the degradation of the incorporated material into amino acids, carbohydrates, fatty acids and nucleotides. Through the recycling of biomolecules, basal autophagy provides new building blocks to support cellular function and homeostasis, as a response to stressful stimuli (e.g. accumulation of damaged organelles, nutrient deprivation, hypoxia). It is therefore a cytoprotective process involved in such diseases as cardiovascular disease, cancer and neurodegenerative disorders. However, little is known about autophagy in arterial ageing and vascular stiffness. Normally, autophagy preserves endothelial function, and age-related impaired autophagy contributes to arterial stiffness and endothelial dysfunction associated with increased levels of oxidative stress and inflammation (De Meyer et al., 2015; Sasaki et al., 2017).

It is well known that reactive oxygen species (ROS) can induce autophagy, which is a defense mechanism against cell death (Perrotta and Aquila, 2015). Recent evidence also indicates that autophagy may exhibit antioxidant properties by taking part in the reduction and repair of oxidative damage through a variety of signaling pathways (Fang et al., 2017), which could have a remarkable impact on cardiovascular health. Figure 3.2 gives an overview of the mechanisms contributing to the pathophysiology of AS.

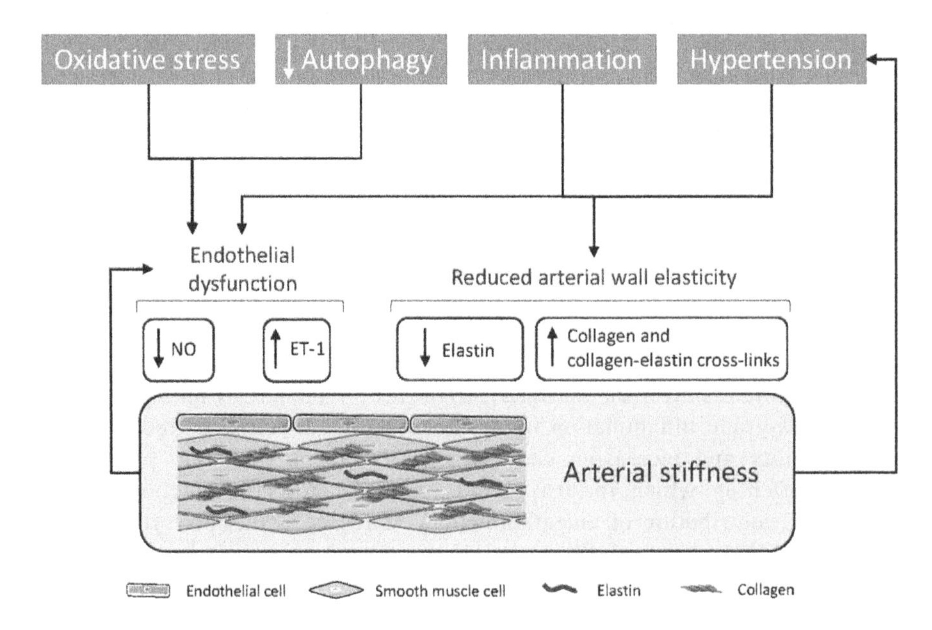

FIGURE 3.2 Mechanisms contributing to the pathophysiology of arterial stiffness. (NO: nitric oxide, ET-1: endothelin 1.)

Heritability of arterial stiffness is about 40%. Associations of **gene** expression levels with arterial stiffening are found on genes involved in extracellular matrix and calcification on one hand and on genes relating to blood-pressure regulation on the other. However, still very little is known about the molecular mechanisms underlying phenotypic variability (Cecelja and Chowienczyk, 2016). Candidate genes (e.g. IGF-1 receptor, IL-6, PACE4) potentially involved in arterial elasticity have been found on chromosomes 2,7,13 and 15 (Wu et al., 2015).

Nutrition and other lifestyle influences are important in the protection against the development of arterial ageing. Generally, diets and nutrients that reduce oxidative stress and inflammation, such as diets emphasizing fruits and vegetables, grains, nuts, seeds, legumes with limited amounts of low-fat meat and fish, and refined foods, are associated with reduced arterial stiffness (LaRocca et al., 2017; Mozos et al., 2017b). Specific nutrient measures like restricted dietary salt seem to have beneficial effects, although it is difficult to differentiate the direct effect on arterial stiffness mechanisms from an effect on blood pressure (LaRocca et al., 2017; Quinn et al., 2012; Wu et al., 2015). Some reports mention specific foods like dairy products, fermented dairy, dark chocolate, tea, soy, olive oil, grains, and nuts, but also here, effects are probably—at least partly—due to a reduction in oxidative stress and inflammation. Energy intake restriction and aerobic exercise also protect against arterial ageing (LaRocca et al., 2017; Mozos and Luca, 2017; Wu et al., 2015). Smoking, on the other hand, has an adverse impact on arterial stiffness (Wu et al., 2015). Extreme (both long and short) sleep duration and poor sleep quality are associated with arterial stiffness and are possibly linked to increased MMP expression.

(Lyle and Raaz, 2017). Mental stress is also seen to contribute to vascular dysfunction involving oxidative stress and inflammation (Daiber et al., 2017).

Hypertension evolves as the haemodynamic response to arterial stiffness. An age-associated increase in aortic stiffness typically is reflected in isolated increased SBP. In reverse, hypertension can also lead to arterial stiffening by upregulation of pathways involved in inflammation, fibrosis and wall hypertrophy (Lyle and Raaz, 2017; Sharman et al., 2017).

Diabetes-accelerated arterial stiffening includes elevated levels of oxidative stress, similar to age-induced stiffness, MMP-mediated elastin fragmentation and calcification (Lyle and Raaz, 2017).

Obesity results in aortic stiffening, at least in part mediated through LOX-downregulation, leading to elastin fragmentation and a significant increase in pulse wave velocity (PWV). Modest weight loss results in the improvement of arterial stiffness (Lyle and Raaz, 2017).

3.1.3 Measurement

Carotid/femoral pulse wave velocity (cf-PWV) (quotient of distance between measuring points to onset delay of pulse wave between those locations) is considered the gold standard and robust surrogate in arterial stiffness measurement, and has the largest amount of epidemiological evidence for its predictive value for cardiovascular (CV) events (Laurent et al., 2006; Avolio, 2013; Van Bortel, 2016). Due to the impaired dampening of pulsative flow, pulse travel is faster and thus PWV is increased in stiffer vessels (Smulyan et al., 2016). It is largely independent of risk factors other than age and blood pressure (Cecelja and Chowienczyk, 2009).

Other arterial stiffness assessment techniques exist and have extensively been used in clinical trials, contributing to different outcomes. Aortic pulse pressure (= central SBP – central DBP), augmentation index AIx (percentage of central pulse pressure attributed to reflected wave overlap in systole; function of other risk factors) and brachial flow mediated dilation (FMD) (relative increase of arterial diameter after temporary vessel occlusion) are common markers (Lilamand et al., 2014).

3.1.4 Targets for Treatment

The majority of treatments target the consequences of the vascular ageing process, rather than the pathophysiology itself (Williams, 2016). For known cardiovascular drugs, it is difficult to evaluate whether improvement of arterial stiffness is of any additional benefit unrelated to the drugs' primary effects (Lyle and Raaz, 2017).

Almost all classes of anti-hypertensive medications could decrease arterial stiffness. Renin-angiotensin-aldosterone system (RAAS) antagonists (angiotensin-converting enzyme (ACE) inhibitors, angiotensin II receptor blockers, aldosterone antagonists) have shown the most promising results (Lyle and Raaz, 2017; Smulyan et al., 2016; Wu et al., 2015). Besides blood-pressure-lowering drugs, potential beneficial effects can be obtained with statins, oral anti-diabetics, AGE and collagen cross-link breakers (e.g. alagebrium) and anti-inflammatory agents. The susceptible population, amount of evidence and degree of success all vary (Laurent et al., 2006; Lyle and Raaz, 2017; Smulyan et al., 2016; Wu et al., 2015).

Long-term intake of food supplements with ω-3 fatty acids (lowest effective daily dosage was 540 mg eicosapentaenoic acid (EPA) and 360 mg docosahexaenoic acid (DHA) in overweight patients with hypertension) improved arterial stiffness, especially in a population with overweight, metabolic syndrome, diabetes or hypertension (Wu et al., 2015). However, results from intervention trials with these agents have been largely inconsistent, at least in the context of arterial aging. Omega-3s seem to improve arterial functions in some populations but not others (LaRocca et al., 2017).

Antioxidant vitamins (A, C, E) generally have the ability to reduce arterial stiffness due to antioxidant and anti-inflammatory effects, and by improvement of the endothelial function. However, results of supplementation studies are still controversial. Care must be taken to correctly select the test group and identify their specific requirements, and to avoid possible pro-oxidant effects as the result of high-dose supplementation (Mozos et al., 2017b).

Vitamin D interferes with several mechanisms involved in arterial stiffening, such as a decrease in RAAS activity and modulation of immunological response and calcification. Arterial stiffness is associated with high levels of parathormone. Nevertheless, up to now, there is only inconsistent evidence on vitamin D levels required; conclusions are hampered by the heterogeneity of the studies (Mozos et al., 2017b; Lyle and Raaz, 2017).

Limited data point to vitamin K, especially K2, for promotion of destiffening primarily by reducing arterial calcification and suppressing inflammation (Mozos et al., 2017b).

Autophagy inducers like spermidine reverse age-associated stiffening of the large arteries, illustrated by an improved endothelial function and a reduction in oxidative stress (De Meyer et al., 2015).

Besides interfering with the consequences of arterial stiffening, or trying to influence general antioxidant and anti-inflammatory status, it might also be important to therapeutically address the intrinsic biomechanical properties of the arterial wall. To this end, epigenetic regulators (such as microRNAs) may provide novel therapeutic targets to counteract structural arterial remodelling (Lyle and Raaz, 2017).

3.2 DIETARY POLYPHENOLS

3.2.1 Introductory Aspects on Dietary Polyphenols

Polyphenols are a widespread class of plant secondary metabolites with a diverse range of biological activities. About 8,000 polyphenolic structures have been identified, and they can be subdivided according to their chemical structure into the following structural classes: phenolic acids, lignans, stilbenes, flavonoids (including isoflavonoids, anthocyanins) and condensed and hydrolysable tannins. They derive from phenylalanine and contain an aromatic ring with at least one reactive hydroxyl group. Polyphenols are widely occurring in plants, often in glycosylated form, and are found in several foods: dietary polyphenols most investigated for their vascular properties include flavonoids (flavanols) and procyanidins in chocolate (*Theobroma cacao*), catechins such as epigallocatechin gallate (EGCG) in green tea (*Camelia sinensis*), isoflavones in soy (*Glycine max*), curcumin from turmeric (*Curcuma longa*), oleuropein and hydroxytyrosol (HT) in olives (*Olea europaea*), anthocyanins in berries and resveratrol and other stilbenes in grapes and wine (*Vitis vinifera*) (Costa et al., 2017). An estimation of average daily total polyphenol intake is hampered by their structural diversity and complexity but has been estimated in a French population to be 1193 ± 510 mg/d, or 820 ± 335 mg/d as aglycone equivalents (Perez-Jimenez et al., 2011). Epidemiological studies demonstrate that polyphenol intake is associated with a reduced risk for cardiovascular events (Arts and Hollman, 2005; Goszcz et al., 2017; Lilamand et al., 2014; McCullough et al., 2012; Tresserra-Rimbau et al., 2014). Figure 3.3 illustrates dietary polyphenol classification.

Polyphenols display pleiotropic effects, and although several biological properties, like radical scavenging activity, are determined by the phenolic nature, and are therefore displayed by all polyphenols, quantitative differences occur according to their specific structural features (e.g. number and position of hydroxyl and catechol groups (Goszcz et al., 2017). Other activities are more specific and only occur in some specific (groups of) polyphenols. However, exact polyphenol composition assessment in trials is often inadequate or insufficient. Data are scarce, especially on polyphenol polymers. Moreover, there may be a non-linear dose-response relationship, as reported for flavonoids and polymers (Kay et al., 2012).

In view of the etiology of arterial stiffness, and the important contribution hereto of oxidative and nitrosative stress and inflammation, among other things, plant polyphenols may be excellent biological agents to interfere with the occurrence and progression of stiffening arteries.

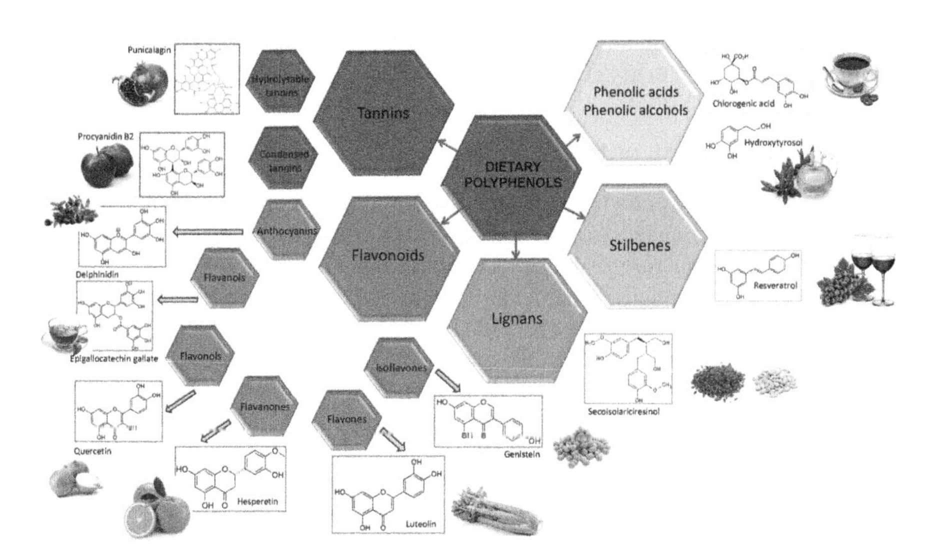

FIGURE 3.3 Classification of dietary polyphenols, with common examples and food sources.

Evidence for the effect of food polyphenols on arterial stiffness is, however, rather limited due to the scarcity and heterogeneity of study design in interventional trials, the complexity of observational trials and the problems related to the translation of observations from animal models to human reality. Moreover, there are important difficulties in assessing polyphenol intake and a lack of uniformity in biomarkers and end points. Nevertheless, limited relevant data exist, and known interference of polyphenols with mechanisms involved in arterial stiffness allows pinpointing some promising interactions.

In general, epidemiological studies and interventional trials suggest an inverse association between dietary polyphenol intake and cardiovascular events both in the general population and in patients (Tresserra-Rimbau et al., 2014; Ludovici et al., 2017b; Rienks et al., 2017).

In this review, the focus will be on the effects of dietary polyphenols on arterial stiffness, although mechanisms involved could also influence other cardiovascular pathologies. Some observational and interventional studies are cited as an illustration of current research, but care should be taken that great variability exists in the compound/food studied, dosage, population, sample sizes, end points and follow-up. Hereafter, rather than discussing effects of separate polyphenolic compounds, or the foods that contain them, effects will be grouped according to the mechanism involved, keeping in mind that the same compound can display diverse activities and that several mechanisms like antioxidative and anti-inflammatory activities are interrelated. Influencing many different targets with lower affinity, like polyphenols do, may result in a combined effect sufficient to provide an overall health benefit (Wang et al., 2017). To our knowledge, this resolves a lack of recent reviews on the effect of dietary polyphenols on arterial stiffness. Besides a review on flavonoids and arterial stiffness, little overview of current evidence is available. Data discussed therein seemed to support an improvement of arterial stiffness with increased flavonoid intake (Lilamand et al., 2014).

3.2.2 Bioavailability of Polyphenols

The biological activity of dietary polyphenols depends on their bioavailability, intestinal absorption and metabolism in the gastrointestinal tract, which itself depends on their chemical structure. Polyphenols can be absorbed from the small intestine, but, more often, since they are frequently present as esters, glycosides or polymers in their food matrix, they cannot be absorbed as such. To be absorbed, these molecules must be previously hydrolyzed by intestinal enzymes or by the colonic microflora. Phase II metabolism then converts them to methylated, sulphated and/or glucuronidated metabolites (Gil-Cardoso et al., 2016). Polyphenols are thus rapidly degraded, metabolized and often poorly absorbed, resulting in limited bioavailability: most native polyphenols are only found in nM to μM ranges, and, in plasma, glucuronidated, sulphated and methylated derivatives are found next to the free phenolic form (Goszcz et al., 2017; Zhang and Tsao, 2016).

On the other hand, large differences exist in bioavailability, and, for example, some flavonoid classes may be absorbed sufficiently to exert cardioprotective effects *in vivo* (Lilamand et al., 2014). Furthermore, intracellular deconjugation metabolism of phase II metabolites, releasing parent polyphenols in cells and tissues and provoking local activity, has to be taken into account.

Non-hydrolyzed polyphenols reach the colon, where they are metabolized by the colonic microbiota, whose phenolic metabolites may act as the true pharmacological agents (Zhang and Tsao, 2016). The microbial composition of the intestine seems to have great relevance for the individual response to these compounds, resulting in a personal specific metabotype or enterotype. This is, for instance, illustrated in the biotransformation of some isoflavones into equol, suggested to have higher efficacy than the parent compound. Only about 30% of the Western population and 60% of Asian subjects can produce equol and have more beneficial health effects from soy consumption, due to the presence of specific bacteria in the gut (Costa et al., 2017).

Moreover, regular consumption of polyphenol-rich foods could in turn influence the colonic bacterial population and their metabolic activities, enlarging inter-individual bioavailability variation. This is illustrated by a significant difference in bacterial metabolite profiles between regular cocoa product consumers and non-consumers after dark chocolate intake (Khan et al., 2014). There is thus a bidirectional phenolic-microbiota interaction. Stratification in clinical trials according to metabotypes is therefore necessary to fully assess the biological activity of polyphenols (Espin et al., 2017; Tomas-Barberan et al., 2016).

The complexity of the metabolic output of the gut microbiota, depending to a large extent on the individual metabolic capacity, emphasizes the need for assessment of functional analysis using metabolomics in conjunction with determinations of gut microbiota compositions (Stevens and Maier, 2016).

Many conflicting results have been reported between *in vitro* and *in vivo* studies Although *in vitro* reports of observed activities are not always reflected in a clinical result, human *in vivo* activities do occur, often by direct interactions with receptors, enzymes and signaling pathways, by modulation of gene expression through activation of various transcription factors or by the activity displayed by degradation products and metabolites.

For flavanols Sies et al. published conclusions reached at the 27th Hohenheim Consensus Conference held on July 11, 2011. Consensus was obtained for flavanol biological functions, which can occur at cellular and systemic levels by modulating cellular signaling and enzyme activities at intakes achievable with a normal diet. Randomized, controlled trials show an effect on blood pressure, LDL cholesterol and FMD (Sies et al., 2012).

The enormous diversity of chemical structures—parent and hydrolyzed polyphenols and phase II and microbial metabolites—hampers the identification of the active compound(s). Often, several metabolites or the whole array of related compounds may be responsible for the observed effect (Feliciano et al., 2016; Mansuri et al., 2014).

3.2.3 Epidemiological Studies with Arterial Stiffness Assessment

Several epidemiological studies indicate a positive correlation between polyphenol-rich food intake (e.g. flavonoid-rich food, cocoa) and several cardiovascular end points like cardiovascular mortality, myocardial infarction, chronic heart disease and heart failure. However, studies specifically linking arterial stiffness outcomes to polyphenols are heterogeneous and limited but seem to indicate a beneficial effect. Next to the already mentioned limitations of dietary polyphenol trials on vascular stiffness, primarily a correct approximation of polyphenol intake seems difficult, especially for polymeric compounds. Usually polyphenol intake is estimated using food frequency questionnaires linked to databases with dietary polyphenol contents. Higher polymers with larger molecular weights will not be bioavailable as such, but their metabolites could contribute to the observed effects. Assessing polyphenol intake should be addressed by the development and measurement of adequate biomarkers in plasma or urine.

Some studies are mentioned here for illustrative purposes.

Higher anthocyanin and flavone intake has been associated with lower PWV in a study on 18–75-year-old women (n = 1898), while higher cocoa intake was linked to lower PWV in another study group of 18–60-year-old persons (n = 198) (Jennings et al., 2012; Lilamand et al., 2014; Vlachopoulos et al., 2007b). Phytoestrogens (isoflavones and lignans) protect against arterial degeneration through an effect on the arterial wall (PWV), especially in older women. (n = 403) (van der Schouw et al., 2002). Epidemiological studies on soy isoflavones in general demonstrated improved arterial compliance, induced nitrite/nitrate levels and decreased endothelin-1 levels in men and postmenopausal women (Upadhyay and Dixit, 2015).

3.2.4 Interventional Human Studies with Arterial Stiffness Assessment

Several, usually small-scale, interventional studies on dietary polyphenols and arterial stiffness have been published. Studies reported are heterogeneous in population, dose, markers and follow-up. The relevance of generally small arterial stiffness effects observed for clinical outcomes remains to be investigated. Often evaluation of arterial stiffness markers is combined with registration of effects on blood pressure and endothelial function, parameters that are also affected by vascular stiffness. Most evidence exists for the beneficial effect of cocoa and its derived products, which has been translated into the approval of a health claim by the European Food Safety Authority (EFSA) on the effect of cocoa polyphenols in maintaining blood vessel elasticity.

3.2.4.1 Cocoa

Cocoa and chocolate are rich in flavonoids and proanthocyanidins. Available literature supports the blood pressure–lowering activity of cocoa (chocolate, cocoa drinks or flavonoid-enriched derivatives); generally, this reduction is more pronounced in systolic blood pressure than in diastolic blood pressure. Cocoa intake has also been associated with decreased cardiovascular risk. Numerous studies also report improvement in vascular function, measured by brachial FMD or by PWV. Effects were best correlated with flavanol intake and plasma concentrations. The positive effect of cocoa flavonoids has been observed in healthy individuals as well as in hypertensive, diabetic, obese, cardiovascular or renal disease patients. Potential mechanisms include activation of NO synthase and increased bioavailability of NO as well as antioxidant and anti-inflammatory properties (Ludovici et al., 2017a; Vlachopoulos et al., 2007a).

As an illustration, a few recent reports on the reduction of arterial stiffness with cocoa or its derived products are listed here.

Vlachopoulos reports increased FMD and reduced aortic AIx for dark chocolate (100g) without changes in PWV in healthy volunteers (n=17) (Vlachopoulos et al., 2005). In another study in healthy individuals (n=20), a simultaneous decrease in arterial stiffness, blood pressure and endothelin-1 levels has been observed (5 times 10g/d during 1 week) (Grassi et al., 2015a). Analogously, dark chocolate increases FMD in healthy and hypertensive subjects with and without glucose intolerance (Borghi and Cicero, 2017; Grassi et al., 2008; Grassi et al., 2012). Similar results were obtained showing decreased PWV, improved FMD and blood pressure after 14 days of intake of cocoa flavanol drinks (450 mg cocoa flavanols/d; n=22 young + n=20 elderly men) (Heiss et al., 2015). Acute cocoa supplementation in diabetes type 2 patients (n=18; 960 mg total polyphenols) only showed an effect in large artery elasticity (but not in small arteries) and on blood pressure or insulin resistance (Basu et al., 2015). In overweight adults, dark chocolate and cocoa enhanced vasodilation generally, but only in women the AIx was decreased (n=30, 4 weeks, 37 g/d dark chocolate and a cocoa beverage (22 g total cocoa/d)) (West et al., 2014).

In addition to several reports on the improvement of arterial stiffness by cocoa flavonols, Sansone et al. observed a more pronounced effect on brachial PWV and on FMD, DBP and circulating angiogenic cells in healthy volunteers (n=47, several studies) when methylxanthines (theobromine, caffeine) were administered simultaneously. The co-ingestion of methylxanthines and cocoa flavonols seems to have an influence on the absorption, resulting in increased plasma concentrations of (–)-epicatechin metabolites (Sansone et al., 2017). Apparently, caffeine does not show comparable effects on coffee polyphenols (Vlachopoulos et al., 2007a).

Recently, correction of vascular impairment (FMD, PWV) after sleep deprivation has been observed after acute intake of flavanol-rich chocolate, while working memory performance was also restored (n=32) (Grassi et al., 2016).

Isolated flavonoids epicatechin (100 mg/d) (present in cocoa) and quercetin (160 mg/d) failed to show any effect on FMD, arterial stiffness, blood pressure, NO or endothelin 1. Epicatechin, however, improved insulin resistance in a four-week crossover trial (n=37) (Dower et al., 2015). In a following study by the same group, however, the effects of pure epicatechin are compared to those of dark chocolate. FMD and AIx were measured. There were no differences between pure epicatechin and dark chocolate, while only dark chocolate influenced NO and endothelin-1 levels. The bioavailability of epicatechin seemed irrespective of the matrix. Epicatechin may thus contribute to the vascular function effects of cocoa (n=20) (Dower et al., 2016).

Although some deviating results have been reported, the bulk of evidence observed in both epidemiological and interventional studies led to the approval of a health claim about the effect of cocoa polyphenols in maintaining blood vessel elasticity by the European Food Safety Authority (EFSA). To achieve this, 200 mg of cocoa flavanols, consumed as 2,5 g high-flavanol cocoa powder or 10g high flavanol dark chocolate, should be ingested daily (Ludovici et al., 2017a).

3.2.4.2 Coffee

Habitual coffee consumption is associated with increased aortic stiffness and wave reflections in healthy subjects. This is not only due, as expected, to the vasoconstrictive caffeine, since de-caffeinated coffee also augments arterial stiffness to a certain extent. Increased inflammatory markers are observed with chronic coffee intake (Vlachopoulos et al., 2007a).

For the acute administration of chlorogenic acids, the main polyphenols in coffee, there were no significant effects observed on BP, nor were there significant effects of the predominant isomer 5-CGA on peak FMD response. However, there were significant improvements in mean post-ischemic FMD response, in healthy individuals (n = 16) (Ward et al., 2016)

In another study, a single consumption of coffee polyphenol extract improved postprandial hyperglycemia and vascular endothelial function (FMD), which was associated with increased glucagon-like peptide 1 (GLP-1) secretion and decreased oxidative stress in healthy humans (n = 19) (Jokura et al., 2015).

3.2.4.3 Tea

Black tea (flavonoids, theaflavins, thearubigins) (800 mg flavonoids/d) intake dose-dependently decreased arterial stiffness as well as blood pressure in healthy volunteers and in hypertensive patients.(n = 19; 5 periods of 1 week) (Grassi et al., 2009). However, a study with green tea (mainly flavanols, like epigallocatechin-3-gallate) (9g; 4 weeks) in type 2 diabetic patients (n = 55) did not show changes in PWV, nor in inflammatory markers (Ryu et al., 2006). Another study with green tea (acute; 836 mg green tea catechins) showed a different influence on the digital volume pulse-stiffness index according to genotype differences with regard to catechol-O-methyl-transferase (COMT) (Miller et al., 2011). The counteracting effect of caffeine in tea reduces the potential beneficial effect of the tea polyphenols but results in a faster and shorter plasma peak level (Vlachopoulos et al., 2007a).

3.2.4.4 Fruit

3.2.4.4.1 Various Fruits

In general, men with increased cardiovascular disease (CVD) risk, consuming flavonoid-rich fruits and vegetables, benefit by an increased endothelium-dependent microvascular reactivity, the prevention of vascular stiffness and reduced NO. Reduction of inflammatory biomarkers was observed (see Section 3.2.5.3) (Grassi et al., 2015b; Macready et al., 2014).

Apples with skin ('flavonoid-rich apples') significantly increased FMD in comparison with apple flesh only ('low flavonoid apple'), without significant differences in blood pressure and arterial stiffness (N = 30; 4 weeks) (Bondonno et al., 2018).

An anthocyanin-rich red orange juice reduced postprandial triglyceride concentration and vascular stiffness (AIx) after intake of a fatty meal. These effects were not observed with the anthocyanin-poor blond orange juice. Both juices significantly prevented white blood cell increase and myeloperoxidase release, thus preventing the low-grade inflammation induced by a fatty meal (n = 18, acute) (Cerletti et al., 2015).

Consumption of grapefruit juice (providing 210 mg naringenin glycosides daily) for six months in a crossover trial in postmenopausal women decreased carotid-femoral PWV without affecting endothelial function, inflammation or oxidative stress (n=48) (Habauzit et al., 2015).

On the other hand, a single intake of orange juice or a matched dose of a hesperidin (320 mg; n=16) supplement did not alter cardiovascular risk biomarkers, including endothelial function, arterial stiffness or blood pressure (Schaer et al., 2015), nor did a single pomegranate drink influence the digital volume pulse-stiffness index (Mathew et al., 2012). In another trial, pomegranate juice supplementation in healthy adults (n=51, 330 ml/d) for four weeks showed no effect on PWV and plasma FRAP while displaying a significant fall in SBP and DBP (Lynn et al., 2012).

Ray and coworkers already concluded in 2014 that several factors limit the assessment of vascular and endothelial dysfunction in nutritional studies with fruit juice intake. Heterogeneity in methodology and study design, limited data, bioavailability and metabolism issues complicate interpretation (Ray et al., 2014), which also holds for other polyphenol research on this topic.

A single dose of a mango (*Mangifera indica*) fruit preparation (100 mg or 300 mg Careless™) provoked an increase in cutaneous blood flow in a small trial on volunteers (n=10) without significant improvement in endothelial function (measured by EndoPAT); activation of eNOS has, however, been observed. The activating properties of Careless towards sirtuin 1 (SIRT1) and AMPK, which have been identified as playing a key role in microcirculation and endothelial function, had already been described (Gerstgrasser et al., 2016).

3.2.4.4.2 Berries

Berry (poly)phenols (primarily flavonoids, isoflavonoids, anthocyanins, proanthocyanidins) have been investigated in a few small-scale studies for their beneficial effects on several surrogate markers of cardiovascular risk, including arterial stiffness. Anthocyanins are probably the main bioactive compounds that characterize berries and are found mainly in the external layer of the pericarp (Vendrame et al., 2016).

Cranberry juice (*Vaccinium macrocarpon;* 500 ml juice/d) failed to show a significant decrease in AIx as a measure of arterial stiffness in abdominally obese men (35 men, 4 weeks crossover against placebo juice). However, a significant within-group decrease in AIx was noted over four weeks (Ruel et al., 2013). In another study, cranberry juice (94 mg anthocyanin—4 weeks) displayed an acute lower PWV in adults (n=44) with coronaropathy (Dohadwala et al., 2011; Lilamand et al., 2014). In nonsmokers and young smokers, a blueberry (*Vaccinium corymbosum*) serving (300 g blueberries, acute) improved peripheral arterial dysfunction measured by the reactive hyperemia index, but failed to display differences in the digital augmentation index as a measure for arterial stiffness. This was also not reflected in modulation of oxidative stress and antioxidant defense markers (Del Bo et al., 2014, 2016, 2017). In postmenopausal women with hypertension (n=48, 8 weeks), daily blueberry powder (22g) did, however, reduce blood pressure and arterial stiffness (brachial-ankle PWV), which may at least partly be due to the observed increase in NO production (Johnson et al., 2015). Addition of freeze-dried strawberry powder (*Fragaria ananassa*) to a high-fat meal did not alter vascular function or postprandial triglycerides, glucose or insulin compared to a control meal (n=30) (Richter et al., 2017b). Blackcurrant drinks (*Ribes nigrum*), rich in anthocyanins, decreased post-prandial glucose and insulin but did not exhibit effects on arterial stiffness or 8-iso-prostane F2α (n=23) (Castro-Acosta et al., 2016).

Black raspberry (*Rubus occidentalis*) extract (750mg/d; 12 weeks) improved arterial stiffness and vascular endothelial function in patients with metabolic syndrome, and inflammatory cytokines IL-6 and tumor necrosis factor-α (TNF-α) were decreased (Jeong et al., 2014, 2016).

Although some promising activities can be noted, the data are nevertheless not sufficient to correlate berry polyphenol intake with improved arterial stiffness. The importance of bioavailability of polyphenols has been stressed (Rodriguez-Mateos et al., 2014). Anti-inflammatory and antioxidant effects are frequently reported in an overview on berry consumption in metabolic syndrome patients (Vendrame et al., 2016).

3.2.4.4.3 Grapes/Wine/Stilbenes

In a dose-response-oriented review of studies published on polyphenols in general and Concord grape (monomeric and oligomeric favan-3-ols) juice, in particular, on cardiovascular risk factors, Blumberg and coworkers concluded that there appeared to be a strong relationship between daily total polyphenol dose and change in FMD observed in chronic beverage intervention studies. Also, for Concord grape juice, clinically significant effects on FMD have been observed (Blumberg et al., 2015).

In healthy smokers, Concord grape juice significantly improved FMD and carotid-femoral PWV (n=26, 2 weeks, crossover) (Siasos et al., 2014).

In another study, on the other hand, grape seed extract reduced SBP and DBP after six weeks of supplementation but failed to show significant changes in FMD or adhesion molecules. Effects were more pronounced in subjects with higher initial BP (n=36) (Park et al., 2016).

Two spray-dried grape extracts (800 mg polyphenols/d), grape-red wine and grape extracts were compared in a four-week study of 60 mildly hypertensive subjects. Only the grape-wine extract, rich in catechins and procyanidins, decreased SBP and DBP, and this effect was independent of alcohol. A reduced plasma endothelin-1 concentration has been proposed as the mechanism of action (Draijer et al., 2015).

Although in postmenopausal women (n=45; 1g polyphenols/d), a six-week consumption of dealcoholized red wine or red wine did not exhibit differences compared to control, arterial stiffness (AIx) improved over time in the dealcoholized red wine group compared to baseline (Naissides et al., 2006).

Resveratrol (stilbene) itself, present in grape skin (100 mg/d; 12 weeks; 50 patients), was able to ameliorate arterial stiffness (cardio-ankle vascular index CAVI) in type 2 diabetes patients, next to a reduction in body weight and body mass index, SBP and oxidative stress (diacron-reactive oxygen metabolites) (Imamura et al., 2017). Both acute (n=19) and chronic (n=28; 75 mg/d; 6 weeks) supplementation with resveratrol also improved FMD in overweight and obese adults (Wong et al., 2011, 2013).

Grape extracts and wine seem to have some blood-pressure-lowering effects, and a few reports point to an improved FMD, although conflicting reports have also been published. An effect on NO production and endothelin-1 synthesis has been postulated (Vlachopoulos et al., 2007a).

3.2.4.5 Soy/Isoflavones

Evidence from earlier trials on soy, summarized by Pase et al., suggested a beneficial effect of soy intakes (and isoflavones herein) on arterial stiffness measured through PWV and arterial compliance, and this could, at least in part, explain the low incidence of heart disease in populations with high soy intake (Pase et al., 2011). This has been confirmed by Lilamand and coworkers, who observed a decrease in PWV in healthy adults after isoflavone supplementation (Lilamand et al., 2014).

Red clover isoflavones (80 mg/d; 6 weeks), enriched in formononetin, had beneficial effects on arterial stiffness and vascular resistance in men and postmenopausal women (n = 80) (Teede et al., 2003). Isoflavone intake (50 mg/d) in black soybean tea effectively reduced the cardio-ankle vascular index (CAVI) among both premenopausal smokers and nonsmokers, but had no effect in postmenopausal smokers or nonsmokers; brachial-ankle PWV, on the other hand, showed no differences, pointing out that results can differ largely, depending on the marker used (n = 55, 8 weeks) (Hoshida et al., 2011).

Treatment with isoflavone-containing soy protein, on the other hand, failed to show differences in AIx and PWV in a patient group with slightly elevated blood pressure in a small randomized crossover trial (25 and 50g soy protein/d; 6 weeks, 20 patients). Authors suggest influence of different bioavailability depending on the food matrix to explain differing results in other studies (Richter et al., 2017a).

A significant decrease of PWV and improvement of pulse pressure variability has been observed in a one-year trial with isoflavones-enriched chocolate (850 mg flavan-3-ols + 100 mg isoflavones) in type 2 diabetic postmenopausal women (n = 93). Equol producers were particularly responsive (Curtis et al., 2013). Also earlier, equol-producing ability had been linked to a decreased arterial stiffness (Gil-Izquierdo et al., 2012). Pasta enriched with soy isoflavone aglycones (33 mg/d) improved arterial stiffness and reduced blood pressure, consistent with an improved oxidative stress status (lower 8-iso-PGF2α, higher reduced glutathione GSH). Improvements were more important in equol producers (n = 26 type 2 diabetics; n = 62 hypercholesterolemia subjects) (Clerici et al., 2007, 2011). Soy isoflavones significantly improved carotid-femoral PWV in male equol producer phenotypes, corresponding to an 11%–12% reduced risk of cardiovascular disease, in contrast to the absence of any effect for commercial equol in non-equol producers, indicating the importance of the equol producer phenotype for vascular protection (n = 28, 80 mg isoflavone aglycones, acute) (Hazim et al., 2016).

However, soy nut snack consumption for a four-week period (101 mg isoflavone aglycones) improved arterial stiffness (AIx) in a small interventional trial (n = 17) in adults with cardiometabolic risk, independent of the investigated inflammatory biomarkers (TNF-α, oxLDL, CRP, interleukins), and independent of equol or O-desmethylangolensin production (Reverri et al., 2015).

In spite of some diverging results, the importance of metabolites in biological activities of polyphenols should thus not be neglected, as illustrated for isoflavones. This is demonstrated by the effect of *trans*-tetrahydrodaidzein (1g/d), a metabolite normally formed after consumption of isoflavones (formononetin, daidzein), which reduces blood pressure and arterial stiffness in obese men and postmenopausal women (n=25; 5 weeks) (Nestel et al., 2007).

3.2.4.6 Olive

Acute ingestion of olive polyphenols oleuropein and hydroxytyrosol in an olive leaf extract (51 mg OLE; 10mg HT) improved vascular function (digital volume pulse-stiffness index) and *ex vivo* IL-8 production. The observed effects were related to analysis of the phenolic metabolites in urine (n=18) (Lockyer et al., 2015). A recent double-blind randomized controlled trial confirmed this antioxidant effect and showed that a combination of red yeast rice and olive extract (containing 9.32 mg HT) decreased SBP and DBP (n=50; 8 weeks) (Hermans et al., 2017; Verhoeven et al., 2015). An olive fruit extract standardized in hydroxytyrosol reduced the cardio-ankle vascular index (CAVI) in subjects with arterial stiffness risk (Pais et al., 2016).

3.2.4.7 Miscellaneous

A short study on 39 subjects investigating the effects of a polyphenol-rich beverage (with 722 mg polyphenols from green tea, grape seed, grape pomace, ruby red grape juice, lemon and apple and 240 mg of vitamin C/d) on PWV and microvascular responses to sodium nitroprusside and acetylcholine laser Doppler iontophoresis did not show any differences. The only effect observed was an increase in IL-6, indicating an increased inflammatory state. Authors postulate a possible pro-oxidant effect of isolated polyphenols and suggest larger and longer intervention studies to evaluate this more thoroughly (n=39; 4 weeks) (Mullan et al., 2016).

A curcuminoid extract (1,5 g/d) significantly reduced PWV in type 2 diabetic subjects in a six-month trial (n=240) (Chuengsamarn et al., 2014), and curcumin (150 mg/d) increased FMD (and therefore endothelial function) in postmenopausal women (n=32; 8 weeks) (Akazawa et al., 2012).

In a crossover trial in an overweight population, daily walnut consumption (56g/d) improved FMD of the brachial artery from baseline together with a reduction in SBP (n=46; 8 weeks) (Katz et al., 2012). The principal polyphenols in walnut are ellagitannins, mainly pedunculagin (Sanchez-Gonzalez et al., 2017).

A quercetin-rich onion-skin extract (162 mg quercetin/d) decreased SBP in a trial on overweight or obese subjects with hypertension. Other variables, however, like endothelial function, adhesion molecules, endothelin-1 and markers of inflammation and oxidative stress, remained unchanged (n=70; 6 weeks) (Bruell et al., 2015)

Melissa officinalis (lemon balm) extract (169 mg polyphenols/d; 123,8 mg rosmarinic acid/d) showed significant reduction in brachial-ankle PWV (28 healthy persons, 6 weeks) in an open label parallel comparative trial (Yui et al., 2017).

3.2.5 Impact on Mechanisms Contributing to Arterial Stiffness

The impact of dietary phenols on the most important mechanisms contributing to the pathophysiology of arterial stiffness are discussed below. Vascular, antioxidant, anti-inflammatory, antiglycation and autophagy-inducing effects are addressed.

However, several of those mechanisms, like oxidative stress and inflammation, are closely interrelated, since oxidative stress can cause inflammation, which in turn can induce oxidative stress. Both oxidative stress and inflammation cause injury to endothelial cells. Endothelial dysfunction in turn promotes a proinflammatory environment, resulting *inter alia* in an increased expression of adhesion molecules. As a positive feedback loop, vascular inflammation leads to endothelial dysfunction (Siti et al., 2015).

Radical scavenging and metal-chelating properties of polyphenols, important for an antioxidative effect, can also contribute to antiglycation. Furthermore, transcription factors and signaling pathways involved in oxidation and inflammation are implicated in autophagy.

3.2.5.1 Vascular Effects

Vasomotor responses of the endothelium are regulated by NO (vasodilating) and endothelin (vasoconstricting). Human trials have demonstrated vasoprotective effects mediated by NO, which is produced by endothelial nitric oxide synthase (eNOS). Adequate production and bioavailability of eNOS-derived NO is necessary for the maintenance of a healthy endothelium; a reduced eNOS-derived NO bioavailability results in endothelial dysfunction (Katz et al., 2011).

On the other hand, blood pressure is determined by the renin-angiotensin-aldosterone system (RAAS), producing the vasoconstrictive angiotensin II (Ang II) from angiotensin I (Ang I) by the angiotensin-converting enzyme (ACE).

Polyphenols display antihypertensive effects by increasing NO production in endothelial cells as well as by a direct inhibition of ACE (Lilamand et al., 2014).

Several dietary polyphenols increase the production or bioavailability of endothelial nitric oxide, as seen for cocoa flavanols. Their vasodilatory response is NO dependent and can be reversed by blocking nitric oxide synthesis (Katz et al., 2011).

Endothelin-1 (ET-1), on the other hand, is a potent vasoconstrictor peptide with pro-oxidant and proinflammatory properties and is of interest in the development of endothelial dysfunction. ET-1 expression and production in endothelial cells is, among others, increased by Ang II-stimulation and ageing. ET-1 overexpression activates NADPH oxidase, and therefore ROS formation, causing oxidative stress and forming a positive feedback loop of oxidative stress-mediated endothelial oxidative injury and dysfunction. Next to that, oxidative stress also causes amplification of the angiotensin-converting enzyme (ACE) activity, subsequently stimulating the angiotensin II receptor type 1 (AT-1) receptor by Ang II, and thus inducing the production of ROS by NADPH oxidase and amplifying the detrimental process (Siti et al., 2015).

Several acute and short-term trials have investigated effects of flavonoid-rich foods and beverages on FMD as marker of endothelial function. indicating an increase in FMD of about 20%–30% (Croft, 2016). Phytoestrogens (isoflavonoids (genistein, daidzein), flavonoids (e.g. artemetin), anthocyanins (e.g. delphinidin) induce vasodilation by binding to estrogen receptors in physiologically relevant concentrations, leading to increased endothelial NOS (eNOS) activity and increased NO synthesis (Goszcz et al., 2017). Inhibition of ET-1 release in human umbilical vein endothelial cells (HUVECs), together with increased eNOS expression, has been demonstrated for delphinidin and cyanidin. For delphinidin glycosides, stimulation of endothelin B receptors and inhibition of ACE have also been reported (Goszcz et al., 2017).

The specific mechanisms by which cocoa flavanols improve vascular function are still under investigation but appear not to be explained by their general antioxidant effects. It is more likely achieved by modulation of NADPH oxidase, which generates superoxide radical anion, to maintain levels low enough to not harm the vascular endothelium. An increased activity of NADPH oxidase is implicated in vascular dysfunction (Katz et al., 2011). Cocoa supplementation decreased both SBP and DBP in several studies. In a 2014 review, Latham et al. concluded that cocoa flavanols' beneficial cardiovascular effects are the result of increased NO bioavailability (Ferri et al., 2015; Latham et al., 2014). Antioxidant activity can contribute to an enhanced endothelial function. NO degradation is related to free radical action, and therefore vascular function is also function of the antioxidant (and anti-inflammatory) actions, as observed for cocoa (Ludovici et al., 2017a).

In a study by Ibero-Baraibar et al., cocoa intake for four weeks significantly decreased postprandial SBP, while bioavailability of cocoa compounds was confirmed by analysis of 14 derived metabolites in plasma (Ibero-Baraibar et al., 2016).

Resveratrol increases endothelial NO production, thereby improving endothelial dysfunction and lowering BP in hypertensive rats, which is explained by calcium-dependent eNOS activation (Li et al., 2016). Morin, a flavonol present in the *Moracaea* family, protects against endothelial dysfunction in diabetes through an Akt (protein kinase B)-dependent activation of eNOS signaling (Taguchi et al., 2016).

Blood pressure reduction by cocoa through stimulation of eNOS activity has been confirmed by Ludovici and coworkers, next to an increase in L-arginine bioavailability by reduced arginase activity. Additionally, inhibition of ET-1 production and of L-NAME has been observed (Ludovici et al., 2017a). Also, ACE is inhibited by procyanidin-rich chocolate (Ferri et al., 2015; Ludovici et al., 2017a). Oligomeric procyanidins are reported to stimulate endothelium-dependent vasodilation, suppress ET-1 synthesis and inhibit the activity of ACE, resulting in blood-pressure-lowering effects. However, bioavailability of those oligomers is an important factor for translation into *in vivo* effects (Hugel et al., 2016).

A similar effect is also observed for tea flavanols (-)-epicatechin, (-)-epigallocatechin and their gallates that dose-dependently inhibit ACE in HUVEC cells (Ferri et al., 2015; Persson et al., 2006). Interference of flavonoids in blood pressure regulation by RAAS results in a lower production of superoxide anion by NADPH oxidase (Ferri et al., 2015).

Adenosine monophosphate-activated protein kinase (AMPK) is an important sensor of cell energy status and can be activated by stressors such as oxidative stress, hypoxia and nutrient deprivation. Targets of AMPK include enzymes of glucose and lipid metabolism, mitochondrial enzymes and eNOS, which is responsible for NO production (Croft, 2016).

EGCG is able to increase cytosolic calcium concentrations, contributing to NO production by binding to calmodulin in the heart and vascular endothelium. Furthermore, it activates AMPK, and consequently reduces ET-1 expression (Kim et al., 2014).

In vivo, resveratrol treatment stimulates the activities of sirtuin 1 (SIRT1) and AMPK, both of which influence the regulation of metabolism (Smoliga et al., 2011).

Increases in media-to-lumen ratios and wall component stiffness were attenuated by resveratrol and the related stilbenoids pterostilbene and gnetol in a spontaneously hypertensive heart failure rat model. However, the authors could not demonstrate a role of AMPK or ERK herein (Lee et al., 2017). In contrast, an earlier report mentions attenuation of the vascular geometry remodeling process and ERK-signaling by resveratrol in spontaneously hypertensive rats, rather than direct effects on arterial wall stiffness (Behbahani et al., 2010). In spontaneous hypertensive rats, whole grape extract tended to reduce arterial wall component stiffness but did not show significance. Reduced blood pressure and improved vascular function and compliance were, however, observed, and due not only to the grape resveratrol but to several grape components (Thandapilly et al., 2012).

A low-molecular grape seed polyphenol extract, rich in flavanols, decreased plasma ET-1, upregulating eNOS and Sirt-1 and downregulating aortic gene expression of ET-1 and NADPH in rats, indicating the vasoprotective effect of grape seed flavanols (Pons et al., 2016).

Curcumin supplementation in young and old mice resulted in amelioration of age-associated large elastic artery stiffening (PWV), NO-mediated vascular endothelial dysfunction, oxidative stress and increases in collagen and AGEs in mice (Fleenor et al., 2013).

3.2.5.2 Oxidant Status

The chemical structure of several polyphenols is ideal for scavenging free radicals and reactive oxygen species. The aromatic feature and highly conjugated system with multiple hydroxyl groups make these compounds excellent electron or hydrogen atom donors, neutralizing free radicals and other ROS. Therefore, dietary phenolics are powerful antioxidants *in vitro* (Zhang and Tsao, 2016). Differences exist depending on the number and location of the free hydroxyl groups—the presence of catechol groups are especially important—and on the electron deficiency in anthocyanins. Besides direct scavenging of ROS and reactive nitrogen species (RNS), polyphenols can react with the peroxidation products of macromolecules such as lipids, proteins, DNA and RNA, or act as metal chelators. Nevertheless, the promising *in vitro* antioxidant capacity cannot easily be extrapolated to an *in vivo* situation due to the limited bioavailability of polyphenols (Goszcz et al., 2017).

The plasma concentration of flavonoids is typically insufficient (less than 1 mmol/L) to exert significant antioxidant activities via direct radical scavenging or reducing power, measurable by the existing *in vitro* assay methods. The complex intrinsic antioxidant system also makes it difficult to validate the systemic antioxidant effects of the poorly absorbed phenolic compounds *in vivo* (Zhang and Tsao, 2016).

Although polyphenols have been linked to a reduced risk for CVD, primarily indicated by altered biomarkers of oxidative stress, a causal link is more difficult to prove (Goszcz et al., 2017). Moreover, high consumption of antioxidant polyphenols has a noxious pro-oxidant effect, thus stimulating oxidation of biomolecules. On the other hand, moderate pro-oxidant effects could also turn out to be beneficial, by stimulation of the intracellular antioxidant defense mechanisms, like antioxidant enzymes (Goszcz et al., 2017). Indeed, it has become clear that the antioxidant effect goes beyond direct interference with ROS. Modulation of ROS production in mitochondria, NADPH oxidases and uncoupled eNOS, together with upregulation of antioxidant enzymes such as glutathione-S-transferase (GST), superoxide dismutase (SOD), glutathione reductase (GR), quinone oxidoreductase 1, heme oxygenase 1 (Hmox-1) and glutamyl-cysteine ligase (GSL) is more relevant (Amiot et al., 2016).

A proposed mechanism for 'nutritional antioxidants,' like polyphenols, involves the paradoxical oxidative activation of the NFE2-related factor 2 (Nrf2) signaling pathway. Nrf2 can be activated by ROS in the cytoplasm, after which it translocates to the nucleus and regulates ARE-mediated transcriptions of various genes encoding the above-mentioned antioxidant enzymes (Goszcz et al., 2017; Zhang and Tsao, 2016). Nrf2 is under constant control of the redox-sensitive repressor protein Keapl (Zenkov et al., 2016).

Low concentrations of phenolic compounds (or their metabolic products) and the quinones formed under the influence of interactions with ROS are electrophiles that can interact with Keapl (Kelch-like ECH-associated protein 1) and thus lead to activation of the redox-sensitive Nrf2 (Croft, 2016; Forman et al., 2014). The consumption or supplementation of dietary polyphenols has indeed been shown to restore the redox homeostasis, inducing an antioxidant response in target cells using the Nrf-2/ARE (Nrf-2/antioxidant responsive element) pathway and thus inducing detoxifying enzymes.

Resveratrol, for instance, demonstrates a wide range of biological effects, of which many are related to the ability to scavenge radicals and activate the Keapl/Nrf2/ARE signaling system. Additionally, it inhibits transcription factors NF-κB, AP-1, p53 and activates kinases MAPK, Akt, AMPK, PI3K as well as SIRT1. Flavonoids, including catechins like EGCG and hydroxytyrosol from olive, also are seen to induce this antioxidant-signaling system (Zenkov et al., 2016). Modulation of such antioxidant-signaling cascades by polyphenols recently has been evidenced extensively *in vitro* and animal models (Goszcz et al., 2017). Flavonoids modulate different signaling cascades such as phosphoinositide 3-kinase (PI 3-kinase), Akt/PKB, tyrosine kinases, protein kinase C (PKC) and MAPK (Ferri et al., 2015). In addition, flavonoids also modulate the expression of various genes through activation of a broad range of transcription factors (Mansuri et al., 2014). Curcumin activates the Keapl/Nrf2/ARE system and induces the expression of antioxidant genes (Zenkov et al., 2016). The vascular protection and antioxidative effect of soy isoflavone diets is attributed mostly to an upregulation of eNOS expression and activity, increased NO bioavailability associated with Nrf2 accumulation and ARE dependent activation of antioxidant defense enzymes (Upadhyay and Dixit, 2015). Cocoa flavanols can directly interact with ROS but exhibit antioxidant effects indirectly through modulation of crucial oxidative stress-related enzymes: induction of antioxidant enzymes and inhibition of pro-oxidant enzymes like NADPH oxidase (Martin and Ramos, 2016). Anthocyanin-rich beverages increased superoxide dismutase and catalase, and decreased malondialdehyde, a biomarker of lipid peroxidation, without affecting inflammatory biomarkers in healthy women (Kuntz et al., 2014).

Moreover, flavonols and isoflavones can regulate aryl hydrocarbon receptor (AhR)-mediated signaling in cells, thus influencing (Nrf)-2 translocation (Zhang and Tsao, 2016).

Chlorogenic acid is seen to protect against HOCl-induced oxidative damage in mouse endothelial cells *ex vivo*, via increased production of NO and induction of heme oxygenase 1 (Hmox-1). Hmox-1 is a key regulator of endothelial function and is involved in vascular protection against ROS-induced oxidative damage, and AMPK activation results in its expression through the Nrf2/ARE pathway (Jiang et al., 2016).

Additionally, dietary polyphenols can also suppress oxidative stress by interfering with inflammatory signaling cascades controlled by nuclear factor kappa B (NF-κB) and mitogen-activated protein kinase (MAPK). Activation of these cellular processes leads to induction of regulatory immune responses. As a result, proinflammatory cytokines, including interleukin (IL)-1β, IL-6, IL-8, tumor necrosis factor (TNF)-α and interferon (IFN)-γ are released (Zhang and Tsao, 2016). Figure 3.4 shows modulation by dietary polyphenols of oxidative/antioxidative pathways involved in arterial stiffness.

3.2.5.3 Anti-Inflammatory Activity

Antioxidant and anti-inflammatory pathways influenced by dietary polyphenols are largely intertwined and can affect similar biomarkers (Zhang and Tsao, 2016).

NF-κB is a central factor in inflammation. It is a transcription factor that stimulates the encoding of several genes, including those responsible for producing cytokines, chemokines, immunoreceptors, cell adhesion molecules and acute-phase proteins. The activation of NF-κB is redox-sensitive. Direct inhibition of NF-κB by polyphenols (e.g. resveratrol and curcumin analogues) is an important mechanism for their anti-inflammatory effect (Goszcz et al., 2017).

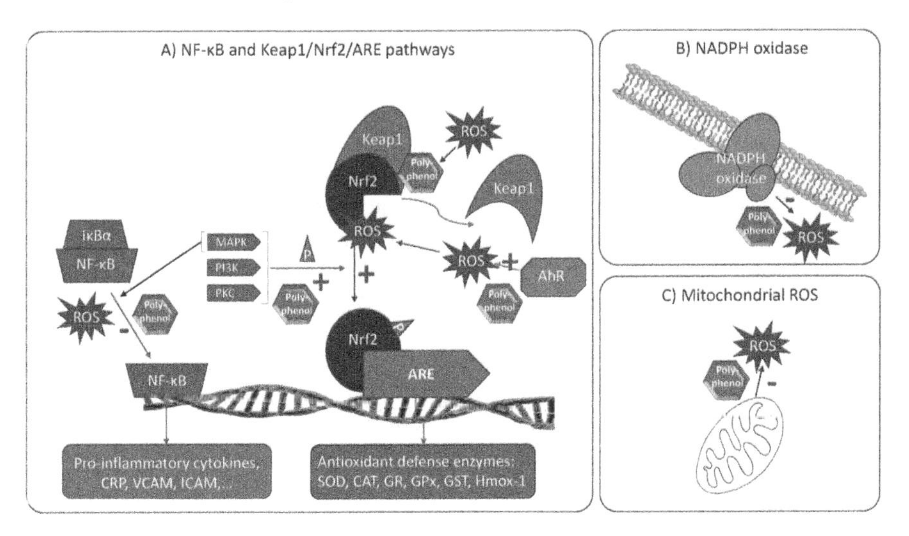

FIGURE 3.4 Overview of oxidative/antioxidative pathways involved in arterial stiffness etiology and interactions of dietary polyphenols. Stimulation is denoted by +; inhibition by –.

The nucleotide-binding oligomerization domain, leucine-rich repeat containing gene family and pyrin-domain containing 3 (NLRP3) inflammasome is a key node that links the signaling cascades between antioxidant response and inflammation and has recently been shown to be modulated by polyphenols. Increased ROS activates NLRP3, which induces IL-1β, and via toll-like receptor (TLR)-1, it triggers NF-κB-activated and MAPK-induced proinflammatory signaling, producing inflammatory cytokines such as IL-1β, IL-6, IL-8, TNF-α and IFN-γ (Zhang and Tsao, 2016). Several reports on the modulation of NLRP3 activation by polyphenols (e.g. resveratrol, procyanidin B2, chlorogenic acid) resulting in an anti-inflammatory effect have been published recently (Misawa et al., 2015; Shi et al., 2018; Yang et al., 2014).

Anthocyanins also target the MAPK pathway (Goszcz et al., 2017). Polyphenols, including flavonoids, have also been reported to stimulate PPAR-γ (peroxisome proliferator-activated receptor γ) or SIRT1-mediated signaling and to interfere with TNF-α-induced MAPKs and NF-κB proinflammatory signaling transductions, resulting in the repression of inflammation (Zhang and Tsao, 2016). In a small study in two different diabetic rat models (insulin-deficient and insulin-resistant), baicalein ameliorated blood pressure elevations and exhibited both antiglycation (AGEs) and anti-inflammatory (NF-κB, TNF-α) mechanisms (El-Bassossy et al., 2014).

Curcumin is a potent multi-targeted polyphenol that modulates multiple cell signaling pathways linked to different chronic diseases. It has been shown to exhibit anti-inflammatory effects by downregulating various cytokines, such as TNF-α, IL-1, IL-6, IL-8, IL-12, monocyte chemoattractant protein (MCP)-1 and IL– 1β, and various inflammatory enzymes and transcription factors (Kunnumakkara et al., 2017).

Dietary polyphenols can also modulate proinflammatory NF-κB signaling through targeting of the RelB/AhR complex, which is also involved in redox management due to binding with the xenobiotic responsive element. Dietary flavonoids are known as AhR modulators (Zhang and Tsao, 2016).

Cell adhesion molecules, including intercellular adhesion molecule-1 (ICAM-1), vascular cell adhesion molecule-1 (VCAM-1) and endothelial selectin (E-selectin) are glycoproteins involved in tissue integrity, cellular communication and interactions, and extracellular matrix contact. They are increased in endothelial dysfunction, vascular remodeling and obesity (Mozos et al., 2017a). Polyphenols also display anti-inflammatory properties by inhibiting adhesion molecule production (VCAM, ICAM-1) by the endothelium. *In vitro*, 14 phenolic acid metabolites and six flavonoids were screened for their relative effects on sVCAM-1 secretion by HUVECs stimulated with tumor necrosis factor alpha (TNF-α). Protocatechuic acid was the most active of the phenolic metabolites, while native flavonoids showed no activity in HUVECs (Warner et al., 2016). *In vivo*, conflicting results exist on modulation of cell adhesion molecules (VCAM and/or ICAM) after interventions with polyphenol containing foods (Amiot et al., 2016). Resveratrol ameliorates aortic stiffness (PWV) in mice with metabolic syndrome by activation of vascular smooth muscle sirtuin-1, associated with a decrease in NF-κB activation and VCAM-1 (Fry et al., 2016).

The expression of other proinflammatory mediators like cyclooxygenase 2 (COX-2) is suppressed by various flavonoids and anthocyanins (Goszcz et al., 2017); for example, the beneficial effect of flavonoids on intestinal inflammation has directly been related to the suppression of proinflammatory enzyme expression, such as COX-2 and iNOS (Gil-Cardoso et al., 2016). Polyphenols from red wine and black tea (quercetin, EGCG, epicatechin gallate and theaflavins) are able to inhibit COX-2 and lipoxygenase in a dose-dependent manner in lipopolysaccharide (LPS)-activated murine macrophages (Upadhyay and Dixit, 2015).

CRP and hsCRP values were associated with arterial stiffness in patients with metabolic syndrome, renal transplant, diabetes mellitus and rheumatoid arthritis (Mozos et al., 2017a).

Resveratrol significantly reduced CRP while increasing Total Antioxidant Status (TAS) values in smokers (Bo et al., 2013). Resveratrol was tested in various studies, sometimes combined with grape extract, and generally showed a decrease in CRP, IL-1β, ICAM, TNF-α expression and IL-10 in different populations (Rangel-Huerta et al., 2015). Red wine phenolics decreased serum levels of ICAM-1, E-selectin and IL-6 (Chiva-Blanch et al., 2012).

In a, 2011 meta-analysis, Dong and coworkers concluded that there was insufficient evidence for the significant reduction of CRP by soy isoflavones in postmenopausal women in general. However, soy isoflavones may reduce CRP significantly among postmenopausal women with elevated CRP (Dong et al., 2011). Several other inflammatory biomarkers have been investigated after isoflavone consumption but without consistent conclusions (Rangel-Huerta et al., 2015). Lignans, however, are not efficient in reducing CRP in the general population but do show a significant decrease in obese subjects (Ren et al., 2016).

Cocoa powder and epicatechin decreased several inflammatory markers like TNF-α, IL-6, IL-10 and CRP. Flavonoid-rich fruit and vegetable intake reduced CRP, E-selectin and VCAM in men with increased CVD risk (Macready et al., 2014).

Pomegranate juice did not exhibit a significant effect on CRP levels in a meta-analysis of five prospective trials (Sahebkar et al., 2016).

Olive oil polyphenols are able to decrease inflammatory markers such as thromboxanes, leukotrienes, cytokines, CRP and soluble adhesion molecules in humans (Scoditti et al., 2014). A traditional Mediterranean diet, including polyphenol-rich virgin olive oil, decreased plasma oxidative and inflammatory status and the corresponding gene expression in peripheral blood mononuclear cells of healthy volunteers, indicating that the benefits associated with a Mediterranean diet and olive oil polyphenol consumption on cardiovascular risk can be mediated, at least in part, through nutrigenomic effects (Konstantinidou et al., 2010).

Catechins and curcumin, for instance, are found to regulate matrix metalloproteinases (MMPs) expression in diverse models, including in VSMC (Upadhyay and Dixit, 2015). Also, red grape skin extracts and their polyphenols (trans-resveratrol, trans-piceid, kaempferol and quercetin) inhibit endothelial invasion as well as the MMP-9 and MMP-2 (gelatinases) release in stimulated endothelial cells and MMP-9 production in monocytes, at concentrations likely to be achieved after moderate red grape skin consumption (Calabriso et al., 2016). Figure 3.5 represents interactions of dietary polyphenols with inflammatory pathways in AS.

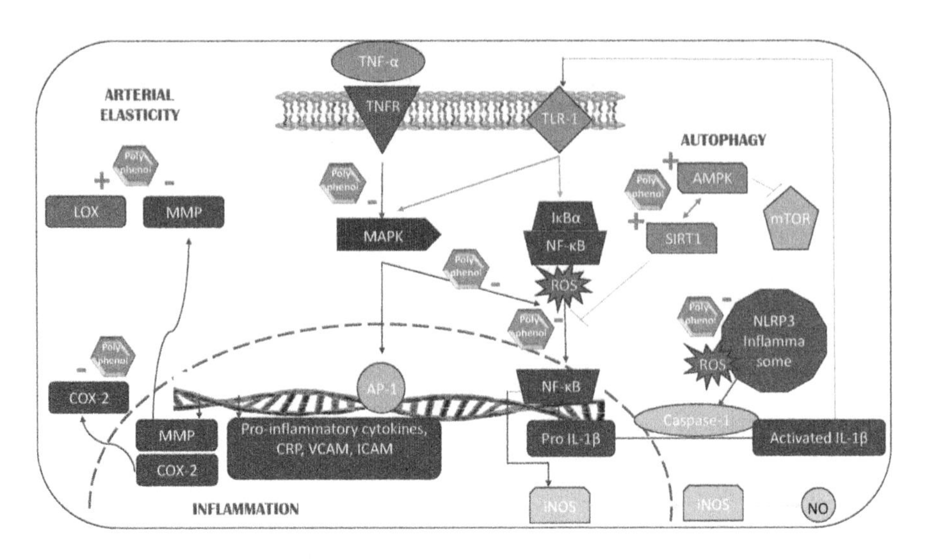

FIGURE 3.5 Overview of inflammatory pathways involved in arterial stiffness etiology, and interactions of dietary polyphenols. Stimulation is denoted by +; inhibition by –.

3.2.5.4 Antiglycation/AGEs

Advanced glycation end products (AGEs) are the end products of the Maillard reaction, in which proteins or amino acids react with reducing sugars. The crosslinking of vascular collagen by AGEs increases arterial stiffness. Soluble receptors for AGE (sRAGE) levels are inversely correlated with the urinary excretion of isoprostanes (8-iso-PGF2α), a biomarker of lipid peroxidation, suggesting a link between vascular stiffness and oxidative stress (Santilli et al., 2015).

Polyphenols can exhibit antiglycation function through their influence on glucose metabolism (aldose reductase inhibition), antioxidant properties (free radical scavenging, metal chelating), protein and receptor binding, modulation of gene expression and dicarbonyl (e.g. methylglyoxal = a highly reactive intermediate) trapping properties (Khangholi et al., 2016).

A bulk of evidence exists for the *in vitro* antiglycation properties of polyphenols. Some polyphenols show to be even more effective than aminoguanidine in inhibiting glycation *in vitro* (Vlassopoulos et al., 2014). Xie and Chen reviewed *in vitro* and animal studies on the antiglycation activity of polyphenols to extract a structure-activity relationship (Xie and Chen, 2013). Lemon balm (*Melissa officinalis*) has been selected as the most active extract from 681 hot water extracts in a pentosidine formation assay, with higher potency than the reference compound aminoguanidine. Rosmarinic acid has been identified as the major active compound herein (Yui et al., 2017).

Redox-active transition metals such as Cu^{2+}, Fe^{2+}, Mn^{2+} and Zn^{2+} catalyze carbonyl formation in proteins. Inhibitors of AGE formation, which have been successful at decreasing arterial stiffness, have known ability to chelate metals (Sell and Monnier, 2012).

Pomegranate (*Punica granatum*) extract and its polyphenolic constituents punicalin, punicagalin and ellagic and gallic acid significantly suppressed AGE formation *in vitro* and in a mouse model (Kumagai et al., 2015). Also rambutan (*Nephelium lappaceum*) extract exhibited antiglycation activity *in vitro*, which correlated to its antioxidant activity. The main compounds were geraniin and ellagic acid (Zhuang et al., 2017). Similarly, glucitol-core containing gallotannins isolated from maple (*Acer*) species inhibit AGEs, mediated by their antioxidant (radical trapping) potential (Ma et al., 2016).

For quercetin, researchers demonstrated that the antiglycation activity *in vitro* was due to its metal-chelating, methylglyoxal-trapping and ROS-trapping properties. It was the most active polyphenol tested (Bhuiyan et al., 2017).

In a recent review on polyphenols with antiglycation activity, Yeh et al. reported on the antiglycation potential of different polyphenol classes. The number of -OH groups seems important for the activity. Simultaneous use of multiple polyphenol types could add to their efficacy (Yeh et al., 2017). This is also illustrated by the *in vitro* antiglycation activity of an olive leaf extract and two characterized fractions. Both inhibition of early and advanced-stage glycation was observed. However, each fraction separately was not able to show the same activity, indicating that compounds from both fractions are necessary for the effect. Hydroxytyrosol in synergy with minor compounds with similar polarity seemed responsible for the activity in a hepatic cell line (Navarro and Morales, 2017; Navarro et al., 2017).

More *in vivo* studies should clarify the relevance of dietary polyphenols in protection against AGE-depending conditions like arterial stiffening.

3.2.5.5 Autophagy

Natural compounds with anti-inflammatory and antioxidant activity like polyphenols are potentially useful to prevent arterial stiffness by promoting autophagy. The mammalian target of rapamycin (mTOR) is a central protein kinase which represses autophagy and is under control of kinase signaling cascades, including the autophagy-activating AMPK pathway (Zenkov et al., 2016).

Limited results are available for resveratrol, activating SIRT1 and AMPK in endothelial and smooth muscle cells *in vitro*, inducing smooth muscle cell differentiation and thus maintaining vascular plasticity (De Meyer et al., 2015). Also, in rhesus monkeys, resveratrol, known to activate endothelial autophagy, reduced arterial ageing (Sasaki et al., 2017). Curcumin and analogues are effective stimulators of autophagy by modulating several transcription factors (NF-κB, Nrf2, AP-1, HIF-1). Green tea polyphenols alleviated autophagy inhibition induced by high glucose, which may be involved in the endothelial protective effects of green tea against hyperglycemia (Zhang et al., 2016). EGCG was able to enhance autophagy-dependent survival through modulation of mTOR-AMPK pathways (Holczer et al., 2018). However, modulation of autophagy by EGCG seems dependent of cell type, stress condition and concentration: low levels of EGCG increase autophagy, while higher levels decrease this process (Kim et al., 2014). EGCG increased formation of LC3-II and autophagosomes and therefore stimulated autophagy in bovine endothelial cells (Kim et al., 2013).

Hydroxytyrosol (HT) is able to induce autophagy, and this is associated with a lower inflammatory response in vascular adventitial fibroblasts (Wang et al., 2018). The activation of the Sirtuin 1 (SIRT1) signaling pathway is thought to play an important role in this process (Cetrullo et al., 2016; Wang et al., 2018; Yang et al., 2017).

HT has been reported to induce autophagosome formation but at the same time to inhibit their degradation by lysosomes in cancer cells; as of yet, there are no reports in endothelial cells or VSMC (Zenkov et al., 2016).

Additionally, oleuropein (OLE)-aglycone has been shown to induce autophagy through AMPK/mTOR-mediated signaling (Rigacci et al., 2015).

3.3 CONCLUSION

Arterial stiffness is an important risk factor for cardiovascular morbidity and mortality, and is reflected by structural and functional changes in the vessel wall. It has received considerable interest as relevant target for patients with increased cardiovascular risk. Nutrition and food-related compounds can offer a suitable strategy in the prevention or reversal of AS.

Considering the different mechanisms involved in the pathophysiology of arterial stiffness, dietary polyphenols offer an array of relevant activities, interfering with vascular, oxidative, inflammatory, glycation and autophagy pathways, and are therefore potentially interesting to counteract or prevent age-induced vascular stiffening. The elucidation of the exact mechanism of action and targets for native polyphenols, but more importantly for their metabolites, stays largely neglected and should receive more attention. *In vivo* beneficial effects of polyphenols on arterial stiffness, and therefore protective effects against cardiovascular risk, have been reported, although findings often have been inconclusive or even inconsistent. Most evidence exists on cocoa and the flavanols therein, isoflavones and anthocyanins.

However, available trials are predominantly small-scale studies with limited duration. Moreover, they profoundly differ in tested compounds or composition of extracts or foods, dosage, intervention schemes, population, endpoints and markers. There is a need for more randomized, placebo-controlled trials using validated biomarkers, and with sufficient follow-up.

Additionally, polyphenol composition of food is a result of cultivation, processing, storage and cooking parameters, implicating a large variability. Often no or inadequate polyphenol composition has been assessed.

Furthermore, it is difficult to identify the bioactivity of a single polyphenol, since clinical effects of foodstuffs are likely to be the result of interactions between different polyphenols and of polyphenols with other food components, interfering with different molecular targets simultaneously.

Additionally, the extensive metabolism of polyphenols adds substantially to the potential bioactive compound array. Most studies do not take into account bioavailability data, and levels and activities of polyphenol phase I and II and microbial metabolites, though this is of major importance. Detailed monitoring of polyphenol bioavailability is therefore required. Apart from the biological activities of those metabolites, an intracellular deconjugation metabolism of phase II metabolites has to be taken into account, releasing parent polyphenols in cells and tissues and provoking local activity. In gut metabolism, the individual microbial community adds variability to the metabolite pool and thus to the biological or clinical effect observed.

Application of metabolomics approaches could identify all polyphenolic metabolites involved in an observed effect, and will help to elucidate mechanisms and targets of their activity in arterial stiffness. However, it should be considered that even the use of valid biomarkers and metabolomics can be biased by several factors, including those influencing microbial metabolism (e.g. by use of antibiotics). Also 24 h urine samples could afford a better reflection of the bioavailable metabolites throughout time, which sampling at a given timepoint cannot provide.

ABBREVIATIONS

5-CGA	5-chlorogenic acid/5-caffeoylquinic acid
ACE	angiotensin-converting enzyme
AGE	advanced glycation endpoint
AhR	aryl hydrocarbon receptor
Aix	augmentation index
Akt	protein kinase B
AMPK	AMP-activated protein kinase
Ang I	angiotensin I
Ang II	angiotensin II
AP-1	activator protein 1
ARE	antioxidant responsive element
AS	arterial stiffness
AT-1	angiotensin II receptor type 1
baPWV	brachial-ankle pulse wave velocity
BP	blood pressure
CAT	catalase
CAVI	cardio-ankle vascular index
COMT	catechol-O-methyltransferase
COX-2	cyclooxygenase 2
CRP	C-reactive protein
CV	cardiovascular
CVD	cardiovascular disease
DHA	docosahexaenoic acid
DNA	deoxyribonucleic acid
DBP	diastolic blood pressure
EFSA	European Food Safety Authority
EGCG	epigallocatechin gallate
eNOS	endothelial nitric oxide synthase
EPA	eicosapentaenoic acid
ERK	extracellular signal-regulated protein kinase
ET-1	Endothelin-1
FMD	flow mediated dilation
FRAP	ferric reducing ability of plasma
GLP-1	glucagon-like peptide 1
GPx	glutathione peroxidase

GR	glutathione reductase
GSH	reduced glutathione
GSL	glutamyl-cysteine ligase
GST	glutathione-S-transferase
HIF-1α	hypoxia-inducible factor 1α
Hmox-1	heme oxygenase 1
hsCRP	high sensitive C-reactive protein
HT	hydroxytyrosol
HUVEC	human umbilical vein endothelial cells
ICAM-1	including intercellular adhesion molecule-1
IFN-γ	interferon-γ
IGF-1	insulin-like growth factor 1
IL	interleukin
iNOS	inducible nitric oxide synthase
Keap1	Kelch-like ECH-associated protein 1
L-NAME	N(ω)-nitro-L-arginine methyl ester
LDL	low-density lipoprotein
LOX	lysyl oxidase
LOX-1	lectin-like oxidized low-density lipoprotein receptor 1
LV	left ventricular
MAPK	mitogen-activated protein kinases
MCP	monocyte chemoattractant protein
MMP	matrix metalloproteinase
mTOR	mammalian target of rapamycin
NAD(P)H	nicotinamide adenine dinucleotide phosphate
NF-κB	nuclear factor kappa B
NLRP3	nucleotide-binding oligomerization domain, leucine-rich repeat containing gene family, and pyrin-domain containing 3
NO	nitric oxide
Nrf2	Nuclear factor (erythroid-derived 2)-like 2/NFE2-related factor 2
OLE	oleuropein
oxLDL	oxidized low-density lipoprotein
p53	tumor protein p53
PACE4	paired basic amino acid-cleaving enzyme 4
PI3K	phosphoinositide 3-kinase
PKC	protein kinase C
PON	paraoxonase
PPAR-γ	peroxisome proliferator-activated receptor γ
PWV	pulse wave velocity
RAAS	Renin-angiotensin-aldosterone system
RAGE	receptor of AGE
RNA	ribonucleic acid
ROS	reactive oxygen species
SBP	systolic blood pressure
SIRT1	sirtuin 1
SOD	superoxide dismutase

TAS	Total Antioxidant Status
TBARS	thiobarbituric acid-reactive substances
TNF-α	tumor necrosis factor-α
TRX	thioredoxin
VCAM-1	vascular cell adhesion molecule-1
VSMC	vascular smooth muscle cells
XO	xanthine oxidase

REFERENCES

Akazawa, N., Y. Choi, A. Miyaki, Y. Tanabe, J. Sugawara, R. Ajisaka, and S. Maeda. 2012. "Curcumin ingestion and exercise training improve vascular endothelial function in postmenopausal women." *Nutrition Research* 32 (10):795–799. doi:10.1016/j.nutres.2012.09.002.

Amiot, M. J., C. Riva, and A. Vinet. 2016. "Effects of dietary polyphenols on metabolic syndrome features in humans: A systematic review." *Obesity Reviews: an Official Journal of the International Association for the Study of Obesity* 17 (7):573–586. doi:10.1111/obr.12409.

Arts, I. C. W., and P. C. H. Hollman. 2005. "Polyphenols and disease risk in epidemiologic studies." *The American Journal of Clinical Nutrition* 81 (1 Suppl):317S–325S.

Avolio, A. 2013. "Arterial stiffness." *Pulse (Basel)* 1 (1):14–28. doi:10.1159/000348620.

Basu, A., N. M. Betts, M. J. Leyva, D. Fu, C. E. Aston, and T. J. Lyons. 2015. "Acute cocoa supplementation increases postprandial HDL cholesterol and insulin in obese adults with Type 2 diabetes after consumption of a high-fat breakfast." *The Journal of Nutrition* 145 (10):2325–2332. doi:10.3945/jn.115.215772.

Behbahani, J., S. J. Thandapilly, X. L. Louis, Y. S. Huang, Z. J. Shao, M. A. Kopilas, P. Wojciechowski, T. Netticadan, and H. D. Anderson. 2010. "Resveratrol and small artery compliance and remodeling in the spontaneously hypertensive rat." *American Journal of Hypertension* 23 (12):1273–1278. doi:10.1038/ajh.2010.161.

Bhuiyan, M. N. I., S. Mitsuhashi, K. Sigetomi, and M. Ubukata. 2017. "Quercetin inhibits advanced glycation end product formation via chelating metal ions, trapping methylglyoxal, and trapping reactive oxygen species." *Bioscience, Biotechnology, and Biochemistry* 81 (5):882–890. doi:10.1080/09168451.2017.1282805.

Blumberg, J. B., J. A. Vita, and C. Y. O. Chen. 2015. "Concord grape juice polyphenols and cardiovascular risk factors: Dose-response relationships." *Nutrients* 7 (12):10032–10052. doi:10.3390/nu7125519.

Bo, S., G. Ciccone, A. Castiglione, R. Gambino, F. De Michieli, P. Villois, M. Durazzo, P. Cavallo-Perin, and M. Cassader. 2013. "Anti-inflammatory and antioxidant effects of resveratrol in healthy smokers: A randomized, double-blind, placebo-controlled, cross-over trial." *Current Medicinal Chemistry* 20 (10):1323–1331. doi:10.2174/0929867311320100009.

Bondonno, N. P., C. P. Bondonno, L. C. Blekkenhorst, M. J. Considine, G. Maghzal, R. Stocker, R. J. Woodman, et al. 2018. "Flavonoid-rich apple improves endothelial function in individuals at risk for cardiovascular disease: A randomised controlled clinical trial." *Molecular Nutrition and Food Research*. doi:10.1002/mnfr.201700674.

Borghi, C., and A. F. G. Cicero. 2017. "Nutraceuticals with a clinically detectable blood pressure-lowering effect: A review of available randomized clinical trials and their meta-analyses." *British Journal of Clinical Pharmacology* 83 (1):163–171. doi:10.1111/bcp.12902.

Bruell, V., C. Burak, B. Stoffel-Wagner, S. Wolffram, G. Nickenig, C. Mueller, P. Langguth, et al. 2015. "Effects of a quercetin-rich onion skin extract on 24 h ambulatory blood pressure and endothelial function in overweight-to-obese patients with (pre-) hypertension: A randomised double-blinded placebo-controlled cross-over trial." *The British Journal of Nutrition* 114 (8):1263–1277. doi:10.1017/S0007114515002950.

Calabriso, N., M. Massaro, E. Scoditti, M. Pellegrino, I. Ingrosso, G. Giovinazzo, and M. A. Carluccio. 2016. "Red grape skin polyphenols blunt matrix metalloproteinase-2 and-9 activity and expression in cell models of vascular inflammation: Protective role in degenerative and inflammatory diseases." *Molecules*:18. doi:10.3390/molecules21091147.

Cardoso, C. R. L., and G. F. Salles. 2016. "Aortic stiffness as a surrogate endpoint to micro- and macrovascular complications in patients with Type 2 diabetes." *International Journal of Molecular Sciences* 17 (12). doi:10.3390/ijms17122044.

Castro-Acosta, M. L., L. Smith, R. J. Miller, D. I. McCarthy, J. A. Farrimond, and W. L. Hall. 2016. "Drinks containing anthocyanin-rich blackcurrant extract decrease postprandial blood glucose, insulin and incretin concentrations." *The Journal of Nutritional Biochemistry* 38:154–161. doi:10.1016/j.jnutbio.2016.09.002.

Cecelja, M., and P. Chowienczyk. 2009. "Dissociation of aortic pulse wave velocity with risk factors for cardiovascular disease other than hypertension A systematic review." *Hypertension* 54 (6):1328–1336. doi:10.1161/HYPERTENSIONAHA.109.137653.

Cecelja, M., and P. Chowienczyk. 2016. "Molecular mechanisms of arterial stiffening." *Pulse* 4 (1):43–48. doi:10.1159/000446399.

Cerletti, C., F. Gianfagna, C. Tamburrelli, A. De Curtis, M. D'Imperio, W. Coletta, L. Giordano, et al. 2015. "Orange juice intake during a fatty meal consumption reduces the postprandial low-grade inflammatory response in healthy subjects." *Thrombosis Research* 135 (2):255–259. doi:10.1016/j.thromres.2014.11.038.

Cetrullo, S., S. D'Adamo, S. Guidotti, R. M. Borzi, and F. Flamigni. 2016. "Hydroxytyrosol prevents chondrocyte death under oxidative stress by inducing autophagy through sirtuin 1-dependent and -independent mechanisms." *Biochimica et Biophysica Acta* 1860 (6):1181–1191. doi:10.1016/j.bbagen.2016.03.002.

Chiva-Blanch, G., M. Urpi-Sarda, R. Llorach, M. Rotches-Ribalta, M. Guillen, R. Casas, S. Arranz, et al. 2012. "Differential effects of polyphenols and alcohol of red wine on the expression of adhesion molecules and inflammatory cytokines related to atherosclerosis: A randomized clinical trial (vol 95, pg 326, 2012)." *American Journal of Clinical Nutrition* 95 (6):1506–1506. doi:10.3945/ajcn.112.038810.

Chuengsamarn, S., S. Rattanamongkolgul, B. Phonrat, R. Tungtrongchitr, and S. Jirawatnotai. 2014. "Reduction of atherogenic risk in patients with type 2 diabetes by curcuminoid extract: A randomized controlled trial." *The Journal of Nutritional Biochemistry* 25 (2):144–150. doi:10.1016/j.jnutbio.2013.09.013.

Clerici, C., K. D. Setchell, P. M. Battezzati, M. Pirro, V. Giuliano, S. Asciutti, D. Castellani, et al. 2007. "Pasta naturally enriched with isoflavone aglycons from soy germ reduces serum lipids and improves markers of cardiovascular risk." *The Journal of Nutrition* 137 (10):2270–2278.

Clerici, C., E. Nardi, P. M. Battezzati, S. Asciutti, D. Castellani, N. Corazzi, V. Giuliano, et al. 2011. "Novel soy germ pasta improves endothelial function, blood pressure, and oxidative stress in patients With Type 2 diabetes." *Diabetes Care* 34 (9):1946–1948. doi:10.2337/dc11-0495.

Costa, C., A. Tsatsakis, C. Mamoulakis, M. Teodoro, G. Briguglio, E. Caruso, D. Tsoukalas, et al. 2017. "Current evidence on the effect of dietary polyphenols intake on chronic diseases." *Food and Chemical Toxicology: an International Journal Published for the British Industrial Biological Research Association* 110:286–299. doi:10.1016/j.fct.2017.10.023.

Croft, K. D. 2016. "Dietary polyphenols: Antioxidants or not?" *Archives of Biochemistry and Biophysics* 595:120–124. doi:10.1016/j.abb.2015.11.014.

Curtis, P. J., J. Potter, P. A. Kroon, P. Wilson, K. Dhatariya, M. Sampson, and A. Cassidy. 2013. "Vascular function and atherosclerosis progression after 1 y of flavonoid intake in statin-treated postmenopausal women with type 2 diabetes: A double-blind randomized controlled trial." *The American Journal of Clinical Nutrition* 97 (5):936–942. doi:10.3945/ajcn.112.043745.

Daiber, A., S. Steven, A. Weber, V. V. Shuvaev, V. R. Muzykantov, I. Laher, H. Li, S. Lamas, and T. Münzel. 2017. "Targeting vascular (endothelial) dysfunction." *British Journal of Pharmacology* 174 (12):1591–1619. doi:10.1111/bph.13517.

De Meyer, G. R. Y., M. O. J. Grootaert, C. F. Michiels, A. Kurdi, D. M. Schrijvers, and W. Martinet. 2015. "Autophagy in vascular disease." *Circulation Research* 116 (3):468–479. doi:10.1161/CIRCRESAHA.116.303804.

Del Bo, C., M. Porrini, D. Fracassetti, J. Campolo, D. Klimis-Zacas, and P. Riso. 2014. "A single serving of blueberry (V. corymbosum) modulates peripheral arterial dysfunction induced by acute cigarette smoking in young volunteers: A randomized-controlled trial." *Food and Function* 5 (12):3107–3116. doi:10.1039/c4fo00570h.

Del Bo, C., M. Porrini, J. Campolo, M. Parolini, C. Lanti, D. Klimis-Zacas, and P. Riso. 2016. "A single blueberry (Vaccinium corymbosum) portion does not affect markers of antioxidant defence and oxidative stress in healthy volunteers following cigarette smoking." *Mutagenesis* 31 (2):215–224. doi:10.1093/mutage/gev079.

Del Bo, C., V. Deon, J. Campolo, C. Lanti, M. Parolini, M. Porrini, D. Klimis-Zacas, and P. Riso. 2017. "A serving of blueberry (V. corymbosum) acutely improves peripheral arterial dysfunction in young smokers and non-smokers: two randomized, controlled, crossover pilot studies." *Food Funct* 8 (11):4108-4117. doi:10.1039/c7fo00861a.

Della Corte, V., A. Tuttolomondo, R. Pecoraro, D. Di Raimondo, V. Vassallo, and A. Pinto. 2016. "Inflammation, endothelial dysfunction and arterial stiffness as therapeutic targets in cardiovascular medicine." *Current Pharmaceutical Design* 22 (30):4658–4668.

Dohadwala, M. M., M. Holbrook, N. M. Hamburg, S. M. Shenouda, W. B. Chung, M. Titas, M. A. Kluge, et al. 2011. "Effects of cranberry juice consumption on vascular function in patients with coronary artery disease." *The American Journal of Clinical Nutrition* 93 (5):934–940. doi:10.3945/ajcn.110.004242.

Dong, J. Y., P. Y. Wang, K. He, and L. Q. Qin. 2011. "Effect of soy isoflavones on circulating C-reactive protein in postmenopausal women: Meta-analysis of randomized controlled trials." *Menopause* 18 (11):1256–1262. doi:10.1097/gme.0b013e31821bfa24.

Dower, J. I., J. M. Geleijnse, L. Gijsbers, P. L. Zock, D. Kromhout, and P. C. Hollman. 2015. "Effects of the pure flavonoids epicatechin and quercetin on vascular function and cardiometabolic health: A randomized, double-blind, placebo-controlled, crossover trial." *The American Journal of Clinical Nutrition* 101 (5):914–921. doi:10.3945/ajcn.114.098590.

Dower, J. I., J. M. Geleijnse, P. A. Kroon, M. Philo, M. Mensink, D. Kromhout, and P. C. Hollman. 2016. "Does epicatechin contribute to the acute vascular function effects of dark chocolate? A randomized, crossover study." *Molecular Nutrition and Food Research* 60 (11):2379–2386. doi:10.1002/mnfr.201600045.

Draijer, R., Y. de Graaf, M. Slettenaar, E. de Groot, and C. I. Wright. 2015. "Consumption of a polyphenol-rich grape-wine extract lowers ambulatory blood pressure in mildly hypertensive subjects." *Nutrients* 7 (5):3138–3153. doi:10.3390/nu7053138.

El-Bassossy, H. M., N. A. Hassan, M. F. Mahmoud, and A. Fahmy. 2014. "Baicalein protects against hypertension associated with diabetes: Effect on vascular reactivity and stiffness." *Phytomedicine: International Journal of Phytotherapy and Phytopharmacology* 21 (12):1742–1745. doi:10.1016/j.phymed.2014.08.012.

Espin, J. C., A. Gonzalez-Sarrias, and F. A. Tomas-Barberan. 2017. "The gut microbiota: A key factor in the therapeutic effects of (poly) phenols." *Biochemical Pharmacology* 139:82–93. doi:10.1016/j.bcp.2017.04.033.

Fang, C., L. Gu, D. Smerin, S. Mao, and X. Xiong. 2017. "The interrelation between reactive oxygen species and autophagy in neurological disorders." *Oxidative Medicine and Cellular Longevity* 2017:8495160. doi:10.1155/2017/8495160.

Feliciano, R. P., G. Istas, C. Heiss, and A. Rodriguez-Mateos. 2016. "Plasma and urinary phenolic profiles after acute and repetitive intake of wild blueberry." *Molecules* 21 (9):15. doi:10.3390/molecules21091120.

Ferri, C., G. Desideri, L. Ferri, I. Proietti, S. Di Agostino, L. Martella, F. Mai, P. Di Giosia, and D. Grassi. 2015. "Cocoa, blood pressure, and cardiovascular health." *Journal of Agricultural and Food Chemistry* 63 (45):9901–9909. doi:10.1021/acs.jafc.5b01064.

Fleenor, B. S., A. L. Sindler, N. K. Marvi, K. L. Howell, M. L. Zigler, M. Yoshizawa, and D. R. Seals. 2013. "Curcumin ameliorates arterial dysfunction and oxidative stress with aging." *Experimental Gerontology* 48 (2):269–276. doi:10.1016/j.exger.2012.10.008.

Forman, H. J., K. J. A. Davies, and F. Ursini. 2014. "How Do Nutritional Antioxidants Really Work: Nucleophilic Tone and Para-hormesis Versus Free radical Scavenging in vivo (vol 66, pg 24, 2014)." *Free Radical Biology and Medicine* 74:307–307. doi:10.1016/j.freeradbiomed.2014.05.012.

Fry, J. L., L. Al. Sayah, R. M. Weisbrod, I. Van Roy, X. Weng, R. A. Cohen, M. M. Bachschmid, and F. Seta. 2016. "Vascular smooth muscle Sirtuin-1 protects Against diet-induced aortic stiffness." *Hypertension* 68 (3):775–784. doi:10.1161/hypertensionaha.116.07622.

Gerstgrasser, A., S. Rochter, D. Dressler, C. Schon, C. Reule, and S. Buchwald-Werner. 2016. "In vitro activation of eNOS by Mangifera indica (Careless (TM)) and determination of an effective dosage in a randomized, double-blind, human pilot study on microcirculation." *Planta Medica* 82 (4):298–304. doi:10.1055/s-0035-1558219.

Gil-Cardoso, K., I. Gines, M. Pinent, A. Ardevol, M. Blay, and X. Terra. 2016. "Effects of flavonoids on intestinal inflammation, barrier integrity and changes in gut microbiota during diet-induced obesity." *Nutrition Research Reviews* 29 (2):234–248. doi:10.1017/S0954422416000159.

Gil-Izquierdo, A., J. L. Penalvo, J. I. Gil, S. Medina, M. N. Horcajada, S. Lafay, M. Silberberg, et al. 2012. "Soy isoflavones and cardiovascular disease epidemiological, clinical and -omics perspectives." *Current Pharmaceutical Biotechnology* 13 (5):624–631.

Goszcz, K., G. G. Duthie, D. Stewart, S. J. Leslie, and I. L. Megson. 2017. "Bioactive polyphenols and cardiovascular disease: Chemical antagonists, pharmacological agents or xenobiotics that drive an adaptive response?" *British Journal of Pharmacology* 174 (11):1209–1225. doi:10.1111/bph.13708.

Grassi, D., G. Desideri, S. Necozione, C. Lippi, R. Casale, G. Properzi, J. B. Blumberg, and C. Ferri. 2008. "Blood pressure is reduced and insulin sensitivity increased in glucose-intolerant, hypertensive subjects after 15 days of consuming high-polyphenol dark chocolate." *The Journal of Nutrition* 138 (9):1671–1676.

Grassi, D., T. P. J. Mulder, R. Draijer, G. Desideri, H. O. F. Molhuizen, and C. Ferri. 2009. "Black tea consumption dose-dependently improves flow-mediated dilation in healthy males." *Journal of Hypertension* 27 (4):774–781. doi:10.1097/HJH.0b013e328326066c.

Grassi, D., G. Desideri, S. Necozione, F. Ruggieri, J. B. Blumberg, M. Stornello, and C. Ferri. 2012. "Protective effects of flavanol-rich dark chocolate on endothelial function and wave reflection During acute hyperglycemia." *Hypertension* 60 (3):827–832. doi:10.1161/hypertensionaha.112.193995.

Grassi, D., G. Desideri, S. Necozione, P. di Giosia, R. Barnabei, L. Allegaert, H. Bernaert, and C. Ferri. 2015a. "Cocoa consumption dose-dependently improves flow-mediated dilation and arterial stiffness decreasing blood pressure in healthy individuals." *Journal of Hypertension* 33 (2):294–303. doi:10.1097/HJH.0000000000000412.

Grassi, D., R. Draijer, G. Desideri, T. Mulder, and C. Ferri. 2015b. "Black tea lowers blood pressure and wave reflections in fasted and postprandial conditions in hypertensive patients: A randomised study." *Nutrients* 7 (2):1037–1051. doi:10.3390/nu7021037.

Grassi, D., V. Socci, D. Tempesta, C. Ferri, L. De Gennaro, G. Desideri, and M. Ferrara. 2016. "Flavanol-rich chocolate acutely improves arterial function and working memory performance counteracting the effects of sleep deprivation in healthy individuals." *Journal of Hypertension* 34 (7):1298–1308. doi:10.1097/HJH.0000000000000926.

Habauzit, V., M. A. Verny, D. Milenkovic, N. Barber-Chamoux, A. Mazur, C. Dubray, and C. Morand. 2015. "Flavanones protect from arterial stiffness in postmenopausal women consuming grapefruit juice for 6 mo: A randomized, controlled, crossover trial." *The American Journal of Clinical Nutrition* 102 (1):66–74. doi:10.3945/ajcn.114.104646.

Hamilton, P. K., C. J. Lockhart, C. E. Quinn, and G. E. McVeigh. 2007. "Arterial stiffness: Clinical relevance, measurement and treatment." *Clinical Science* 113 (4):157–170. doi:10.1042/CS20070080.

Hazim, S., P. J. Curtis, M. Y. Schär, L. M. Ostertag, C. D. Kay, A. M. Minihane, and A. Cassidy. 2016. "Acute benefits of the microbial-derived isoflavone metabolite equol on arterial stiffness in men prospectively recruited according to equol producer phenotype: A double-blind randomized controlled trial." *The American Journal of Clinical Nutrition* 103 (3):694–702. doi:10.3945/ajcn.115.125690.

Heiss, C., R. Sansone, H. Karimi, M. Krabbe, D. Schuler, A. Rodriguez-Mateos, T. Kraemer, et al. 2015. "Impact of cocoa flavanol intake on age-dependent vascular stiffness in healthy men: A randomized, controlled, double-masked trial." *Age (Dordr)* 37 (3). doi:10.1007/s11357-015-9794-9.

Hermans, N., A. Van der Auwera, A. Breynaert, A. Verlaet, T. De Bruyne, L. Van Gaal, L. Pieters, and V. Verhoeven. 2017. "A red yeast rice-olive extract supplement reduces biomarkers of oxidative stress, OxLDL and Lp-PLA(2), in subjects with metabolic syndrome: A randomised, double-blind, placebo-controlled trial." *Trials* 18:8. doi:10.1186/s13063-017-2058-5.

Holczer, M., B. Besze, V. Zambo, M. Csala, G. Banhegyi, and O. Kapuy. 2018. "Epigallocatechin-3-Gallate (EGCG) promotes autophagy-dependent survival via influencing the balance of mTOR-AMPK pathways upon endoplasmic reticulum stress." *Oxidative Medicine and Cellular Longevity* 2018. doi:10.1155/2018/6721530.

Hoshida, S., T. Miki, T. Nakagawa, Y. Shinoda, N. Inoshiro, K. Terada, and T. Adachi. 2011. "Different effects of isoflavones on vascular function in premenopausal and postmenopausal smokers and nonsmokers: NYMPH study." *Heart and Vessels* 26 (6):590–595. doi:10.1007/s00380-010-0103-3.

Hugel, H. M., N. Jackson, B. May, A. L. Zhang, and C. C. Xue. 2016. "Polyphenol protection and treatment of hypertension." *Phytomedicine: International Journal of Phytotherapy and Phytopharmacology* 23 (2):220–231. doi:10.1016/j.phymed.2015.12.012.

Ibero-Baraibar, I., M. Suarez, A. Arola-Arnal, M. A. Zulet, and J. A. Martinez. 2016. "Cocoa extract intake for 4 weeks reduces postprandial systolic blood pressure response of obese subjects, even after following an energy-restricted diet." *Food & Nutrition Research* 60. doi:10.3402/fnr.v60.30449.

Imamura, H., T. Yamaguchi, D. Nagayama, A. Saiki, K. Shirai, and I. Tatsuno. 2017. "Resveratrol ameliorates arterial stiffness assessed by cardio-ankle vascular index in patients With type 2 diabetes mellitus." *International Heart Journal* 58 (4):577–583. doi:10.1536/ihj.16-373.

Jennings, A., A. A. Welch, S. J. Fairweather-Tait, C. Kay, A. M. Minihane, P. Chowienczyk, B. Y. Jiang, et al. 2012. "Higher anthocyanin intake is associated with lower arterial stiffness and central blood pressure in women." *The American Journal of Clinical Nutrition* 96 (4):781–788. doi:10.3945/ajcn.112.042036.

Jeong, H. S., S. J. Hong, T. B. Lee, J. W. Kwon, J. T. Jeong, H. J. Joo, J. H. Park, et al. 2014. "Effects of black raspberry on lipid profiles and vascular endothelial function in patients with metabolic syndrome." *Phytotherapy Research: PTR* 28 (10):1492–1498. doi:10.1002/ptr.5154.

Jeong, H. S., S. Kim, S. J. Hong, S. C. Choi, J. H. Choi, J. H. Kim, C. Y. Park, et al. 2016. "Black raspberry extract increased circulating endothelial progenitor cells and improved arterial stiffness in patients with metabolic syndrome: A randomized controlled trial." *Journal of Medicinal Food* 19 (4):346–352. doi:10.1089/jmf.2015.3563.

Jiang, R. J., J. M. Hodgson, E. Mas, K. D. Croft, and N. C. Ward. 2016. "Chlorogenic acid improves ex vivo vessel function and protects endothelial cells against HOCl-induced oxidative damage, via increased production of nitric oxide and induction of Hmox-1." *The Journal of Nutritional Biochemistry* 27:53–60. doi:10.1016/j.jnutbio.2015.08.017.

Johnson, S. A., A. Figueroa, N. Navaei, A. Wong, R. Kalfon, L. T. Ormsbee, R. G. Feresin, et al. 2015. "Daily blueberry consumption improves blood pressure and arterial stiffness in postmenopausal women with Pre- and Stage 1-hypertension: A randomized, double-blind, placebo-controlled clinical trial." *Journal of the Academy of Nutrition and Dietetics* 115 (3):369–377. doi:10.1016/j.jand.2014.11.001.

Jokura, H., I. Watanabe, M. Umeda, T. Hase, and A. Shimotoyodome. 2015. "Coffee polyphenol consumption improves postprandial hyperglycemia associated with impaired vascular endothelial function in healthy male adults." *Nutrition Research* 35 (10):873–881. doi:10.1016/j.nutres.2015.07.005.

Katz, D. L., K. Doughty, and A. Ali. 2011. "Cocoa and chocolate in human health and disease." *Antioxidants & Redox Signaling* 15 (10):2779–2811. doi:10.1089/ars.2010.3697.

Katz, D. L., A. Davidhi, Y. Y. Ma, Y. Kavak, L. Bifulco, and V. Y. Njike. 2012. "Effects of walnuts on endothelial function in overweight adults with visceral obesity: A randomized, controlled, crossover trial." *Journal of the American College of Nutrition* 31 (6):415–423. doi:10.1080/07315724.2012.10720468.

Kay, C. D., L. Hooper, P. A. Kroon, E. B. Rimm, and A. Cassidy. 2012. "Relative impact of flavonoid composition, dose and structure on vascular function: A systematic review of randomised controlled trials of flavonoid-rich food products." *Molecular Nutrition and Food Research* 56 (11):1605–1616. doi:10.1002/mnfr.201200363.

Khan, N., O. Khymenets, M. Urpi-Sarda, S. Tulipani, M. Garcia-Aloy, M. Monagas, X. Mora-Cubillos, R. Llorach, and C. Andres-Lacueva. 2014. "Cocoa polyphenols and inflammatory markers of cardiovascular disease." *Nutrients* 6 (2):844–880. doi:10.3390/nu6020844.

Khangholi, S., F. A. A. Majid, N. J. A. Berwary, F. Ahmad, and R. B. Aziz. 2016. "The mechanisms of inhibition of advanced glycation end products formation through polyphenols in hyperglycemic condition." *Planta Medica* 82 (1–2):32–45. doi:10.1055/s-0035-1558086.

Kim, H. S., V. Montana, H. J. Jang, V. Parpura, and J. A. Kim. 2013. "Epigallocatechin gallate (EGCG) stimulates autophagy in vascular endothelial cells a POTENTIAL ROLE FOR REDUCING LIPID ACCUMULATION." *The Journal of Biological Chemistry* 288 (31):22693–22705. doi:10.1074/jbc.M113.477505.

Kim, H. S., M. J. Quon, and J. A. Kim. 2014. "New insights into the mechanisms of polyphenols beyond antioxidant properties; lessons from the green tea polyphenol, epigallocatechin 3-gallate." *Redox Biology* 2:187–195. doi:10.1016/j.redox.2013.12.022.

Konstantinidou, V., M. I. Covas, D. Munoz-Aguayo, O. Khymenets, R. de la Torre, G. Saez, Mdel C. Tormos, et al. 2010. "In vivo nutrigenomic effects of virgin olive oil polyphenols within the frame of the Mediterranean diet: A randomized controlled trial." *FASEB Journal: Official Publication of the Federation of American Societies for Experimental Biology* 24 (7):2546–2557. doi:10.1096/fj.09-148452.

Kumagai, Y., S. Nakatani, H. Onodera, A. Nagatomo, N. Nishida, Y. Matsuura, K. Kobata, and M. Wada. 2015. "Anti-glycation effects of pomegranate (Punica granatum L.) fruit extract and its components in vivo and in vitro." *Journal of Agricultural and Food Chemistry* 63 (35):7760–7764. doi:10.1021/acs.jafc.5b02766.

Kunnumakkara, A. B., D. Bordoloi, G. Padmavathi, J. Monisha, N. K. Roy, S. Prasad, and B. B. Aggarwal. 2017. "Curcumin, the golden nutraceutical: Multitargeting for multiple chronic diseases." *British Journal of Pharmacology* 174 (11):1325–1348. doi:10.1111/bph.13621.

Kuntz, S., C. Kunz, J. Herrmann, C. H. Borsch, G. Abel, B. Frohling, H. Dietrich, and S. Rudloff. 2014. "Anthocyanins from fruit juices improve the antioxidant status of healthy young female volunteers without affecting anti-inflammatory parameters: Results from the randomised, double-blind, placebo-controlled, cross-over ANTHONIA (ANTHOcyanins in Nutrition Investigation Alliance) study." *The British Journal of Nutrition* 112 (6):925–936. doi:10.1017/S0007114514001482.

LaRocca, T. J., C. R. Martens, and D. R. Seals. 2017. "Nutrition and other lifestyle influences on arterial aging." *Ageing Research Reviews* 39:106–119. doi:10.1016/j.arr.2016.09.002.

Latham, L. S., Z. K. Hensen, and D. S. Minor. 2014. "Chocolate–Guilty pleasure or healthy supplement?" *Journal of Clinical Hypertension* 16 (2):101–106.

Laurent, S., J. Cockcroft, L. Van Bortel, P. Boutouyrie, C. Giannattasio, D. Hayoz, B. Pannier, et al. 2006. "Invasive Investigation of Large arteries." *European Heart Journal* 27 (21):2588–2605. doi:10.1093/eurheartj/ehl254.

Lee, D. I., C. Acosta, C. M. Anderson, and H. D. Anderson. 2017. "Peripheral and cerebral resistance arteries in the spontaneously hypertensive heart failure rat: Effects of stilbenoid polyphenols." *Molecules* 22 (3). doi:10.3390/molecules22030380.

Li, X., Y. N. Dai, S. J. Yan, Y. L. Shi, J. X. Li, J. L. Liu, L. Cha, and J. J. Mu. 2016. "Resveratrol lowers blood pressure in spontaneously hypertensive rats via calcium-dependent endothelial NO production." *Clinical and Experimental Hypertension* 38 (3):287–293. doi: 10.3109/10641963.2015.1089882.

Li, P., L. Wang, and C. Liu. 2017a. "Overweightness, obesity and arterial stiffness in healthy subjects: A systematic review and meta-analysis of literature studies." *Postgraduate Medicine* 129 (2):224–230. doi:10.1080/00325481.2017.1268903.

Li, X. X., P. Y. Lyu, Y. Y. Ren, J. An, and Y. H. Dong. 2017b. "Arterial stiffness and cognitive impairment." *Journal of the Neurological Sciences* 380:1–10. doi:10.1016/j.jns.2017.06.018.

Lilamand, M., E. Kelaiditi, S. Guyonnet, R. A. Incalzi, A. Raynaud-Simon, B. Vellas, and M. Cesari. 2014. "Flavonoids and arterial stiffness: Promising perspectives." *Nutrition, Metabolism, and Cardiovascular Diseases: NMCD* 24 (7):698–704. doi:10.1016/j.numecd.2014.01.015.

Lockyer, S., G. Corona, P. Yaqoob, J. P. E. Spencer, and I. Rowland. 2015. "Secoiridoids delivered as olive leaf extract induce acute improvements in human vascular function and reduction of an inflammatory cytokine: A randomised, double-blind, placebo-controlled, cross-over trial." *The British Journal of Nutrition* 114 (1):75–83. doi:10.1017/S0007114515001269.

Ludovici, V., J. Barthelmes, M. P. Nägele, F. Enseleit, C. Ferri, A. J. Flammer, F. Ruschitzka, and I. Sudano. 2017a. "Cocoa, blood pressure, and vascular function." *Frontiers in Nutrition* 4. doi:10.3389/fnut.2017.00036.

Ludovici, V., J. Barthelmes, M. P. Nägele, F. Enseleit, C. Ferri, A. J. Flammer, F. Ruschitzka, and I. Sudano. 2017b. "Cocoa, blood pressure, and vascular function." *Frontiers in Nutrition* 4:36. doi:10.3389/fnut.2017.00036.

Lyle, A. N., and U. Raaz. 2017. "Killing me unsoftly: Causes and mechanisms of arterial stiffness." *Arteriosclerosis, Thrombosis, and Vascular Biology* 37 (2):e1–e11. doi:10.1161/ATVBAHA.116.308563.

Lynn, A., H. Hamadeh, W. C. Leung, J. M. Russell, and M. E. Barker. 2012. "Effects of pomegranate juice supplementation on pulse wave velocity and blood pressure in healthy young and middle-aged men and women." *Plant Foods for Human Nutrition* 67 (3):309–314. doi:10.1007/s11130-012-0295-z.

Ma, H., W. X. Liu, L. Frost, L. J. Kirschenbaum, J. A. Dain, and N. P. Seeram. 2016. "Glucitol-core containing gallotannins inhibit the formation of advanced glycation end-products mediated by their antioxidant potential." *Food & Function* 7 (5):2213–2222. doi:10.1039/c6fo00169f.

Macready, A. L., T. W. George, M. F. Chong, D. S. Alimbetov, Y. Jin, A. Vidal, J. P. Spencer, et al. 2014. "Flavonoid-rich fruit and vegetables improve microvascular reactivity and inflammatory status in men at risk of cardiovascular disease--FLAVURS: A randomized controlled trial." *The American Journal of Clinical Nutrition* 99 (3):479–489. doi:10.3945/ajcn.113.074237.

Mansuri, M. L., P. Parihar, I. Solanki, and M. S. Parihar. 2014. "Flavonoids in modulation of cell survival signalling pathways." *Genes and Nutrition* 9 (3). doi:10.1007/s12263-014-0400-z.

Martin, M. A., and S. Ramos. 2016. "Cocoa polyphenols in oxidative stress: Potential health implications." *Journal of Functional Foods* 27:570–588. doi:10.1016/j.jff.2016.10.008.

Mathew, A. S., G. M. Capel-Williams, S. E. Berry, and W. L. Hall. 2012. "Acute effects of pomegranate extract on postprandial lipaemia, vascular function and blood pressure." *Plant Foods for Human Nutrition* 67 (4):351–357. doi:10.1007/s11130-012-0318-9.

McCullough, M. L., J. J. Peterson, R. Patel, P. F. Jacques, R. Shah, and J. T. Dwyer. 2012. "Flavonoid intake and cardiovascular disease mortality in a prospective cohort of US adults." *The American Journal of Clinical Nutrition* 95 (2):454–464. doi:10.3945/ajcn.111.016634.

Miller, R. J., K. G. Jackson, T. Dadd, A. E. Mayes, A. L. Brown, A. M. Minihane. 2011. "The impact of the catechol-O-methyltransferase genotype on the acute responsiveness of vascular reactivity to a green tea extract." *The British Journal of Nutrition* 105 (8):1138–1144. doi:10.1017/S0007114510004836.

Misawa, T., T. Saitoh, T. Kozaki, S. Park, M. Takahama, and S. Akira. 2015. "Resveratrol inhibits the acetylated alpha-tubulin-mediated assembly of the NLRP3-inflammasome." *International Immunology* 27 (9):425–434. doi:10.1093/intimm/dxv018.

Mozos, I., and C. T. Luca. 2017. "Crosstalk between oxidative and nitrosative stress and arterial stiffness." *Current Vascular Pharmacology* 15 (5):446–456. doi:10.2174/1570161115666170201115428.

Mozos, I., C. Malainer, J. Horbanczuk, C. Gug, D. Stoian, C. T. Luca, and A. G. Atanasov. 2017a. "Inflammatory markers for arterial stiffness in cardiovascular diseases." *Frontiers in Immunology* 8:1058. doi:10.3389/fimmu.2017.01058.

Mozos, I., D. Stoian, and C. T. Luca. 2017b. "Crosstalk between vitamins A, B12, D, K, C, and E status and arterial stiffness." *Disease Markers*. doi:10.1155/2017/8784971.

Mullan, A., C. Delles, W. Ferrell, W. Mullen, C. A. Edwards, J. H. McColl, S. A. Roberts, M. E. Lean, and N. Sattar. 2016. "Effects of a beverage rich in (poly)phenols on established and novel risk markers for vascular disease in medically uncomplicated overweight or obese subjects: A four week randomized placebo-controlled trial." *Atherosclerosis* 246:169–176. doi:10.1016/j.atherosclerosis.2016.01.004.

Naissides, M., S. Pal, J. C. L. Mamo, A. P. James, and S. Dhaliwal. 2006. "The effect of chronic consumption of red wine polyphenols on vascular function in postmenopausal women." *European Journal of Clinical Nutrition* 60 (6):740–745. doi:10.1038/sj.ejcn.1602377.

Navarro, M., and F. J. Morales. 2017. "Evaluation of an olive leaf extract as a natural source of antiglycative compounds." *Food Research International* 92:56–63. doi:10.1016/j.foodres.2016.12.017.

Navarro, M., F. J. Morales, and S. Ramos. 2017. "Olive leaf extract concentrated in hydroxytyrosol attenuates protein carbonylation and the formation of advanced glycation end products in a hepatic cell line (HepG2)." *Food & Function* 8 (3):944–953. doi:10.1039/c6fo01738j.

Nestel, P., A. Fujii, and L. Zhang. 2007. "An isoflavone metabolite reduces arterial stiffness and blood pressure in overweight men and postmenopausal women." *Atherosclerosis* 192 (1):184–189. doi:10.1016/j.atherosclerosis.2006.04.033.

O'Rourke, M. F., and J. Hashimoto. 2007. "Mechanical factors in arterial aging: A clinical perspective." *Journal of the American College of Cardiology* 50 (1):1–13. doi:10.1016/j.jacc.2006.12.050.

Pais, P., S. Rull, and A. Villar. 2016. "Impact of a proprietary standardized olive fruit extract (Proliva (R)) on CAVI assessments in subjects with arterial stiffness risk." *Planta Medica* 82:2. doi:10.1055/s-0036-1596929.

Palombo, C., and M. Kozakova. 2016. "Arterial stiffness, atherosclerosis and cardiovascular risk: Pathophysiologic mechanisms and emerging clinical indications." *Vascular Pharmacology* 77:1–7. doi:10.1016/j.vph.2015.11.083.

Papaioannou, T. G., K. Karatzi, T. Psaltopoulou, and D. Tousoulis. 2017. "Arterial ageing: Major nutritional and life-style effects." *Ageing Research Reviews* 37:162–163. doi:10.1016/j.arr.2016.10.004.

Park, E., I. Edirisinghe, Y. Y. Choy, A. Waterhouse, and B. Burton-Freeman. 2016. "Effects of grape seed extract beverage on blood pressure and metabolic indices in individuals with pre-hypertension: A randomised, double-blinded, two-arm, parallel, placebo-controlled trial." *The British Journal of Nutrition* 115 (2):226–238. doi:10.1017/S0007114515004328.

Pase, M. P., N. A. Grima, and J. Sarris. 2011. "The effects of dietary and nutrient interventions on arterial stiffness: A systematic review." *The American Journal of Clinical Nutrition* 93 (2):446–454. doi:10.3945/ajcn.110.002725.

Perez-Jimenez, J., L. Fezeu, M. Touvier, N. Arnault, C. Manach, S. Hercberg, P. Galan, and A. Scalbert. 2011. "Dietary intake of 337 polyphenols in French adults." *The American Journal of Clinical Nutrition* 93 (6):1220–1228. doi:10.3945/ajcn.110.007096.

Perrotta, I., and S. Aquila. 2015. "The role of oxidative stress and autophagy in atherosclerosis." *Oxidative Medicine and Cellular Longevity* 2015:130315. doi:10.1155/2015/130315.

Persson, I. A. L., M. Josefsson, K. Persson, and R. G. G. Andersson. 2006. "Tea flavanols inhibit angiotensin-converting enzyme activity and increase nitric oxide production in human endothelial cells." *The Journal of Pharmacy and Pharmacology* 58 (8):1139–1144. doi:10.1211/jpp.58.8.0016.

Pons, Z., M. Margalef, F. I. Bravo, A. Arola-Arnal, and B. Muguerza. 2016. "Grape seed flavanols decrease blood pressure via Sirt-1 and confer a vasoprotective pattern in rats." *Journal of Functional Foods* 24:164–172. doi:10.1016/j.jff.2016.03.030.

Quinn, U., L. A. Tomlinson, and J. R. Cockcroft. 2012. "Arterial stiffness." *JRSM Cardiovascular Disease* 1 (6). doi:10.1258/cvd.2012.012024.

Rangel-Huerta, O. D., B. Pastor-Villaescusa, C. M. Aguilera, and A. Gil. 2015. "A systematic review of the efficacy of bioactive compounds in cardiovascular disease: Phenolic compounds." *Nutrients* 7 (7):5177–5216. doi:10.3390/nu7075177.

Ray, S., C. Miglio, T. Eden, and D. Del Rio. 2014. "Assessment of vascular and endothelial dysfunction in nutritional studies." *Nutrition, Metabolism, and Cardiovascular Diseases: NMCD* 24 (9):940–946. doi:10.1016/j.numecd.2014.03.011.

Ren, G. Y., C. Y. Chen, G. C. Chen, W. G. Chen, A. Pan, C. W. Pan, Y. H. Zhang, L. Q. Qin, and L. H. Chen. 2016. "Effect of flaxseed intervention on inflammatory marker C-reactive protein: A systematic review and meta-analysis of randomized controlled trials." *Nutrients* 8 (3):136. doi:10.3390/nu8030136.

Reverri, E. J., C. D. LaSalle, A. A. Franke, and F. M. Steinberg. 2015. "Soy provides modest benefits on endothelial function without affecting inflammatory biomarkers in adults at cardiometabolic risk." *Molecular Nutrition and Food Research* 59 (2):323–333. doi:10.1002/mnfr.201400270.

Richter, C. K., A. C. Skulas-Ray, J. A. Fleming, C. J. Link, R. Mukherjea, E. S. Krul, and P. M. Kris-Etherton. 2017a. "Effects of isoflavone-containing soya protein on ex vivo cholesterol efflux, vascular function and blood markers of CVD risk in adults with moderately elevated blood pressure: A dose-response randomised controlled trial." *The British Journal of Nutrition* 117 (10):1403–1413. doi:10.1017/S000711451700143X.

Richter, C. K., A. C. Skulas-Ray, T. L. Gaugler, J. D. Lambert, D. N. Proctor, and P. M. Kris-Etherton. 2017b. "Incorporating freeze-dried strawberry powder into a high-fat meal does not alter postprandial vascular function or blood markers of cardiovascular disease risk: A randomized controlled trial." *The American Journal of Clinical Nutrition* 105 (2):313–322. doi:10.3945/ajcn.116.141804.

Rienks, J., J. Barbaresko, and U. Nöthlings. 2017. "Association of polyphenol biomarkers with cardiovascular disease and mortality risk: A systematic review and meta-analysis of observational studies." *Nutrients* 9 (4). doi:10.3390/nu9040415.

Rigacci, S., C. Miceli, C. Nediani, A. Berti, R. Cascella, D. Pantano, P. Nardiello, et al. 2015. "Oleuropein aglycone induces autophagy via the AMPK/mTOR signalling pathway: A mechanistic insight." *Oncotarget* 6 (34):35344–35357. doi:10.18632/oncotarget.6119.

Rodriguez-Mateos, A., C. Heiss, G. Borges, and A. Crozier. 2014. "Berry (poly)phenols and cardiovascular health." *Journal of Agricultural and Food Chemistry* 62 (18):3842–3851. doi:10.1021/jf403757g.

Ruel, G., A. Lapointe, S. Pomerleau, P. Couture, S. Lemieux, B. Lamarche, and C. Couillard. 2013. "Evidence that cranberry juice may improve augmentation index in overweight men." *Nutrition Research* 33 (1):41–49. doi:10.1016/j.nutres.2012.11.002.

Ryu, O. H., J. Lee, K. W. Lee, H. Y. Kim, J. A. Seo, S. G. Kim, N. H. Kim, et al. 2006. "Effects of green tea consumption on inflammation, insulin resistance and pulse wave velocity in type 2 diabetes patients." *Diabetes Research and Clinical Practice* 71 (3):356–358. doi:10.1016/j.diabres.2005.08.001.

Sahebkar, A., C. Gurban, A. Serban, F. Andrica, and M. C. Serban. 2016. "Effects of supplementation with pomegranate juice on plasma C-reactive protein concentrations: A systematic review and meta-analysis of randomized controlled trials." *Phytomedicine: International Journal of Phytotherapy and Phytopharmacology* 23 (11):1095–1102. doi:10.1016/j.phymed.2015.12.008.

Sanchez-Gonzalez, C., C. J. Ciudad, V. Noe, and M. Izquierdo-Pulido. 2017. "Health benefits of walnut polyphenols: An exploration beyond their lipid profile." *Critical Reviews in Food Science and Nutrition* 57 (16):3373–3383. doi:10.1080/10408398.2015.1126218.

Sansone, R., J. I. Ottaviani, A. Rodriguez-Mateos, Y. Heinen, D. Noske, J. P. Spencer, A. Crozier, et al. 2017. "Methylxanthines enhance the effects of cocoa flavanols on cardiovascular function: Randomized, double-masked controlled studies." *The American Journal of Clinical Nutrition* 105 (2):352–360. doi:10.3945/ajcn.116.140046.

Santilli, F., D. D'Ardes, and G. Davi. 2015. "Oxidative stress in chronic vascular disease: From prediction to prevention." *Vascular Pharmacology* 74:23–37. doi:10.1016/j.vph.2015.09.003.

Sasaki, Y., Y. Ikeda, M. Iwabayashi, Y. Akasaki, and M. Ohishi. 2017. "The impact of autophagy on cardiovascular senescence and diseases." *International Heart Journal* 58 (5):666–673. doi:10.1536/ihj.17-246.

Schaer, M. Y., P. J. Curtis, S. Hazim, L. M. Ostertag, C. D. Kay, J. F. Potter, and A. Cassidy. 2015. "Orange juice-derived flavanone and phenolic metabolites do not acutely affect cardiovascular risk biomarkers: A randomized, placebo-controlled, crossover trial in men at moderate risk of cardiovascular disease." *The American Journal of Clinical Nutrition* 101 (5):931–938. doi:10.3945/ajcn.114.104364.

Scoditti, E., C. Capurso, A. Capurso, and M. Massaro. 2014. "Vascular effects of the Mediterranean diet-Part II: Role of omega-3 fatty acids and olive oil polyphenols." *Vascular Pharmacology* 63 (3):127–134. doi:10.1016/j.vph.2014.07.001.

Sell, D. R., and V. M. Monnier. 2012. "Molecular basis of arterial stiffening: Role of glycation - A mini-review." *Gerontology* 58 (3):227–237. doi:10.1159/000334668.

Sharman, J. E., P. Boutouyrie, and S. Laurent. 2017. "Arterial (aortic) stiffness in patients with resistant hypertension: from assessment to treatment." *Current Hypertension Reports* 19 (1). doi:10.1007/s11906-017-0704-7.

Shi, A. M., H. T. Shi, Y. Wang, X. Liu, Y. Cheng, H. Li, H. L. Zhao, S. N. Wang, and L. Dong. 2018. "Activation of Nrf2 pathway and inhibition of NLRP3 inflammasome activation contribute to the protective effect of chlorogenic acid on acute liver injury." *International Immunopharmacology* 54:125–130. doi:10.1016/j.intimp.2017.11.007.

Siasos, G., D. Tousoulis, E. Kokkou, E. Oikonomou, M. E. Kollia, A. Verveniotis, N. Gouliopoulos, et al. 2014. "Favorable effects of Concord grape juice on endothelial function and arterial stiffness in healthy smokers." *American Journal of Hypertension* 27 (1):38–45. doi:10.1093/ajh/hpt176.

Sies, H., P. C. H. Hollman, T. Grune, W. Stahl, H. K. Biesalski, and G. Williamson. 2012. "Protection by flavanol-rich foods Against vascular dysfunction and oxidative damage: 27th Hohenheim Consensus Conference." *Advances in Nutrition* 3 (2):217–221. doi:10.3945/an.111.001578.

Siti, H. N., Y. Kamisah, and J. Kamsiah. 2015. "The role of oxidative stress, antioxidants and vascular inflammation in cardiovascular disease (a review)." *Vascular Pharmacology* 71:40–56. doi:10.1016/j.vph.2015.03.005.

Smoliga, J. M., J. A. Baur, and H. A. Hausenblas. 2011. "Resveratrol and health - A comprehensive review of human clinical trials." *Molecular Nutrition & Food Research* 55 (8):1129–1141. doi:10.1002/mnfr.201100143.

Smulyan, H., S. Mookherjee, and M. E. Safar. 2016. "The two faces of hypertension: Role of aortic stiffness." *Journal of the American Society of Hypertension: JASH* 10 (2):175–183. doi:10.1016/j.jash.2015.11.012.

Stevens, J. F., and C. S. Maier. 2016. "The chemistry of gut microbial metabolism of polyphenols." *Phytochemistry Reviews: Proceedings of the Phytochemical Society of Europe* 15 (3):425–444. doi:10.1007/s11101-016-9459-z.

Taguchi, K., M. Hida, M. Hasegawa, T. Matsumoto, and T. Kobayashi. 2016. "Dietary polyphenol morin rescues endothelial dysfunction in a diabetic mouse model by activating the Akt/eNOS pathway." *Molecular Nutrition & Food Research* 60 (3):580–588. doi:10.1002/mnfr.201500618.

Teede, H. J., B. P. McGrath, L. DeSilva, M. Cehun, A. Fassoulakis, and P. J. Nestel. 2003. "Isoflavones reduce arterial stiffness: A placebo-controlled study in men and postmenopausal women." *Arteriosclerosis, Thrombosis, and Vascular Biology* 23 (6):1066–1071. doi:10.1161/01.ATV.0000072967.97296.4A.

Thandapilly, S. J., J. L. LeMaistre, X. L. Louis, C. M. Anderson, T. Netticadan, and H. D. Anderson. 2012. "Vascular and cardiac effects of grape powder in the spontaneously hypertensive rat." *American Journal of Hypertension* 25 (10):1070–1076. doi:10.1038/ajh.2012.98.

Tomas-Barberan, F. A., M. V. Selma, and J. C. Espin. 2016. "Interactions of gut microbiota with dietary polyphenols and consequences to human health." *Current Opinion in Clinical Nutrition and Metabolic Care* 19 (6):471–476. doi:10.1097/MCO.0000000000000314.

Tresserra-Rimbau, A., E. B. Rimm, A. Medina-Remon, M. A. Martinez-Gonzalez, R. de la Torre, D. Corella, J. Salas-Salvado, et al. 2014. "Inverse association between habitual polyphenol intake and incidence of cardiovascular events in the PREDIMED study." *Nutrition Metabolism and Cardiovascular Diseases* 24 (6):639–647. doi:10.1016/j.numecd.2013.12.014.

Upadhyay, S., and M. Dixit. 2015. "Role of polyphenols and other phytochemicals on molecular signaling." *Oxidative Medicine and Cellular Longevity*. doi:10.1155/2015/504253.

Van Bortel, L. 2016. "Arterial stiffness: From surrogate marker to therapeutic target." *Artery Research* 14:10–14. doi:10.1016/j.artres.2016.01.001.

van der Schouw, Y. T., A. Pijpe, C. E. Lebrun, M. L. Bots, P. H. Peeters, W. A. van Staveren, S. W. Lamberts, and D. E. Grobbee. 2002. "Higher usual dietary intake of phytoestrogens is associated with lower aortic stiffness in postmenopausal women." *Arteriosclerosis, Thrombosis, and Vascular Biology* 22 (8):1316–1322.

Vendrame, S., C. Del Bo, S. Ciappellano, P. Riso, and D. Klimis-Zacas. 2016. "Berry fruit consumption and metabolic syndrome." *Antioxidants* 5 (4). doi:10.3390/antiox5040034.

Verhoeven, V., A. Van der Auwera, L. Van Gaal, R. Remmen, S. Apers, M. Stalpaert, J. Wens, and N. Hermans. 2015. "Can red yeast rice and olive extract improve lipid profile and cardiovascular risk in metabolic syndrome?: A double blind, placebo controlled randomized trial." *BMC Complementary and Alternative Medicine* 15:52. doi:10.1186/s12906-015-0576-9.

Vlachopoulos, C., K. Aznaouridis, N. Alexopoulos, E. Economou, I. Andreadou, and C. Stefanadis. 2005. "Effect of dark chocolate on arterial function in healthy individuals." *American Journal of Hypertension* 18 (6):785–791. doi:10.1016/j.amjhyper.2004.12.008.

Vlachopoulos, C., N. Alexopoulos, and C. Stefanadis. 2007a. "Effects of nutrition on arterial rigidity and reflected waves." *Sang Thrombose Vaisseaux* 19 (9):479–486.

Vlachopoulos, C. V., N. A. Alexopoulos, K. A. Aznaouridis, N. C. Ioakeimidis, I. A. Dima, A. Dagre, C. Vasiliadou, E. C. Stefanadi, and C. I. Stefanadis. 2007b. "Relation of habitual cocoa consumption to aortic stiffness and wave reflections, and to central hemodynamics in healthy individuals." *The American Journal of Cardiology* 99 (10):1473–1475. doi:10.1016/j.amjcard.2006.12.081.

Vlassopoulos, A., M. E. J. Lean, and E. Combet. 2014. "Oxidative stress, protein glycation and nutrition - Interactions relevant to health and disease throughout the lifecycle." *The Proceedings of the Nutrition Society* 73 (3):430–438. doi:10.1017/S0029665114000603.

Wang, J. N., Z. H. Guo, Y. X. Fu, Z. Y. Wu, C. Huang, C. L. Zheng, P. A. Shar, et al. 2017. "Weak-binding molecules are not drugs?-toward a systematic strategy for finding effective weak-binding drugs." *Briefings in Bioinformatics* 18 (2):321–332. doi:10.1093/bib/bbw018.

Wang, W., T. Jing, X. Yang, Y. He, B. Wang, Y. Xiao, C. Shang, J. Zhang, and R. Lin. 2018. "Hydroxytyrosol regulates the autophagy of vascular adventitial fibroblasts through the SIRT1-mediated signaling pathway." *Canadian Journal of Physiology and Pharmacology* 96 (1):88–96. doi:10.1139/cjpp-2016-0676.

Ward, N. C., J. M. Hodgson, R. J. Woodman, D. Zimmermann, L. Poquet, A. Leveques, L. Actis-Goretta, I. B. Puddey, and K. D. Croft. 2016. "Acute effects of chlorogenic acids on endothelial function and blood pressure in healthy men and women." *Food & Function* 7 (5):2197–2203. doi:10.1039/c6fo00248j.

Warner, E. F., Q. Z. Zhang, K. S. Raheem, D. O'Hagan, M. A. O'Connell, and C. D. Kay. 2016. "Common phenolic metabolites of flavonoids, but not their unmetabolized precursors, reduce the secretion of vascular cellular adhesion molecules by human endothelial cells." *The Journal of Nutrition* 146 (3):465–473. doi:10.3945/jn.115.217943.

West, S. G., M. D. McIntyre, M. J. Piotrowski, N. Poupin, D. L. Miller, A. G. Preston, P. Wagner, L. F. Groves, and A. C. Skulas-Ray. 2014. "Effects of dark chocolate and cocoa consumption on endothelial function and arterial stiffness in overweight adults." *The British Journal of Nutrition* 111 (4):653–661. doi:10.1017/S0007114513002912.

Williams, B. 2016. "Vascular ageing and interventions: Lessons and learnings." *Therapeutic Advances in Cardiovascular Disease* 10 (3):126–132. doi:10.1177/1753944716642681.

Willum-Hansen, T., J. A. Staessen, C. Torp-Pedersen, S. Rasmussen, L. Thijs, H. Ibsen, and J. Jeppesen. 2006. "Prognostic value of aortic pulse wave velocity as index of arterial stiffness in the general population." *Circulation* 113 (5):664–670. doi:10.1161/CIRCULATIONAHA.105.579342.

Wong, R. H. X., P. R. C. Howe, J. D. Buckley, A. M. Coates, I. Kunz, and N. M. Berry. 2011. "Acute resveratrol supplementation improves flow-mediated dilatation in overweight/obese individuals with mildly elevated blood pressure." *Nutrition, Metabolism, and Cardiovascular Diseases: NMCD* 21 (11):851–856. doi:10.1016/j.numecd.2010.03.003.

Wong, R. H. X., N. M. Berry, A. M. Coates, J. D. Buckley, J. Bryan, I. Kunz, and P. R. C. Howe. 2013. "Chronic resveratrol consumption improves brachial flow-mediated dilatation in healthy obese adults." *Journal of Hypertension* 31 (9):1819–1827. doi:10.1097/HJH.0b013e328362b9d6.

Wu, C. F., P. Y. Liu, T. J. Wu, Y. Hung, S. P. Yang, and G. M. Lin. 2015. "Therapeutic modification of arterial stiffness: An update and comprehensive review." *World Journal of Cardiology* 7 (11):742–753. doi:10.4330/wjc.v7.i11.742.

Xie, Y. X., and X. Q. Chen. 2013. "Structures required of polyphenols for inhibiting advanced glycation end products formation." *Current Drug Metabolism* 14 (4):414–431. doi:10.2174/1389200211314040005.

Yang, H. X., L. Xiao, Y. Yuan, X. Q. Luo, M. L. Jiang, J. H. Ni, and N. P. Wang. 2014. "Procyanidin B2 inhibits NLRP3 inflammasome activation in human vascular endothelial cells." *Biochemical Pharmacology* 92 (4):599–606. doi:10.1016/j.bcp.2014.10.001.

Yang, X., T. Jing, Y. Li, Y. He, W. Zhang, B. Wang, Y. Xiao, et al. 2017. "Hydroxytyrosol attenuates LPS-induced acute lung injury in mice by regulating autophagy and sirtuin expression." *Current Molecular Medicine* 17 (2):149–159. doi:10.2174/1566524017666170421151940.

Yeh, W. J., S. M. Hsia, W. H. Lee, and C. H. Wu. 2017. "Polyphenols with antiglycation activity and mechanisms of action: A review of recent findings." *Journal of Food and Drug Analysis* 25 (1):84–92. doi:10.1016/j.jfda.2016.10.017.

Yui, S., S. Fujiwara, K. Harada, M. Motoike-Hamura, M. Sakai, S. Matsubara, and K. Miyazaki. 2017. "Beneficial effects of lemon balm leaf extract on in vitro glycation of proteins, arterial stiffness, and skin elasticity in healthy adults." *Journal of Nutritional Science and Vitaminology* 63 (1):59–68. doi:10.3177/jnsv.63.59.

Zenkov, N. K., A. V. Chechushkov, P. M. Kozhin, N. V. Kandalintseva, G. G. Martinovich, and E. B. Menshchikova. 2016. "Plant phenols and autophagy." *Biochemistry. Biokhimiia* 81 (4):297–314. doi:10.1134/S0006297916040015.

Zhang, H., and R. Tsao. 2016. "Dietary polyphenols, oxidative stress and antioxidant and anti-inflammatory effects." *Current Opinion in Food Science* 8:33–42. doi:10.1016/j.cofs.2016.02.002.

Zhang, P. W., C. Tian, F. Y. Xu, Z. Chen, R. Burnside, W. J. Yi, S. Y. Xiang, et al. 2016. "Green tea polyphenols alleviate autophagy inhibition induced by high glucose in endothelial cells." *Biomedical and Environmental Sciences: BES* 29 (7):524–528. doi:10.3967/bes2016.069.

Zhuang, Y. L., Q. Y. Ma, Y. Guo, and L. P. Sun. 2017. "Purification and identification of rambutan (Nephelium lappaceum) peel phenolics with evaluation of antioxidant and antiglycation activities invitro." *International Journal of Food Science and Technology* 52 (8):1810–1819. doi:10.1111/ijfs.13455.

4 Potential Therapeutic Uses of the Genus *Cecropia* as an Antihypertensive Herbal Medicinal Product

Mahabir Prashad Gupta, Orlando O. Ortíz, Andrés Rivera-Mondragón and Catherina Caballero-George

CONTENTS

4.1 INTRODUCTION

In several Latin-American countries, many species of the genus *Cecropia* (Urticaceae) have been widely used in folk medicine as diuretic, cardiotonic, antioxidant, antitussive and expectorant, as well as for the treatment of cough, asthma, diabetes, inflammation, anxiety, depression and hypertension (Costa et al., 2011a; Gazal et al., 2014; Pacheco et al., 2014; Pio-Corrêa, 1978; Matos, 1989; Gupta, 1995; Caballero-George et al., 2001). Souccar et al. (2008) have also made claims of the efficacy of plant-derived material from *Cecropia glaziovii* in wound healing, analgesic and antimicrobial activities.

Even though the exact mechanism by which these plants reduce blood pressure is still unclear, there is scientific evidence that support the use of *Cecropia glaziovii, C. hololeuca, C. obtusifolia, C. pachystachya (Syn.: C. adenopus, C. lyratiloba, C. catarinensis)* and *C. peltata* as antihypertensive agents. This evidence will be discussed in this chapter.

4.2 TAXONOMY AND GEOGRAPHICAL DISTRIBUTION OF THE GENUS *CECROPIA*

Trees of the genus *Cecropia* are generally dioecious, with few branches (usually with a candelabrum-like branching system) and a hollow trunk. These trees can have stilt roots, fully amplexicaul stipules, peltate blades with one to two trichilia at the base of the petioles and inflorescences arranged in digitate clusters (or a single inflorescence) often enveloped by a spathe until anthesis. In such trees, the interfloral bracts are generally absent, flowers have two stamens and trees have small dry fruits enveloped by a tubular greenish perianth (Berg et al., 1990; Berg and Franco-Rosselli, 2005).

These generally fast-growing trees are widespread and abundant. They are distributed across the tropical and subtropical rainforests of Mexico, Central America and South America at elevations below 2,600 m (Franco-Rosselli and Berg, 1997). The genus *Cecropia* includes 61 species (Berg and Franco-Rosselli, 2005), including species popularly known as "*yarumo*," "*guarumo*," "*guarumbo*," "*embaúba*," "*ambay*," "*torém*," and "trumpet tree." (Luengas-Caicedo et al., 2007; Costa et al., 2011a; Ospina Chávez et al., 2013; Hernández-Carvajal and Luengas-Caicedo, 2013; Montoya Peláez et al., 2013).

The ecological significance of the trees within the genus *Cecropia* is associated with their rapid growth rate, which makes them the primary colonizers of deforested tropical areas (Monro, 2009) and invasive species in non-native regions. (Conn et al., 2012). In addition, most species within the genus *Cecropia* are ant-plants or Myrmecophytes (Gutiérrez-Valencia et al., 2017). They may live in a symbiotic relationship with a colony of symbiotic ants, especially ants of the genus *Azteca* (Treiber et al., 2016). They possess specialized structures for offering shelter and food to ants in exchange for protection against natural enemies (Dejean et al., 2010; Oliveira et al., 2015).

4.2.1 *CECROPIA GLAZIOVII* SNETHL.

This species is characterized by trees of 20 m height. Their petioles have fused trichilia, are 20–55 cm long, usually puberulous to sub-hispidulous or sub-glabrous; their stipules usually are 15–27 cm long, bright to dark red to purplish or sometimes greenish. Their lamina is subcoriaceous to chartaceous with 8–12 segments, the upper surface scabrous, hispidulous to strigose, the lower surface usually minutely puberulous on the main veins; with lateral veins in the free part of the medial segment usually 12–16 pairs. This species has staminate inflorescences with a dark to pale red or greenish spathes, 9–22 cm long; with spikes usually 8–12 per inflorescence, yellow, sometimes pink. Their pistillate inflorescences have red or green spathes, which are 10–15 cm long; with 4–9 spikes per inflorescence. Their fruit is ellipsoid to oblongoid, 2–2.5 mm long, smooth, pale or dark brown (Figures 4.1 and 4.2) (Berg and Franco-Rosselli, 2005).

4.2.2 *Cecropia hololeuca* Miq.

Cecropia hololeuca species are trees up to 25 m high. They have petioles without trichilia, 30–95 cm long, glabrous, with arachnoid indumentum or hirsute to subvillous at the base, sometimes only with brown pluricellular hairs; and yellowish-white to brownish stipules of 10–40 cm long. Their coriaceous lamina has 8–10 segments, the upper surface is glabrous, with dense arachnoid indumentum, the lower surface with arachnoid indumentum in the areoles and on the smaller veins or also on the main veins. There are 12–15 pairs of lateral veins in the free part of the medial

FIGURE 4.1 Inflorescences of *Cecropia glaziovii* Snethl. (Urticaceae); embaúba. Location: Praia Laranjeiras; Balneário Camboriú, Santa Catarina, Brazil, June 26, 2018. (Courtesy of Oscar Benigno Iza.)

FIGURE 4.2 Habitat of *Cecropia glaziovii* Snethl. (Urticaceae); embaúba. Location: Praia Laranjeiras; Balneário Camboriú, Santa Catarina, Brazil, June 26, 2018. (Courtesy of Oscar Benigno Iza.)

segment. This species has staminate inflorescences without spathes (sometimes with bracts on the upper part of the peduncle); with 9–17 red-wine to dark purple or black spikes per inflorescence. It also has pistillate inflorescences without spathes, but with 1 or 2 bracts on the upper part or the apex of the peduncle; 1 or 2 (rarely 3) spikes per inflorescence, red to purplish, turning blackish. Fruit ellipsoid to oblongoid, 3–4 mm long, tuberculate, dark brown (Figure 4.3) (Berg and Franco-Rosselli, 2005).

FIGURE 4.3 Leaves and inflorescences of *Cecropia hololeuca* Miq. (Urticaceae). (Courtesy of Harari Lorenzi, Instituto Plantarum, Brazil.)

4.2.3 *Cecropia obtusifolia* Bertol.

The trees of this species are up to 12 m high. Their petioles have a fused trichilia, usually 20–80 cm long, puberulous to hirtellous, occasionally with arachnoid indumentum or sub-glabrous; red or yellowish stipules of 5–15 cm long. They also have a chartaceous to subcoriaceous lamina, usually with 10–13 segments, the upper surface scabrous, hispidulous, the lower surface puberulous to hirtellous to subtomentose on the veins, with arachnoid indumentum in the areoles; with usually 15–25 pairs of lateral veins in the free part of the medial segment. Their staminate inflorescences have red, pink, purple or yellowish 5–20 cm long spathes; with usually 8–18 spikes per inflorescence. Their pistillate inflorescences have reddish to yellowish 12–22 cm long spathes; spikes usually 2–4 per inflorescence. Fruit ellipsoid, 2–2.5 mm long, smooth, dark brown (Figures 4.4 and 4.5) (Berg and Franco-Rosselli, 2005).

4.2.4 *Cecropia pachystachya* Trécul (Syn.: *C. adenopus* Mart. ex Miq., *C. lyratiloba* Miq. and *C. catarinensis* Cuatrec.)

This species is characterized by trees of 20 m height. Their petioles have fused trichilia, 10–55 cm long, puberulous to hirtellous, mostly with sparse to dense arachnoid indumentum; stipules usually 10–22 cm long, white to pale green. Lamina subcoriaceous to chartaceous with 9–13 segments, the upper surfaces scabrous, hispidulous or partly hirtellous, the lower surfaces puberulous to sub-hirtellous to sub-tomentose, with arachnoid indumentum in the areoles; with 10–20 pairs of lateral veins in the free part of the medial segment. Their staminate inflorescences have a white

FIGURE 4.4 Pistillate inflorescences of *Cecropia obtusifolia* Bertol (Urticaceae). Location: Panamá Oeste Province; Cerro Campana, Panama, June 2016. (Courtesy of Orlando O. Ortíz.)

FIGURE 4.5 Habitat of *Cecropia obtusifolia* Bertol (Urticaceae). Location: Eastern Panama; Cerro Azul, Panama, June 2016. (Courtesy of Orlando O. Ortíz.)

to pale green 3–18 cm long spathes; 5–20 spikes per inflorescence. Their pistillate inflorescences have greenish-white spathes, 3–10 cm long; usually 3–6 spikes per inflorescence. Fruit oblongoid, 2–2.2 cm long, tuberculate, dark brown (Figure 4.6) (Berg and Franco-Rosselli, 2005).

4.2.5 *Cecropia peltata* L.

Cecropia peltata are trees usually of 15 m height. Their petiole has fused trichilia, usually 20–50 cm long, puberlous or also with arachnoid indumentum; with pinkish or reddish stipules 3–10 cm long. This psecies has a chartaceous to subcoriaceous lamina, usually with 8–10 segments; the upper surfaces is scabrous and hispidulous, the lower surface minutely puberulous, intermixed with sparse longer hairs, with arachnoid indumentum in the areoles; with usually 10–15 pairs of lateral veins in the free part of the medial segment. Their staminate inflorescences have pinkish, greenish or whitish 2.5–7 cm long spathes; usually with 15–25 spikes per inflorescence. Their pistillate inflorescences have pinkish or greenish-white spathes, 3.5–6 cm long; usually with 3–4 spikes per inflorescence. Their fruit is ovoid to ellipsoid, 2 mm long, tuberculate, dark brown (Figures 4.7 and 4.8) (Berg and Franco-Rosselli, 2005).

FIGURE 4.6 Inflorescences of *Cecropia pachystachya* Trécul (Urticaceae); embaúba. Location: Praia Laranjeiras; Balneário Camboriú, Santa Catarina, Brazil, June 26, 2018. (Courtesy of Oscar Benigno Iza/UNIVALI.)

4.3 BIOLOGICAL ACTIVITIES AND PHYTOCONSTITUENTS THAT SUPPORT ETHNOMEDICAL USES OF THE GENUS *CECROPIA* AS ANTIHYPERTENSIVE AGENT

With an objective of understanding the mechanism by which different extracts of *Cecropia* species have the capacity to reduce blood pressure, numerous studies have been carried out and are summarized *vide infra* in this review.

4.3.1 *Cecropia glaziovii*.

Early pioneer experiments on *C. glaziovii* from Brazil by Lapa and collaborators showed that oral administration of the water decoction of its leaves (0.5 g/kg/*bid*) and the corresponding *n*-butanol fraction (0.1 g/kg/day) induced hypotension in normotensive rats, decreased hypertension in spontaneous hyptertensive rats (SHR) and induced (by L-NAME and unilateral reduction of renal blood flow). These effects were established after 4–15 days treatment (Lapa et al., 1999; Lima-Landman et al., 2007).

FIGURE 4.7 Leaves and pistillate inflorescences of *Cecropia peltata* L. (Urticaceae). Location: Eastern Panama; Cerro Azul, Panama, June 2016. (Courtsey of Orlando O. Ortíz.)

FIGURE 4.8 Young leaves and pistillate inflorescences of *Cecropia peltata* L. (Urticaceae). Location: Panamá Oeste Province; Cerro Campana, Panama, June 2016. (Courtesy of Orlando O. Ortíz.)

Additionally, the *n*-butanol fraction (100 μg/ml) blocked the peak Ca^{++} current in cromaffin cells of a PC12 cell line after five minutes and decreased smooth muscle contraction (Lapa et al., 1999). Moreover, intravenous administration of the *n*-butanol fraction to anesthetized rats induced a transient hypotension and inhibited the pressor responses to angiotensin II, noradrenaline and angiotensin I by 42.8 ± 11.7%, 45.6 ± 10.8% and 47.6 ± 12.2%, respectively (Lima-Landman et al., 2007).

Additional studies showed that hot methanol extract (0.33 mg/ml) of the spitules of *C. glaziovii* inhibited 91 ± 9% the activity of angiotensin converting enzyme

(ACE), while a similar type of extract of its leaves inhibited only half of this activity (Lacaille-Dubois et al., 2001). In contrast, *in vivo* experiments suggested that hypotension induced by the aqueous extract of the leaves of *C. glaziovii* was unrelated to inhibition of ACE (Ninahuaman et al., 2007).

The above described active *n*-butanol fraction contained catechin, epicatechin, procyanidin B_2, procyanidin B_5, procyanidin C_1, isoorientin and isovitexin (Tanae et al., 2007), while the active methanol extract from the spitules contained, in addition, isoquercitrin but did not contain isovitexin nor procyanidin B_5 (Lacaille-Dubois et al., 2001).

4.3.2 *Cecropia hololeuca* Miq.

At a concentration of 0.33 mg/ml, the methanolic extracts of the leaves and the bark of *C. hololeuca* inhibited 40 ± 4% and 33 ± 3% the activity of angiotensin converting enzyme (ACE), respectively. Major components within this extract were identified as orientin, isoorientin, (+)-catechin, (–)-*epi*catechin and two (–)-*epi*catechin-derived oligomeric procyanidins (procyanidin B_2, procyanidin C_1) chlorogenic and protocatechic acids (Lacaille-Dubois et al., 2001).

4.3.3 *Cecropia obtusifolia* Bertol.

Evidence for the blood-pressure-lowering effect of the leaves of *C. obtusifolia* was seen after intravenous administration of 10 mg/kg of lyophilized ethanol extract to normotensive anesthetized male rats (Vidrio et al., 1982). Additionally, 50 mg/kg of an aqueous extract administered intravenously produced a fall in mean arterial blood pressure of 23.5% when it was administered to conscious SHR (Salas et al., 1987). Both acid and neutral methanol and dichloromethane fractions of the aerial parts of this plant caused relaxation of endothelium-intact aorta pre-contracted with phenylephrine. Methanol fractions showed a greater effect (>53.3 ± 3.3%) when compared to dichloromethane fractions (<42.2 ± 3.4%) (Guerrero et al., 2010).

A preliminary biological screening by radioligand-binding techniques revealed that both methanol/dichloromethane (1:1) and ethanol extracts from stems and leaves of *C. obtusifolia* at 100 µg/ml inhibited angiotensin II binding to the angiotensin II type 1 receptor by 80 ± 16% (stems) and 58 ± 2% (leaves), and endothelin-1 binding to the endothelin-1 type A receptor by 51 ± 12% (stems) and 63 ± 7 (leaves) (Caballero-George et al., 2001).

Chlorogenic acid and isoorientin are the main compounds identified in the aqueous extract and the butanolic fraction of the leaves of *C. obtusifolia* (Andrade-Cetto and Wiedenfeld, 2001). The leaves of this plant contain five anthraquinones (aloe-emodin, emodin, rhein, chrysophanol and physcion) (Yan et al., 2013), two saturated fatty acids (palmitic and stearic acid) (Guerrero et al., 2010), five steroids (*β*-sitosterol, stigmasterol (Andrade-Cetto and Heinrich, 2005), stigmast-4-en-3-one, 4-cholestene-3,24-dione and 4,22-cholestadien-3-one (Guerrero et al., 2010), and vanillic acid (Guerrero et al., 2010) have been identified in a small amount and only for the plant in these studies. Besides these, 4-vinyl-2-methoxy-phenol, 2-methylbenzaldehyde, 2,3-dihydrobenzofuran, 3′-methoxyacetophenone (Guerrero et al., 2010),

1-(2-methyl-1-nonen-8-il)-aziridine and 4-ethyl-5-(n-3-valeroil)-6-hexahydrocoumarin (Andrade-Cetto and Heinrich, 2005) have also been described.

4.3.4 *Cecropia pachystachya* Trécul (Syn.: *C. adenopus* Mart. ex Miq., *C. lyratiloba* Miq. and *C. catarinensis* Cuatrec.)

The ethanol extract (0.33 mg/ml) of the leaves of *C. pachystachya* inhibited 42 ± 2% the activity of ACE (Lacaille-Dubois et al., 2001).

The aqueous extract of the leaves (300 mg/kg) of this plant decreased blood pressure up to 46.2% of basal and increased heart rate up to 133% of basal on Wistar rats (Consolini and Migliori, 2005). These results may be explained by central blockage of sympathetic nerves of vessels and central cholinergic inhibition of the heart, respectively. In this study, the activities were not associated to specific compounds. Furthermore, a 500 μg/ml flavonoid fraction of the methanol extract of leaves induced cardiac depression (reduction to 56.7 ± 5.1%) and inhibited adrenaline-induced contractions of the aorta (34.2 ± 6.9%) in Wistar rats (Oliveira et al., 2003; Ramos Almeida et al., 2006).

The main flavonoids isolated from the flavonoid fraction were orientin, isoorientin, isovitexin and apigenin-6-galactosyl-6″-*O*-galactopyranoside (Oliveira et al., 2003; Ramos Almeida et al., 2006).

The main compounds identified in the aqueous and ethanol extracts of the leaves were chlorogenic acid and orientin (Maquiaveli et al., 2014).

The potential use of this plant in the treatment of renal chronic diseases was demonstrated by reducing the inflammation and renal lesions in male Wistar rats submitted to 5/6 nephrectomy. These effects were associated with ACE inhibition (67%), reduction of macrophage (ED-1 positive cells) infiltration, angiotensin II (AII) and c-Jun N-terminal kinase (p-JNK) expression and arginase activity in renal cortex of rats. (Maquiaveli et al., 2014).

Compounds found in the methanol extract of the leaves of *C. pachystachya* included isoorientin, isovitexin, protocatechic acid, chlorogenic acid, catechin, epicatechin and isoquercitrin (Lacaille-Dubois et al., 2001).

4.3.5 *Cecropia peltata* L.

Even though traditional medicine in Brazil and Trinidad and Tobago claim *C. peltata* is useful to treat heart diseases and hypertension (Lans, 2006; Agra et al., 2007) there are no studies available supporting its antihypertensive properties. Nevertheless, since its chemistry is similar to that of other *Cecropia* species (Costa et al., 2011b), it can be suggested to also have antihypertensive potential.

4.4 PROPOSED MECHANISMS OF ACTION

The overall composition of aqueous and alcoholic extracts of the aerial parts of these plants mainly contains procyanidins, phenolic acids, triterpenes and *C*-glycosylflavones. Literature describing the effect of these groups of compounds evaluated individually on different targets involved in blood pressure regulation is discussed below.

4.4.1 ACE Inhibition

The renin-angiotensin system is a bioenzymatic cascade that plays a role in cardiovascular homeostasis by influencing vascular tone, fluid and electrolyte balance (Mirabito Colafella and Danser, 2017). ACE is the key enzyme responsible for converting angiotensin I into the potent vasoconstrictor octapeptide angiotensin II (Studdy et al., 1983). The inhibition of this enzyme is a key factor for potential therapeutic interventions and, therefore, a target of selection when studying the antihypertensive potential of drugs.

Major components of extracts from *Cecropia hololeuca* were identified as orientin, isoorientin, (+)-catechin, (–)-epicatechin and two (–)-epicatechin-derived oligomeric procyanidins (procyanidin B_2, procyanidin C_1) chlorogenic and protocatechic acids (Lacaille-Dubois et al., 2001). All these compounds were tested individually at a concentration of 0.33 mg/ml, showing very little inhibition of ACE ($< 48 \pm 1$). Interestingly, a fraction containing mainly procyanidins inhibited ACE activity by $94 \pm 4\%$ at the same concentration (Lacaille-Dubois et al., 2001).

4.4.2 Interaction with Angiotensin and Endothelin Receptors

Angiotensin II AT_1 and endothelin-1 ET_A receptors are G-protein coupled receptors involved in blood pressure regulation and two of the most common targets for the discovery of antihypertensive drugs (Timmermans et al., 1992; Takigawa et al., 1995).

Procyanidins have previously been shown to inhibit the binding of specific ligands to angiotensin II AT_1 and endothelin-1 ET_A receptors. Their effect depends on their degree of polymerization; thus, monomer and dimers showed less inhibition of binding than tetramers and procyanidins with higher polymerization degrees. Catechin, epicatechin, procyanidins B2 and B5 showed very little inhibitory activity over [^3H] angiotensin binding to the AT_1 receptor ($Ki = 286 \pm 1 \mu M$) and less than 56% binding inhibition to the ET_A receptor at 100 μM. In contrast, procyaniding C1 inhibited 76% the binding of the selective ET_A receptor antagonist [^3H] BQ123 but showed very little effect on the binding of [^3H] angiotensin ($Ki = 184 \pm 19$) (Caballero-George et al., 2002; Caballero-George, 2002).

Other compounds like ursolic acid and α-amyrin have also been studied for their ability to inhibit these two ligands. While 10 μg/ml ursolic acid inhibited $52 \pm 2\%$ [^3H] BQ123 binding to the ET_A receptor and $65 \pm 27\%$ [^3H] angiotensin binding to the AT1 receptor, no activity was found with 10 μg/ml α-amyrin (Caballero-George et al., 2004).

4.4.3 Vascular Smooth-Muscle Relaxation

Vasodilation produced by methanol and flavonoid fraction of *Cecropia pachystachya* is due to an endothelin-dependent effect probably by stimulation of NO production. Although a complex mixture like the crude extract showed activity, their main flavonoids isoorientin and a mixture of orientin/isovitexin did not seem to be active when tested individually (Ramos Almeida et al., 2006).

4.5 CONCLUSIONS

The therapeutic properties of the plants in this genus have generally been attributed to its chemical composition, consisting mainly in terpenoids, steroids (Ospina Chávez et al., 2013), chlorogenic and caffeic acids (Lacaille-Dubois et al., 2001), proanthocyanidins, flavonoids (Luengas-Caicedo et al., 2007) and other phenolic compounds (Gazal et al., 2014).

A mixture of phytoconstituents may be responsible for complex pharmacological effects, and this may be the reason it is difficult to correlate individual compounds with specific biological activities. In several studies, synergism has been suggested as the mechanism of action for these extracts (Ramos Almeida et al., 2006; Lacaille-Dubois et al., 2001).

Caution must be used when comparing results from extracts obtained from different *Cecropia* species, since chemical composition may vary among species and geographical location.

Variation in chemical composition of extracts can depend on the method of extraction and on the species and the collection site. Although, some species of this genus have enough scientific support to be recommended as a phytomedicine to treat hypertension, additional studies regarding the methods to assure its quality need to be carried out.

Even though there are no studies available to support the use of *Cecropia peltata* in the treatment of hypertension, its chemical composition is similar to the well-studied plants of this genus, thus it is possible to consider this plant as a valid option to this treatment.

While the medicinal use of the plants of this genus is officially recognized in National Pharmacopoeias and Formularies of several Latin-American countries, it is important to recognize that these phytomedicines are complex mixtures requiring a better understanding of their chemical composition and their correlation with the biological activities to assure their quality, safety and efficacy.

REFERENCES

Agra M de F, De Freitas PF, Barbosa-Filho JM. 2007. Synopsis of the plants known as medicinal and poisonous in northeast of Brazil. *Brazilian J Pharmacogn.* 17:114–140.

Andrade-Cetto A, Heinrich M. 2005. Mexican plants with hypoglycaemic effect used in the treatment of diabetes. *J Ethnopharmacol.* 99:325–348.

Andrade-Cetto A, Wiedenfeld H. 2001. Hypoglycemic effect of *Cecropia obtusifolia* on streptozotocin diabetic rats. *J Ethnopharmacol.* 78:145–149.

Berg C, Franco-Rosselli P. 2005. *Cecropia. Flora Neotrop.* 94:1–230.

Berg C, Akkermans R, van Heusden E. 1990. Cecropiaceae: *Coussapoa* and *Pourouma*, with an introduction to the family. *Flora Neotrop.* 51:1–208.

Caballero-George C. 2002. Application of ligand-binding techniques in the Discovery of bioactive natural products: chemical and biological evaluation of *Bocconia frutescens* L. and *Guazuma ulmifolia* Lam. Doctoral thesis. Faculteit Farmaceutische, Biomedische en Diergeneeskundige Wetenschappen, Departement Farmaceutische Wetenschappen, Universiteit Antwerpen, Belgium.

Caballero-George C, Vanderheyden PM, Solis PN, Pieters L, Shahat AA, Gupta MP, Vauquelin G, Vlietinck AJ. 2001. Biological screening of selected medicinal panamanian plants by radioligand-binding techniques. *Phytomedicine.* 8:59–70.

Caballero-George C, Vanderheyden PM, De Bruyne T, Shahat AA, Van den Heuvel H, Solis PN, Gupta MP, Claeys M, Pieters L, Vauquelin G, Vlietinck AJ. 2002. In vitro inhibition of [3H]-angiotensin II binding on the human AT1 receptor by proanthocyanidins from Guazuma ulmifolia bark. *Planta Med.* 68(12):1066–1071.

Caballero-George C, Vanderheyden PM, Okamoto Y, Masaki T, Mbwambo Z, Apers S, Gupta MP, Pieters L, Vauquelin G, Vlietinck A. 2004. Evaluation of bioactive saponins and triterpenoidal aglycons for their binding properties on human endothelin ETA and angiotensin AT1 receptors. *Phytother Res.* 18(9):729–736.

Conn BJ, Hadiah JT, Webber BL. 2012. The status of *Cecropia* (Urticaceae) introductions in Malesia: addressing the confusion. *Blumea J Plant Taxon Plant Geogr.* 57:136–142.

Consolini AE, Migliori GN. 2005. Cardiovascular effects of the South American medicinal plant *Cecropia pachystachya* (ambay) on rats. *J Ethnopharmacol.* 96(3):417–22. Epub 2004 Nov 21. 96:417–422.

Costa GM, Ortmann CF, Schenkel EP, Reginatto FH. 2011a. An HPLC-DAD method to quantification of main phenolic compounds from leaves of *Cecropia* species. *J Braz Chem Soc.* 22:1096–1102.

Costa GM, Schenkel EP, Reginatto FH. 2011b. Chemical and pharmacological aspects of the genus *Cecropia. Nat Prod Comun.* 6:913–920.

Dejean A, Leroy C, Corbara B, Céréghino R, Roux O, Hérault B, Rossi V, Guerrero RJ, Delabie JH, Orivel J, Boulay R. 2010. A temporary social parasite of tropical plant-ants improves the fitness of a myrmecophyte. *Naturwissenschaften.* 97:925–934.

Franco-Rosselli P, Berg CC. 1997. Distributional patterns of *Cecropia* (Cecropiaceae): a pan-biogeographic analysis. *Caldasia.* 19:285–296.

Gazal M, Ortmann CF, Martins FA, Streck EL, Quevedo J, de Campos AM, Stefanello FM, Kaster MP, Ghisleni G, Reginatto FH, Lencina CL. 2014. Antidepressant-like effects of aqueous extract from *Cecropia pachystachya* leaves in a mouse model of chronic unpredictable stress. *Brain Res Bull.* 108:10–17.

Guerrero EI, Morán-Pinzón JA, Gabriel L, Olmedo D, López-Pérez JL, San Feliciano A, Gupta MP. 2010. Vasoactive effects of different fractions from two Panamanians plants used in amerindian traditional medicine. *J Ethnopharmacol.* 131:497–501.

Gupta M. 1995. *Cecropia pathystachya* Mart. In: Gupta, M. (Ed.), 270 *Plantas medicinales iberoamericanas.* Convenio Andrés Bello, CYTED, Santafé de Bogotá, DC, Colombia, pp. 407–408.

Gutiérrez-Valencia J, Chomicki G, Renner SS. 2017. Recurrent breakdowns of mutualisms with ants in the neotropical ant-plant genus *Cecropia* (Urticaceae). *Mol Phylogenet Evol.* 111:196–205.

Hernández-Carvajal JF, Luengas-Caicedo PE. 2013. Preliminary phytochemical study of *Cecropia membranacea* Trécul. and *Cecropia metensis* Cuatrec. *Revista Cubana de Plantas Medicinales.* 18(4):586–595.

Lacaille-Dubois MA, Franck U, Wagner H. 2001. Search for potential angiotensin converting enzyme (ACE)-inhibitors from plants. *Phytomedicine.* 8:47–52.

Lans CA. 2006. Ethnomedicines used in Trinidad and Tobago for urinary problems and diabetes mellitus. *J Ethnobiol Ethnomed.* 2:45.

Lapa AJ, Lima-Landman MTR, Cysneiros RM, Borges ACR, Souccar C, Barreta IP, Lima TCM. 1999. Chemistry, biological and pharmacological properties of medicinal plants from the Americas: *Proceedings of the IOCD/CYTED Symposium*, Panama City, Panama, February 23–26, 1997. Amsterdam: Harwood Academic Publishers. Chapter 10, The Brazilian folk medicine program to validate medicinal plants – a topic in new antihypertensive drug research; pp. 185–196.

Lima-Landman MT, Borges AC, Cysneiros RM, De Lima TC, Souccar C, Lapa AJ. 2007. Antihypertensive effect of a standardized aqueous extract of *Cecropia glaziovii* Sneth in rats: an *in vivo* approach to the hypotensive mechanism. *Phytomedicine.* 14:314–320.

Luengas-Caicedo PE, Braga FC, Brandão GC, De Oliveira AB. 2007. Seasonal and intraspecific variation of flavonoids and proanthocyanidins in *Cecropia glaziovi* Sneth. leaves from native and cultivated specimens. *Z Naturforsch C.* 62:701–709.

Maquiaveli CC, da Silva ER, Rosa LC, Francescato HD, Lucon Júnior JF, Silva CG, Casarini DE, Ronchi FA, Coimbra TM. 2014. *Cecropia pachystachya* extract attenuated the renal lesion in 5/6 nephrectomized rats by reducing inflammation and renal arginase activity. *J Ethnopharmacol.* 158:49–57.

Matos FJ. 1989. *Plantas Medicinais: Guia de Seleção e Emprego de Plantas Usadas em Fitoterapia no Nordeste do Brasil.* IOCE, Fortaleza.

Mirabito Colafella KM, Danser AHJ. 2017. Recent advances in angiotensin research. *Hypertension.* 69(6):994–999.

Monro A. 2009. Neotropical urticaceae. In: Milliken W, Klitgård B, Baracat, A. Neotropikey - Interactive key and information resources for flowering plants of the Neotropics. [Internet]. Available from: http://www.kew.org/science/tropamerica/neotropikey/families/Urticaceae.htm. Accessed on July 20, 2016.

Montoya Peláez GL, Sierra JA, Alzate F, Holzgrabe U, Ramirez-Pineda JR. 2013. Pentacyclic triterpenes from *Cecropia telenitida* with immunomodulatory activity on dendritic cells. *Brazilian J Pharmacogn.* 23:754–761.

Ninahuaman MFML, Souccar C, Lapa AJ, Lima-Landman MTR. 2007. ACE activity during the hypotension produced by standardized aqueous extract of Cecropia glaziovii Sneth: a comparative study to captopril effects in rats. *Phytomedicine.* 14(5), 321–327.

Oliveira RR, Moraes MC, Castilho RO, Valente AP, Carauta JP, Lopes D, Kaplan MA. 2003. High-speed countercurrent chromatography as valuable tool to isolate C-glycosylflavones from *Cecropia lyratiloba* Miquel. *Phytochem Anal.* 14:96–99.

Oliveira KN, Coley PD, Kursar TA, Kaminski LA, Moreira MZ, Campos RI. 2015. The effect of symbiotic ant colonies on plant growth: a test using an Azteca-*Cecropia* system. *PLoS ONE.* 10(3):e0120351.

Ospina Chávez J, Rincón Velanda J, Guerrero Pabón M. 2013. Perfil neurofarmacológico de la fracción butanólica de las hojas de *Cecropia peltata* L [Neuropharmacological profile of the butanolic fraction obtained from leaves of *Cecropia peltata* L.]. *Rev Colomb Cienc Quím Farm.* 42:244–259. Spanish.

Pacheco NR, Pinto NC, da Silva JM, Mendes RF, da Costa JC, Aragão DM, Castañon MC, Scio E. 2014. *Cecropia pachystachya*: a species with expressive *in vivo* topical anti-inflammatory and *in vitro* antioxidant effects. *Biomed Res Int.* 2014:1–10.

Pio-Corrêa M. 1978. Diccionário das Plantas Úteis do Brasil e das Exóticas Cultivadas. Imprensa Nacional, Rio de Janeiro.

Ramos Almeida R, Montani Raimundo J, Rodrigues Oliveira R, Coelho Kaplan MA, Rocah Gattass CR, Takashi Sudo R, Zapata-Sudo G. 2006. Activity of *Cecropia lyratiloba* extract on contractility of cardiac and smooth muscles in Wistar rats. *Clin Exp Pharmacol Physiol.* 33:109–113.

Salas I, Brenes JR, Morales OM. 1987. Antihypertensive effect of *Cecropia obtusifolia* (Moraceae) leaf extract on rats. *Rev Biol Trop.* 35:127–130.

Souccar C, Cysneiros RM, Tanae MM, Torres LM, Lima-Landman MT, Lapa AJ. 2008. Inhibition of gastric acid secretion by a standardized aqueous extract of *Cecropia glaziovii* Sneth and underlying mechanism. *Phytomedicine.* 15:462–469.

Studdy PR, Lapworth R, Bird R. 1983. Angiotensin-converting enzyme and its clinical significance--a review. *J Clin Pathol.* 36(8):938–947.

Takigawa M, Sakurai T, Kasuya Y, Abe Y, Masaki T, Goto K. 1995. Molecular identification of guanine-nucleotide-binding regulatory proteins which couple to endothelin receptors. *Eur J Biochem.* 228(1):102–108.

Tanae MM, Lima-Landman MT, De Lima TC, Souccar C, Lapa AJ. 2007. Chemical standardization of the aqueous extract of *Cecropia glaziovii* Sneth endowed with antihypertensive, bronchodilator, antiacid secretion and antidepressant-like activities. *Phytomedicine*. 14:309–313.

Timmermans PB, Benfield P, Chiu AT, Herblin WF, Wong PC, Smith RD. 1992. Angiotensin II receptors and functional correlates. *Am J Hypertens*. 5 (12 Pt 2):221S–235S. Review.

Treiber EL, Gaglioti AL, Romaniuc-Neto S, Madriñán S, Weiblen GD. 2016. Phylogeny of the Cecropieae (Urticaceae) and the evolution of an ant-plant mutualism. *Systematic Botany*. 41(1): 56–66.

Vidrio H, García-Márquez F, Reyes J, Soto RM. 1982. Hypotensive activity of *Cecropia obtusifolia*. *J Pharm Sci*. 71:475–476.

Yan Y, Hao Y, Hu S, Chen X, Bai X. 2013. Hollow fibre cell fishing with high performance liquid chromatography for screening bioactive anthraquinones from traditional Chinese medicines. *J Chromatogr A*. 1322:8–17.

5 Berries and Lipids in Cardiovascular Health

Arpita Basu, Nancy Betts, Paramita Basu and Timothy J. Lyons

CONTENTS

5.1 INTRODUCTION

Cardiovascular disease (CVD) is the leading cause of global mortality and a growing worldwide public health problem. Among the multiple traditional risk factors of CVD, blood lipids, especially blood cholesterol level, is an established predictor of CVD risks and subsequent complications. Low-density lipoprotein cholesterol (LDL-C) is a well-established risk factor for CVD and was recognized more than 30 years ago by the National Heart, Lung, and Blood Institute when it created the National Cholesterol Education Program (NCEP) to educate both the medical community and the public about the need to lower levels of blood cholesterol to reduce the risk of major vascular events (NCEP, 1988). In a meta-regression analysis of 49 clinical trials with 312,175 participants, each 1-mmol/L reduction in LDL-C level was associated with a 23% risk reduction of major vascular events for statins, and a 25% risk reduction for non-statin interventions, including dietary approaches to lower LDL-C (Silverman et al., 2016). Studies further reveal that residual cardiovascular risk remains after LDL-C goals are achieved with lipid-lowering treatments, especially in high-risk patients such as those with type 2 diabetes or the metabolic syndrome. This residual risk can be attributed to low high-density lipoprotein (HDL) and high triglyceride-rich lipoprotein levels routinely measured in clinical care. Thus, in addition to LDL-C, triglyceride, HDL-C, non-HDL-C, total apolipoprotein (apo) B, apoB/A-I and TC/HDL-C levels are considered significant predictors of cardiovascular events and have been validated in large prospective studies in cardiovascular and diabetes epidemiology (Arsenault et al., 2009; Kastelein et al., 2008; Ray et al., 2009). Further, qualitative changes in lipids, such as shifts among low, medium and large lipid particle size and molar concentrations, have been associated with CVD risks and events beyond conventional lipid profiles (Basu et al., 2016; Garvey et al., 2003). Thus, lipid and lipoprotein subclasses based on size and density have been widely examined in lipid-lowering interventions.

Dietary and food-based approaches of lowering lipids have been widely practiced as a secondary or adjunct strategy, especially in achieving weight loss and healthy lifestyle goals associated with favorable lipid profiles. Among these dietary approaches, several foods and supplements containing phytosterols, fiber and polyphenols have gained recognition for their lipid-lowering effects. In this context, berry fruits and their products have shown much promise in offering several cardiovascular benefits, including those related to favorable lipid profiles. In a meta-analysis of clinical studies on *Vaccinium* berries, significant lipid-lowering effects were observed based on a pooled analysis of 16 clinical studies in 1,109 participants (Zhu et al., 2015). This brief review aims to provide further insights into the role of dietary berries in blood lipid management, with special emphasis on the reported clinical studies.

5.2 BERRIES: COMPOSITION AND NUTRITIONAL VALUE

Berries have a wide range of nutrients and bioactive compounds that have been extensively reviewed for their protective effects against CVD (Basu et al., 2010c). The commonly consumed berries in the United States include blackberry, black raspberry, blueberry, cranberry, red raspberry and strawberry. Less commonly consumed berries include acai, black currant, chokeberry and mulberry. In general, berries are low in calories and are high in moisture and fiber, and contain natural antioxidants, such as vitamins C and E, and micronutrients such as folic acid, calcium, selenium, alpha and beta carotene and lutein. For example, 100g frozen unsweetened strawberries provide only 35 kcal and thus can be included as a low-calorie fat-free nutrient-dense snack in the dietary management of CVD (Basu et al., 2014a). Berries, including commonly consumed blueberries and strawberries, are naturally rich in polyphenolic flavonoids, with high proportions of flavonoids, including anthocyanins and ellagitannins. Anthocyanins comprise the largest group of natural, water-soluble, plant pigments and impart the bright colors to berry fruits. Approximately 400 individual anthocyanins have been determined, and these are generally more concentrated in the skins of fruits, especially berry fruits. However, red berry fruits, such as strawberries and cherries, also have anthocyanins in their flesh. Studies suggest that Americans consume an average of 12.5–215 mg of anthocyanins per day (Manach et al., 2004).

5.3 BERRIES: MECHANISMS OF ACTION OF BIOACTIVE CHEMICALS

Dietary berries are a rich source of several categories of phytochemicals with structural properties that confer antioxidant functions that have been associated with health benefits of berries. The bioactive compounds in berries are constituted by mainly phenolic compounds (phenolic acids, flavonoids, such as anthocyanins and flavonols, and tannins) and ascorbic acid. These compounds, either individually or combined, are responsible for various health benefits of berries, such as prevention of inflammation disorders, CVD or protective effects to lower the risk of various cancers. Phenolics represent a large group of secondary metabolites, consisting of one or more aromatic rings with variable degrees of hydroxylation, methoxylation and glycosylation, contributing to fruit color, astringency and bitterness. Phenolic compounds in berries especially include flavonoids, such as anthocyanins (i.e., cyanidin glucosides and pelargonidin glucosides), flavonols (quercetin, kaempferol, myricetin) and flavanols (catechins and epicatechin). Furthermore, phenolic acids (hydroxybenzoic acids and hydroxycinnamic acids) and hydrolysable tannins, such as ellagitannins, act as important bioactive compounds (de Souza et al., 2014). These components, either individually or combined, are mainly responsible for berry health benefits and are also associated with their antioxidant properties. Among the commonly consumed dietary berries, blueberries have one of the highest antioxidant capacities, and this antioxidant activity of blueberries depends on their phytochemical complex, being mainly represented by anthocyanins, procyanidins, chlorogenic acid and other flavonoid compounds. It is supposed that the major contributors to their antioxidant activity are mainly anthocyanins, responsible for about 84% of total antioxidant capacity (TAC), and not ascorbic acid. Ascorbic acid, which is present in blueberries in a significant amount, was found to contribute to antioxidant capacity only with a small portion up to 10% (Moyer et al., 2002). In another detailed comparative study of anthocyanin composition of dark-colored berries, especially blackberries, black currants and blueberries, cyanidin-3-O-glucoside accounted for 94% of blackberry anthocyanins, and as one of the strongest antioxidants present in these three berries, it contributed to approximately 96% TAC of blackberries. This was followed by delphinidin-3-O-rutinoside and cyanidin-3-O-rutinoside in black currants and blueberries (Lee et al., 2015). Hydroxyl groups on these anthocyanidins have been shown to be the main functional groups for radical scavenging and thus their antioxidant activities. Data further show that flavonoids with free phenolic hydroxyl groups showed effective radical scavenging activity, while compounds having only a methoxyl group did not exhibit such effects (Lee et al., 2015). As a result, anthocyanidins with B ring *o*-diphenyl patterns, such as cyanidin and delphinidin in blueberries, have higher antioxidant activity than malvidin, pelargonidin, petunidin and peonidin, for example, found in strawberries (Manganaris et al., 2014). Thus, based on these differences in chemical structures and corresponding antioxidant activities, it is largely recommended that a combination of different colored berries be consumed in the human diet for their protective functions against chronic diseases.

5.4 BERRIES AND LIPIDS: ANIMAL MODELS

Animal models provide mechanistic insights into the role of berries in lipid and lipoprotein metabolism. Berries and their bioactive constituents, such as polyphenolic flavonoids and phenolic acids, have been shown to increase paraoxonase (PON) activity associated with antioxidant function of HDL-cholesterol, and to increase hepatic synthesis of apolipoprotein A-I. For example, acai berry native to the Amazon region has been reported to increase serum activity of PON1 and levels of HDL-cholesterol and to increase fecal cholesterol excretion in rat models of hyperlipidemia and hepatic steatosis (de Souza et al., 2010; de Souza et al., 2012; Pereira et al., 2016). Commonly consumed dietary berries, such as blueberries and strawberry extracts, were also demonstrated to improve lipid profiles by downregulating the activity of genes related to fatty acid synthesis, causing regression of aortic lesions and decreasing inflammation and oxidative damage in these animals (Mandave et al., 2017; Ströher et al., 2015). Strawberry and blueberry pomace, a by-product from industrial fruit processing, has emerged as a promising source of antioxidants and micronutrients and has been tested for its health benefits by many researchers mainly as an alternative and cost-effective way of supplementing human diet with berry polyphenols. In two separate studies, blueberry and strawberry pomace supplementation improved multiple metabolic parameters, including plasma and liver cholesterol, insulin resistance and abdominal fat content in fructose-fed animals (Jaroslawska et al., 2011; Khanal et al., 2012). As expected, the polyphenol content of the berry pomace was shown to play an important role in its lipid-lowering effects based on larger decreases in liver cholesterol content in the high-polyphenol pomace group when compared to the regular pomace-fed animals (Jaroslawska et al., 2011). Thus, these studies provide evidence for the protective action of polyphenols and fiber mediated by berries in rat models of hyperlipidemia. Further studies are needed to assess the possible role of these compounds in the prevention of metabolic and lipid disorders in humans.

5.5 BERRIES AND LIPIDS: EPIDEMIOLOGICAL STUDIES

More than a decade ago, Djoussé et al. reported a large epidemiological study on the inverse association of fruit and vegetable intake with serum LDL cholesterol concentration in participants from the National Heart, Lung, and Blood Institute (NHLBI) Family Heart Study (Djoussé et al., 2004). Since then, accumulating epidemiological evidence further reveals the inverse association of dietary berries with chronic disease outcomes linked to dyslipidemia and hyperlipidemia, especially insulin resistance, type 2 diabetes, coronary artery disease and non-fatal myocardial infarction (MI) in large prospective studies of adult men and women (Cassidy et al., 2013; Jennings et al., 2014; Wedick et al., 2012). Among the bioactive compounds in berries, anthocyanins responsible for the red/blue hue in these fruits have been mainly linked to their protective effects. These compounds have been shown to directly improve dyslipidemia and lipoprotein profiles and to decrease surrogate markers of atherosclerosis in reported clinical trials (Qin et al., 2009; Zhu et al., 2011). In one of the epidemiological studies, food-based analyses revealed a trend toward a reduction in risk of MI with increasing intake of the two main sources of anthocyanins from strawberries and blueberries, with a 34% decrease in risk for those who consumed >3 portions per week compared to those who consumed these berries less than once a month. These data are important from a public health perspective because these fruits can be readily incorporated into the habitual diet (Cassidy et al., 2013). In another longitudinal study, higher intakes of red/purple fruits and vegetables were associated with significantly lower serum total cholesterol in adults (Mirmiran et al., 2015). In the Women's Health Study, strawberry intake was specifically associated with marginally significant but lower levels of C-reactive protein, a stable marker of inflammation shown to be elevated with higher serum lipids (Sesso et al., 2007). Of practical relevance was that these significant associations were observed at intakes as low as two servings/week of strawberry in women. Thus, on the basis of this epidemiological evidence and the related mechanistic insights, adding red and purple berries may be considered a prudent dietary choice in combating lipid abnormalities that explain much of the underlying causes of cardiovascular complications in adults.

5.6 BERRIES AND LIPIDS: CLINICAL STUDIES

As summarized in Tables 5.1 and 5.2, dietary berries as whole fruits and in different processed forms have been shown to decrease circulating levels of conventional lipids and shift lipid and lipoprotein subclasses to a less atherogenic profile in adults with one or more cardiovascular risks. In most of these studies, berry intervention was consistently shown to decrease total and LDL-cholesterol (Basu et al., 2014b; Broncel et al., 2010; Kianbakht et al., 2014; Lee et al., 2008; Qin et al., 2009; Soltani et al., 2014; Zhu et al., 2011; Zunino et al., 2012) and increase HDL-cholesterol (Kianbakht et al., 2014; Qin et al., 2009; Ruel et al., 2006; Zhu et al., 2011), thereby lowering cardiometabolic risks in these adults. While lipid-lowering effects of whole berries may be supported by several bioactive compounds in the fruit, including their fiber and phytosterol content, studies using commercially available berry juice products, especially the low-calorie cranberry juice (Novotny et al., 2015; Ruel et al., 2006), also showed similar beneficial effects of lowering triglycerides and increasing HDL-cholesterol. These clinical findings have been supported by epidemiological data from the NHANES (2005–2008) survey, which reports a higher proportion of cranberry beverage consumers (approximately 221mL/day) were predicted to be normal weight (BMI < 25 kg/m^2) with lower waist circumference and had significantly lower triglycerides and CRP than non-consumers (Duffey and Sutherland, 2013). These data provide evidence on the hypolipidemic and anti-atherosclerotic actions of berry fruits and juices in humans.

In addition to measuring lipid outcomes related to conventional lipids, findings reported by Qin et al. (2009) and Zhu et al. (2015) provide further mechanistic insights on the hypolipidemic effects of purified berry extracts in adults with independent cardiovascular risks such as dyslipidemia/hyperlipidemia and type 2 diabetes. In a 12-week randomized placebo-controlled trial in adults with dyslipidemia, purified anthocyanins derived from bilberries and black currants were shown to decrease the mass and activity of cholesteryl ester transfer protein (CETP) (Qin et al., 2009). CETP is a plasma protein that mediates the removal of cholesteryl esters from HDL in exchange for a triglyceride molecule derived primarily from either LDL, VLDL or chylomicrons. Thus, CETP inhibition has been shown to be a possible mechanism for the elevation of HDL cholesterol and decrease of LDL cholesterol (Inazu et al., 1990). In another study in participants with type 2 diabetes, supplementation of a similar berry anthocyanin extract was shown to decrease specific plasma apolipoproteins, especially apolipoprotein B

TABLE 5.1
Effects of Dietary Berries (Whole Fruit, Juice or Freeze-Dried Berries) on Serum Lipids: Clinical Studies

Author (year)	Study Design	Participants	Intervention	Significant Effects on Conventional Lipids	Significant Effects on Lipid Subclasses/Apolipoproteins
Ruel et al. (2006)	Four-week successive periods of intervention with increasing doses of CJC	Obese men (n=30)	Three doses of CJC (125, 250 and 500mL) vs. placebo juice/day	Increases in plasma HDL-cholesterol (46±5 to 49±6mg/dL) and decreases in ratio of total and HDL-cholesterol with increasing doses of CJC vs. placebo	Plasma Apo-A-I increased in the CJC phase, but did not reach significance
Burton-Freeman et al. (2010)	Randomized crossover trial; 12 wks	Hyperlipidemic adults (n=24)	Strawberry beverage (10g FDS) vs. matched placebo with or without high-fat meal challenge	Decrease in postprandial triglycerides in the strawberry (131±2) vs. placebo (136±2) group	Not reported
Zunino et al. (2012)	Randomized crossover trial; 7 wks	Obese adults (n=20)	Strawberry beverage (4 servings strawberries) vs. strawberry-flavored control beverage	Decreases in total cholesterol (182±38 to 169±37 mg/dL) in the strawberry vs. control group	Decreases in NMR-derived small HDL particle concentrations (18.3±4.4 to 17.2±3.8 μmol/L), and increase in LDL size (21±0.7 to 21.22±0.6 nm) in the strawberry vs. control group
Basu et al. (2014a)	Randomized parallel trial; 12 wks	Adults with above optimal serum lipids (n=60)	High-dose FDS (50g/d), low-dose FDS (25g/d) vs. fiber and calorie-matched control beverages	Decreases in total (214±7 to 181±5 mg/dL) and LDL-C (130±7 to 103±5 mg/dL) in high-dose vs. low-dose FDS & controls at 12 wks vs. baseline	Decreases in NMR-derived small LDL particle concentrations (697±106 to 396±69 nmol/L) in high-dose vs. low-dose FDS & controls at 12 wks vs. baseline

(Continued)

TABLE 5.1 (CONTINUED)
Effects of Dietary Berries (Whole Fruit, Juice or Freeze-Dried Berries) on Serum Lipids: Clinical Studies

Author (year)	Study Design	Participants	Intervention	Significant Effects on Conventional Lipids	Significant Effects on Lipid Subclasses/Apolipoproteins
Lankinen et al. (2014)	Randomized parallel trial; 12 wks	Overweight/obese adults with the metabolic syndrome (n=131)	Healthy Diet (whole grains, fish, bilberries (300g/day), Whole Grain diet (whole grains), or Control diet (refined grains)	Increase in large HDL particle concentrations and particle size in the Healthy Diet group vs. Control	No changes in Apo-A-I and Apo-B 100 among diets
Novotny et al. (2015)	Randomized parallel trial; 8 wks	Overweight adults (n=56)	LCCJ or matched placebo beverage (480mL)/day	Decreases in serum triglycerides (113±9 to 102±4 mg/dL) in the LCCJ vs. placebo at 8 wks	No changes in Apo-A-I, A-II and Apo-B

Apo: apolipoprotein; CJC: cranberry juice cocktail; FDS: freeze-dried strawberries; LCCJ: low-calorie cranberry juice; NMR: nuclear magnetic resonance.

TABLE 5.2
Effects of Berry Bioactive Compound Extracts on Serum Lipids: Clinical Studies

Author (year)	Study Design	Participants	Intervention	Significant Effects on Conventional Lipids	Significant Effects on Lipid Subclasses/Apo Lipoproteins
Lee et al. (2008)	Randomized parallel trial; 12 wks	Adults with type 2 diabetes (n=30)	Cranberry extracts (3 capsules/day ~ 1500mg extracts) vs. placebo	Decreases in total (-16±4mg/d/L) and LDL-cholesterol (-15±4mg/dL) in the cranberry vs. placebo	Not reported
Qin et al. (2009)	Randomized parallel trial; 12 wks	Adults with dyslipidemia (n=120)	Anthocyanin extracts (4 capsules/day ~ 320 mg extracts) vs. placebo	Decreases in LDL (159±34 to 140±35 mg/dL) and increases in HDL-cholesterol (46±8 to 51±9 mg/dL) in anthocyanin vs. placebo	Decreases in CETP mass and activity in anthocyanin vs. placebo; no changes in Apo-A-I and Apo-B
Broncel et al. (2010)	Uncontrolled study; 2 months vs. baseline	Healthy adults (n=22) and adults with MS (n=25)	Chokeberry extracts (300mg/day)	Decreases in total (243±35 to 228±33 mg/dL), LDL-cholesterol (159±36 to 146±35 mg/dL), and triglycerides (216±67 to 188±90 mg/dL) in berry group vs. baseline	Not reported
Zhu et al. (2011)	Randomized parallel trial; 12 wks	Adults with hypercholesterolemia (n=150)	Anthocyanin extracts (4 capsules/day ~ 320 mg extracts) vs. placebo	Decreases in LDL (130±22 to 117±16 mg/dL) and increases in HDL-cholesterol (47±9 to 53±8 mg/dL) in anthocyanin vs. placebo	No changes in Apo-A-I and Apo-B

(*Continued*)

TABLE 5.2 (CONTINUED)
Effects of Berry Bioactive Compound Extracts on Serum Lipids: Clinical Studies

Author (year)	Study Design	Participants	Intervention	Significant Effects on Conventional Lipids	Significant Effects on Lipid Subclasses/Apo Lipoproteins
Soltani et al. (2014)	Randomized parallel trial; 4 wks	Adults with hyperlipidemia (n=50)	Whortleberry extracts (90mg anthocyanins) vs. placebo	Decreases in total (225±32 to 192±29 mg/dL), LDL-cholesterol (133±24 to 122±27 mg/dL) and triglycerides (226±97 to 156±47 mg/dL) in the berry group vs. placebo	Not reported
Kianbakht et al. (2014)	Randomized parallel trial; 8 wks	Adults with hyperlipidemia (n=80)	Whortleberry extracts (1050 mg fruit extracts) vs. placebo	Decreases in total (282±38 to 202±37 mg/dL), LDL-cholesterol (172±48 to 117±35 mg/dL) and triglycerides (305±23 to 248±19 mg/dL), and increases in HDL-cholesterol (44±5 to 59±7 mg/dL) in the berry group vs. placebo	Not reported

Apo: apolipoprotein; CETP: cholesteryl ester transfer protein; MS: metabolic syndrome.

and CIII that have been associated with increased risks of atherosclerotic CVD in epidemiological observations (Jiang et al., 2004; Wyler von Ballmoos et al., 2015; Zhu et al., 2015).

Table 5.3 highlights clinical studies examining the role of dietary berries in decreasing biomarkers of inflammation related to CVD and atherosclerosis. Our group has previously reported the role of blueberries, strawberries and low-calorie cranberry juice in lowering biomarkers of oxidative stress and adhesion molecules following supplementation in free-living adults with the metabolic syndrome (Basu et al., 2010a,b, 2011). However, few studies using berry supplementation have shown an effect on key biomarkers of inflammation, such as C-reactive protein (CRP) and interleukin-6 (IL-6), associated with increased CVD risks. Inflammation, demonstrated primarily by the elevated levels of serum CRP, has been associated with insulin resistance and the metabolic syndrome (Festa et al., 2000). Adipose tissue also secretes adiponectin, a protein showing anti-inflammatory activity, which inhibits tumor necrosis factor-α, adhesion molecule expression and nuclear transcriptional factor kB signaling, a pivotal pathway in inflammatory reactions in endothelial cells and in the propagation of atherosclerosis (Ouchi et al., 1999, 2000). In a few reported studies, cranberry juice has been shown to increase levels of circulating adiponectin and to decrease CRP in adults with features of the metabolic syndrome (Novotny et al., 2015; Simão et al., 2013), while biomarkers of inflammation were not altered in other reported studies, as shown in Table 5.3. Thus, based on the known differences in polyphenol composition among different dietary berries, we may expect to see differential effects in modulating biomarkers of CVD risks. Nonetheless, all berry products show consistent evidence of lowering one or more biomarkers of conventional lipids in reported studies assessing lipid outcomes.

5.7 CONCLUSIONS

Reported studies show a consistent role of dietary whole berries, berry juices and extracts in decreasing blood total and LDL-C and triglycerides, and/or increasing HDL-C in subjects with one or more elevations of lipid biomarkers. A few studies further show that dietary supplementation of freeze-dried whole berries can also improve qualitative changes in lipids, such as by decreasing small HDL and LDL particle concentrations and increasing LDL size, thus conferring less atherogenicity and reducing CVD risks. These clinical observations have been explained by mechanistic studies demonstrating the role of dietary berries in modulating lipid metabolism, mainly by increasing hepatic synthesis of apolipoprotein A-I, downregulating the activity of genes related to fatty acid synthesis, causing regression of aortic lesions and decreasing inflammation and oxidative damage in experimental animals. In addition to lowering lipids, clinical studies also provide evidence on the role of dietary berries in decreasing surrogate markers of atherosclerosis, including biomarkers of oxidative stress and inflammation. However, there is a large heterogeneity in the biomarkers reported by each study and thus further investigation is needed on the role of berries in markers of atherosclerosis and endothelial function in clinical settings.

TABLE 5.3
Effects of Dietary Berries and Markers of Atherosclerosis and Inflammation: Clinical Studies

Author (year)	Study Design	Participants	Intervention	Significant Effects on Surrogate Markers of Atherosclerosis	Significant Effects on Biomarkers of Inflammation
Basu et al. (2010a)	Randomized parallel trial; 8 wks	Adults with MS (n=27)	FDS (50g/day) vs. control beverage	Decreases in VCAM-1	Not reported
Basu et al. (2010b)	Randomized parallel trial; 8 wks	Adults with MS (n=48)	FDB (50g/day) vs. control beverage	Decreases in plasma oxidized LDL and MDA; no effects on adhesion molecules	No effects on CRP and IL-6
Basu et al. (2011)	Randomized parallel trial; 8 wks	Adults with MS (n=31)	LCCJ (480mL/day) vs. matched placebo	Decreases in plasma oxidized LDL and MDA	No effects on CRP and IL-6
Simão et al. (2013)	Randomized parallel trial; 60d	Adults with MS (n=56)	Cranberry juice vs. no juice (usual diet as control group)	Decrease in homocysteine, hydroperoxides and AOPP	Increase in adiponectin; no change in CRP, IL-1,6 & TNF-α
Novotny et al. (2015)	Randomized parallel trial; 8 wks	Overweight adults (n=56)	LCCJ or matched placebo beverage (480mL)/day	No changes in adhesion molecules	Decrease in CRP
Johnson et al. (2015)	Randomized parallel trial; 8 wks	Postmenopausal women with hypertension (n=48)	FDB (22g/day) vs. control powder	Increase in nitric oxide	No effects on CRP

AOPP: Advanced oxidation; CRP: C-reactive protein; FDB: freeze-dried blueberries; FDS: freeze-dried strawberries; protein products; IL-6: interleukin-6; LCCJ: low-calorie cranberry juice; MDA: malondialdehyde; MS: metabolic syndrome; TNF-α: tumor necrosis factor-alpha; VCAM-1: vascular cell adhesion molecule-1.

5.8 FUTURE RESEARCH AND RECOMMENDATIONS

Based on the emerging evidence, dietary berries and berry products hold promise as a natural and alternative means to lower blood lipids in adults with CVD risks. On the basis of epidemiological and clinical findings, including two cups of low-calorie cranberry juice or half to one cup of whole berries in the daily diet may provide these health benefits in adults. However, it is important to note that these recommendations are made in the context of the existing diet and the presence of the magnitude of CVD risks in specific individuals. Most of the studies reviewed herein were in overweight/obese but otherwise healthy adults and thus recommendations might differ for those with advanced CVD and type 2 diabetes. Also, some of the clinical studies used a large dose of berries not feasible to consume on a daily basis. Thus, future studies must address the dose-response effects of berry products in modulating lipids and related cardiometabolic variables with or without changes in background diet and lifestyle. Studies comparing the effects of berry products with conventional lipid-lowering medications, especially targeting LDL-C, will be useful in assessing the magnitude of absolute risk reduction of CVD among different therapies.

In addition to the determination of blood lipids and lipoproteins, future studies must also examine the effects of berry supplementation on changes in the gut microbiome that have been significantly associated with risks of chronic diseases, including blood lipids (Fu et al., 2015; Madeeha et al., 2016). Interindividual variations in diet, lifestyle factors and gut microbiome may further explain and modify lipid and metabolic responses to dietary berry interventions, and such studies will help identify personalized approaches for optimal lipid management in lowering CVD risks.

REFERENCES

Arsenault, BJ, Rana, JS, Stroes, ES, Després, JP, Shah, PK, Kastelein, JJ, Wareham, NJ, Boekholdt, SM; Khaw, KT 2009. Beyond low-density lipoprotein cholesterol: Respective contributions of non-high-density lipoprotein cholesterol levels, triglycerides, and the total cholesterol/high-density lipoprotein cholesterol ratio to coronary heart disease risk in apparently healthy men and women. *J. Am. Coll. Cardiol.* 55(1):35–41.

Basu, A, Du, M, Leyva, MJ, Sanchez, K, Betts, NM, Wu, M, Aston, CE; Lyons, TJ 2010a. Blueberries decrease cardiovascular risk factors in obese men and women with metabolic syndrome. *J. Nutr.* 140(9):1582–7.

Basu, A, Fu, DX, Wilkinson, M, Simmons, B, Wu, M, Betts, NM, Du, M; Lyons, TJ 2010b. Strawberries decrease atherosclerotic markers in subjects with metabolic syndrome. *Nutr. Res.* 30(7):462–9.

Basu, A, Rhone, M; Lyons, TJ 2010c. Berries: Emerging impact on cardiovascular health. *Nutr. Rev.* 68(3):168–77.

Basu, A, Betts, NM, Ortiz, J, Simmons, B, Wu, M; Lyons, TJ 2011. Low-energy cranberry juice decreases lipid oxidation and increases plasma antioxidant capacity in women with metabolic syndrome. *Nutr. Res.* 31(3):190–6.

Basu, A, Betts, NM, Nguyen, A, Newman, ED, Fu, D; Lyons, TJ 2014a. Freeze-dried strawberries lower serum cholesterol and lipid peroxidation in adults with abdominal adiposity and elevated serum lipids. *J. Nutr.* 144(6):830–7.

Basu, A, Nguyen, A, Betts, NM; Lyons, TJ 2014b. Strawberry as a functional food: An evidence-based review. *Crit. Rev. Food Sci. Nutr.* 54(6):790–806.

Basu, A, Jenkins, AJ, Zhang, Y, Stoner, JA, Klein, RL, Lopes-Virella, MF, Garvey, WT, Lyons, TJ; DCCT/EDIC Research Group 2016. Nuclear magnetic resonance-determined lipoprotein subclasses and carotid intima-media thickness in type 1 diabetes. *Atherosclerosis* 244:93–100.

Broncel, M, Kozirog, M, Duchnowicz, P, Koter-Michalak, M, Sikora, J; Chojnowska-Jezierska, J 2010. Aronia melanocarpa extract reduces blood pressure, serum endothelin, lipid, and oxidative stress marker levels in patients with metabolic syndrome. *Med. Sci. Monit.* 16(1):CR28–34.

Burton-Freeman, B, Linares, A, Hyson, D; Kappagoda, T 2010. Strawberry modulates LDL oxidation and postprandial lipemia in response to high-fat meal in overweight hyperlipidemic men and women. *J. Am. Coll. Nutr.* 29(1):46–54.

Cassidy, A, Mukamal, KJ, Liu, L, Franz, M, Eliassen, AH; Rimm, EB 2013. High anthocyanin intake is associated with a reduced risk of myocardial infarction in young and middle-aged women. *Circulation* 127(2):188–96.

de Souza, MO, Silva, M, Silva, ME, Oliveira, Rde P; Pedrosa, ML 2010. Diet supplementation with acai (Euterpe oleracea Mart.) pulp improves biomarkers of oxidative stress and the serum lipid profile in rats. *Nutrition* 26(7–8):804–10.

de Souza, MO, Souza E Silva, L, de Brito Magalhães, CL, de Figueiredo, BB, Costa, DC, Silva, ME; Pedrosa, ML 2012. The hypocholesterolemic activity of açaí (Euterpe oleracea Mart.) is mediated by the enhanced expression of the ATP-binding cassette, subfamily G transporters 5 and 8 and low-density lipoprotein receptor genes in the rat. *Nutr. Res.* 32(12):976–84.

de Souza, VR, Pereira, PA, da Silva, TL, de Oliveira Lima, LC, Pio, R; Queiroz, F 2014. Determination of the bioactive compounds, antioxidant activity and chemical composition of Brazilian blackberry, red raspberry, strawberry, blueberry and sweet cherry fruits. *Food Chem.* 156:362–8.

Djoussé, L, Arnett, DK, Coon, H, Province, MA, Moore, LL; Ellison, RC 2004. Fruit and vegetable consumption and LDL cholesterol: The National Heart, Lung, and Blood Institute Family Heart Study. *Am. J. Clin. Nutr.* 79(2):213–7.

Duffey, KJ; Sutherland, LA 2013. Adult cranberry beverage consumers have healthier macronutrient intakes and measures of body composition compared to non-consumers: National Health and Nutrition Examination Survey (NHANES) 2005–2008. *Nutrients* 5(12):4938–49.

Festa, A, D'Agostino, R. Jr, Howard, G, Mykkänen, L, Tracy, RP; Haffner, SM 2000. Chronic subclinical inflammation as part of the insulin resistance syndrome: The Insulin Resistance Atherosclerosis Study (IRAS). *Circulation* 102(1):42–7.

Fu, J, Bonder, MJ, Cenit, MC, Tigchelaar, EF, Maatman, A, Dekens, JA, Brandsma, E, et al. 2015. The gut microbiome contributes to a substantial proportion of the variation in blood lipids. *Circ. Res.* 117(9):817–24.

Garvey, WT, Kwon, S, Zheng, D, Shaughnessy, S, Wallace, P, Hutto, A, Pugh, K, et al. 2003. Effects of insulin resistance and type 2 diabetes on lipoprotein subclass particle size and concentration determined by nuclear magnetic resonance. *Diabetes* 52(2):453–62.

Inazu, A, Brown, ML, Hesler, CB, Agellon, LB, Koizumi, J, Takata, K, Maruhama, Y, Mabuchi, H; Tall, AR 1990. Increased high-density lipoprotein levels caused by a common cholesteryl-ester transfer protein gene mutation. *N. Engl. J. Med.* 323(18):1234–8.

Jaroslawska, J, Juskiewicz, J, Wroblewska, M, Jurgonski, A, Krol, B; Zdunczyk, Z 2011. Polyphenol-rich strawberry pomace reduces serum and liver lipids and alters gastrointestinal metabolite formation in fructose-fed rats. *J. Nutr.* 141(10):1777–83.

Jennings, A, Welch, AA, Spector, T, Macgregor, A; Cassidy, A 2014. Intakes of anthocyanins and flavones are associated with biomarkers of insulin resistance and inflammation in women. *J. Nutr.* 144(2):202–8.

Jiang, R, Schulze, MB, Li, T, Rifai, N, Stampfer, MJ, Rimm, EB; Hu, FB 2004. Non-HDL cholesterol and apolipoprotein B predict cardiovascular disease events among men with type 2 diabetes. *Diabetes Care* 27(8):1991–7.

Johnson, SA, Figueroa, A, Navaei, N, Wong, A, Kalfon, R, Ormsbee, LT, Feresin, RG, et al. 2015. Daily blueberry consumption improves blood pressure and arterial stiffness in postmenopausal women with pre- and stage 1-hypertension: A randomized, double-blind, placebo-controlled clinical trial. *J. Acad. Nutr. Diet.* 115(3):369–77.

Kastelein, JJ, van der Steeg, WA, Holme, I, Gaffney, M, Cater, NB, Barter, P, Deedwania, P, et al. 2008. Lipids, apolipoproteins, and their ratios in relation to cardiovascular events with statin treatment. *Circulation* 117(23):3002–9.

Khanal, RC, Howard, LR, Wilkes, SE, Rogers, TJ; Prior, RL 2012. Effect of dietary blueberry pomace on selected metabolic factors associated with high fructose feeding in growing Sprague-Dawley rats. *J. Med. Food* 15(9):802–10.

Kianbakht, S, Abasi, B; Hashem Dabaghian, F 2014. Improved lipid profile in hyperlipidemic patients taking Vaccinium arctostaphylos fruit hydroalcoholic extract: A randomized double-blind placebo-controlled clinical trial. *Phytother. Res.* 28(3):432–6.

Lankinen, M, Kolehmainen, M, Jääskeläinen, T, Paananen, J, Joukamo, L, Kangas, AJ, Soininen, P, et al. 2014. Effects of whole grain, fish and bilberries on serum metabolic profile and lipid transfer protein activities: A randomized trial (Sysdimet). *PLOS ONE* 9(2):e90352.

Lee, IT, Chan, YC, Lin, CW, Lee, WJ; Sheu, WH 2008. Effect of cranberry extracts on lipid profiles in subjects with Type 2 diabetes. *Diabet. Med.* 25(12):1473–7.

Lee, SG, Vance, TM, Nam, TG, Kim, DO, Koo, SI; Chun, OK 2015. Contribution of anthocyanin composition to total antioxidant capacity of berries. *Plant Foods Hum. Nutr.* 70(4):427–32.

Madeeha, IR, Ikram, A; Imran, M 2016. A preliminary insight of correlation between human fecal microbial diversity and blood lipid profile. *Int. J. Food Sci. Nutr.* 67(7):865–71.

Manach, C, Scalbert, A, Morand, C, Rémésy, C; Jiménez, L 2004. Polyphenols: Food sources and bioavailability. *Am. J. Clin. Nutr.* 79(5):727–47.

Mandave, P, Khadke, S, Karandikar, M, Pandit, V, Ranjekar, P, Kuvalekar, A; Mantri, N 2017. Antidiabetic, lipid normalizing, and nephroprotective actions of the strawberry: A potent supplementary fruit. *Int. J. Mol. Sci.* 18(1). pii:E124.

Manganaris, GA, Goulas, V, Vicente, AR; Terry, LA 2014. Berry antioxidants: Small fruits providing large benefits. *J. Sci. Food Agric.* 94(5):825–33.

Mirmiran, P, Bahadoran, Z, Moslehi, N, Bastan, S; Azizi, F 2015. Colors of fruits and vegetables and 3-year changes of cardiometabolic risk factors in adults: Tehran lipid and glucose study. *Eur. J. Clin. Nutr.* 69(11):1215–9.

Moyer, RA, Hummer, KE, Finn, CE, Frei, B; Wrolstad, RE 2002. Anthocyanins, phenolics, and antioxidant capacity in diverse small fruits: Vaccinium, rubus, and ribes. *J. Agric. Food Chem.* 50(3):519–25.

NCEP, 1988. Report of the National Cholesterol Education Program expert panel on detection, evaluation, and treatment of high blood cholesterol in adults. The Expert Panel. *Arch. Intern. Med.* 148(1):36–69.

Novotny, JA, Baer, DJ, Khoo, C, Gebauer, SK; Charron, CS 2015. Cranberry juice consumption lowers markers of cardiometabolic risk, including blood pressure and circulating C-reactive protein, triglyceride, and glucose concentrations in adults. *J. Nutr.* 145(6):1185–93.

Ouchi, N, Kihara, S, Arita, Y, Maeda, K, Kuriyama, H, Okamoto, Y, Hotta, K, et al. 1999. Novel modulator for endothelial adhesion molecules: Adipocyte-derived plasma protein adiponectin. *Circulation* 100(25):2473–6.

Ouchi, N, Kihara, S, Arita, Y, Okamoto, Y, Maeda, K, Kuriyama, H, Hotta, K, et al. 2000. Adiponectin, an adipocyte-derived plasma protein, inhibits endothelial NF-kappaB signaling through a cAMP-dependent pathway. *Circulation* 102(11):1296–301.

Pereira, RR, de Abreu, IC, Guerra, JF, et al. 2016. Açai (Euterpe oleracea Mart.) upregulates paraoxonase 1Gene expression and activity with concomitant reduction of hepatic steatosis in high-fat diet-fed rats. *Oxid. Med. Cell. Longev.* 2016:8379105.

Qin, Y, Xia, M, Ma, J, Hao, Y, Liu, J, Mou, H, Cao, L; Ling, W 2009. Anthocyanin supplementation improves serum LDL- and HDL-cholesterol concentrations associated with the inhibition of cholesteryl ester transfer protein in dyslipidemic subjects. *Am. J. Clin. Nutr.* 90(3):485–92.

Ray, KK, Cannon, CP, Cairns, R, Morrow, DA, Ridker, PM; Braunwald, E 2009. Prognostic utility of apoB/AI, total cholesterol/HDL, non-HDL cholesterol, or hs-CRP as predictors of clinical risk in patients receiving statin therapy after acute coronary syndromes: Results from PROVE IT-TIMI 22. *Arterioscler. Thromb. Vasc. Biol.* 29(3):424–30.

Ruel, G, Pomerleau, S, Couture, P, Lemieux, S, Lamarche, B; Couillard, C 2006. Favourable impact of low-calorie cranberry juice consumption on plasma HDL-cholesterol concentrations in men. *Br. J. Nutr.* 96(2):357–64.

Sesso, HD, Gaziano, JM, Jenkins, DJ; Buring, JE 2007. Strawberry intake, lipids, C-reactive protein, and the risk of cardiovascular disease in women. *J. Am. Coll. Nutr.* 26(4):303–10.

Silverman, MG, Ference, BA, Im, K, Wiviott, SD, Giugliano, RP, Grundy, SM, Braunwald, E; Sabatine, MS 2016. Association between lowering LDL-C and cardiovascular risk reduction Among different therapeutic interventions: A systematic review and meta-analysis. *JAMA* 316(12):1289–97.

Simão, TN, Lozovoy, MA, Simão, AN, Oliveira, SR, Venturini, D, Morimoto, HK, Miglioranza, LH; Dichi, I 2013. Reduced-energy cranberry juice increases folic acid and adiponectin and reduces homocysteine and oxidative stress in patients with the metabolic syndrome. *Br. J. Nutr.* 110(10):1885–94.

Soltani, R, Hakimi, M, Asgary, S, Ghanadian, SM, Keshvari, M; Sarrafzadegan, N 2014. Evaluation of the Effects of Vaccinium Arctostaphylos L. fruit Extract on Serum Lipids and hs-CRP Levels and Oxidative Stress in Adult Patients with hyperlipidemia: A Randomized, Double-Blind, Placebo-Controlled Clinical Trial. *Evid. Based Complement. Alternat. Med.* 2014:217451.

Ströher, DJ, Escobar Piccoli Jda, Jda C, Güllich, AA, Pilar, BC, Coelho, RP, Bruno, JB, Faoro, D; Manfredini, V 2015. 14 days of supplementation with blueberry extract shows anti-atherogenic properties and improves oxidative parameters in hypercholesterolemic rats model. *Int. J. Food Sci. Nutr.* 66(5):559–68.

Wedick, NM, Pan, A, Cassidy, A, Rimm, EB, Sampson, L, Rosner, B, Willett, W, et al. 2012. Dietary flavonoid intakes and risk of type 2 diabetes in US men and women. *Am. J. Clin. Nutr.* 95(4):925–33.

Wyler von Ballmoos, MC, Haring, B; Sacks, FM 2015. The risk of cardiovascular events with increased apolipoprotein CIII: A systematic review and meta-analysis. *J. Clin. Lipidol.* 9(4):498–510.

Zhu, Y, Xia, M, Yang, Y, Liu, F, Li, Z, Hao, Y, Mi, M, Jin, T; Ling, W 2011. Purified anthocyanin supplementation improves endothelial function via NO-cGMP activation in hypercholesterolemic individuals. *Clin. Chem.* 57(11):1524–33.

Zhu, Y, Miao, Y, Meng, Z; Zhong, Y 2015. Effects of Vaccinium berries on serum lipids: A meta-analysis of randomized controlled trials. *Evid. Based Complement. Alternat. Med.* 2015:790329.

Zunino, SJ, Parelman, MA, Freytag, TL, Stephensen, CB, Kelley, DS, Mackey, BE, Woodhouse, LR; Bonnel, EL 2012. Effects of dietary strawberry powder on blood lipids and inflammatory markers in obese human subjects. *Br. J. Nutr.* 108(5):900–9.

6 Cardiovascular Protection Effects of Proanthocyanidins

Graham C. Llivina, Megan M. Waguespack and Angela I. Calderón

CONTENTS

6.1 INTRODUCTION

The leading cause of death worldwide is cardiovascular disease (Deaton et al., 2011). Cardiovascular diseases include myocardial infarction, stroke, cardiomyopathy, aortic aneurysms, hypertension and heart failure (Mensah et al., 2014). Because of its enormous societal burden, there is growing interest in dietary intervention to mitigate the severity of cardiovascular disease. Proanthocyanidins–polyphenolic compounds found in extracts from grape skins, grape seeds, chocolate, pomegranates, bilberries, cranberries and other plant-derived sources—have shown promise in preventing cardiovascular disease. The information cited here was compiled using research-specific search engines such as PubMed and SciFinder, filtered for content and used on the basis of the research strength.

6.2 BIOACTIVE PROANTHOCYANIDINS

6.2.1 Chemistry

6.2.1.1 A-Type Proanthocyanidins

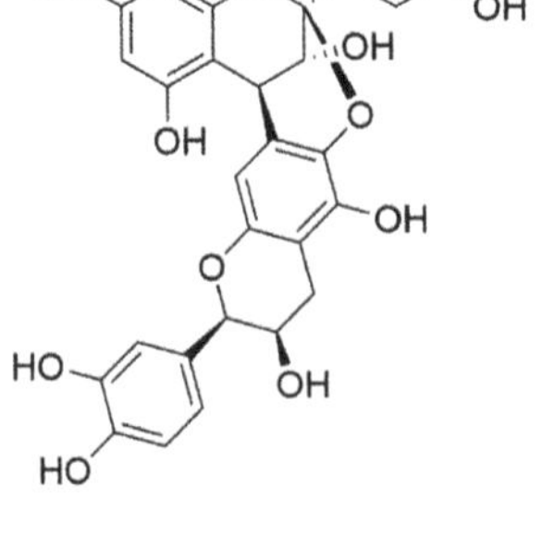

HO O OH OH OH O OH OH O HO OH HO

- Proanthocyanidin A1

  Proanthocyanidin A1 is an epicatechin-(2β→7,4β→8)-epicatechin dimer commonly found in cranberries, bilberries, hawthorn and peanut skins. A-type linkage is a relatively uncommon formation in proanthocyanidins with both 4β→8 and 2β→O→7 bonds. Proanthocyanidin A1 is the most common formation of A-type proanthocyanidins (Neto et al., 2007).

6.2.1.2 Proanthocyanidins with 4→8 Bonds

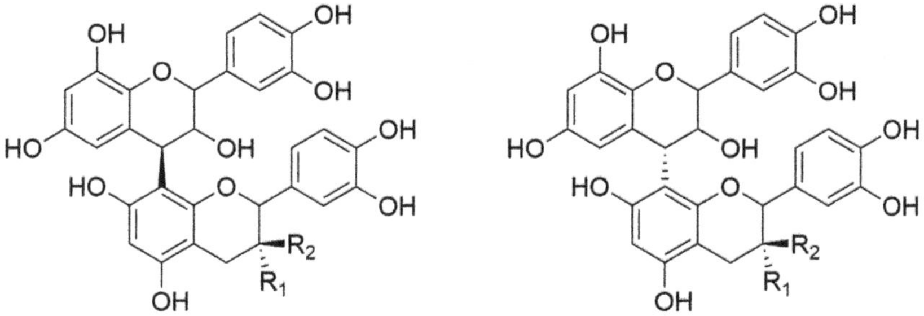

Compound	R1	R2
Procyanidin B1	OH	H
Procyanidin B2	H	OH
Procyanidin B3	OH	H
Procyanidin B4	H	OH

- Proanthocyanidin B1

  Proanthocyanidin B1 is a B-type proanthocyanidin represented by the formula epicatechin-(4β→8)-catechin. It is found in the common grape vine, *Vitis vinifera*, and the fruit of peeled pomegranate, *Punica granatum* (de Pascual-Teresa et al., 2000).
- Proanthocyanidin B2

  Proanthocyanidin B2 is a B-type proanthocyanidin structurally represented as (–)-epicatechin-(4β→8)-(–)-epicatechin. It is the most common subclass of proanthocyanidin molecular groups with 4→8 bonding and occurs in the leaves and stems of *Vitis vinifera.*
- Proanthocyanidin B3

  Proanthocyanidin B3 is a B-type proanthocyanidin with a catechin-(4α→8)-catechin structure. Proanthocyanidin B3, a molecular dimer, is mainly found in red wine. Proan-thocyanidin dimer B3 is also found in pomegranate juice at a rate of 0.16 mg/100 g (Eren et al., 2014).

- Proanthocyanidin B4

  Proanthocyanidin B4 is a B-type proanthocyanidin whose molecular structure is represented by catechin-(4α→8)-epicatechin. It is isolated primarily from grape seeds rather than from the skin.

6.2.1.3 Proanthocyanidins with 4→6 Bonds

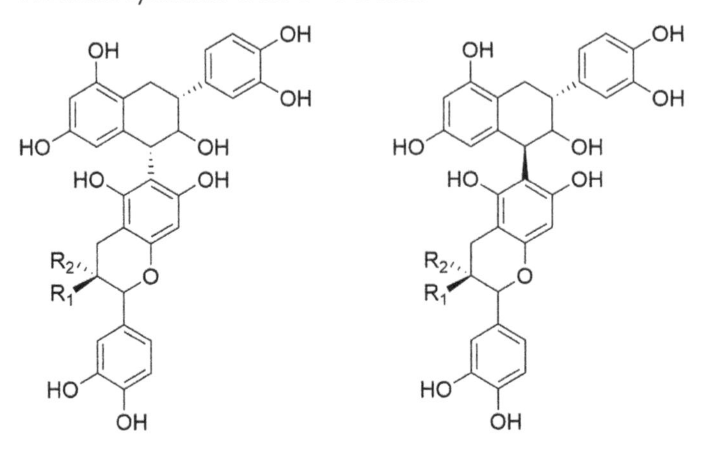

Compound	R1	R2
Procyanidin B5	H	OH
Procyanidin B6	OH	H
Procyanidin B7	OH	H
Procyanidin B8	H	OH

- Proanthocyanidin B5

  Proanthocyanidin B5—or epicatechin-(4β→6)-epicatechin—is an epicatechin dimer encountered in grape seeds, cocoa and chocolate.
- Proanthocyanidin B6

  Proanthocyanidin B6 is a catechin-(4α→6)-catechin dimer detected in grape seeds.

These bioactive proanthocyanidin groups exhibit cardioprotective effects. Unfortunately, most research projects do not discriminate between these groups, so the current understanding of the relative protections offered by specific compounds is limited.

6.3 CARDIOPROTECTIVE MECHANISMS

The following are confirmed mechanisms of action of proanthocyanidins in human subjects:

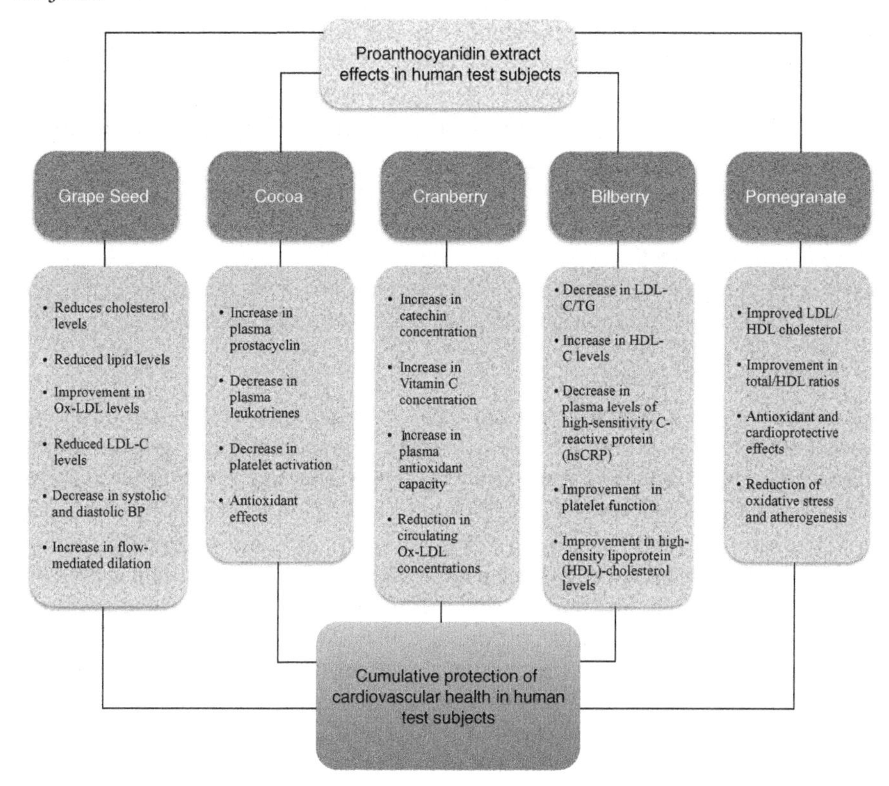

6.3.1 Effects on Total Cholesterol, Lipids and Low-Density Lipoprotein (LDL-C) and/or Oxidized Low-Density Lipoprotein (OX-LDL)

6.3.1.1 Grapes (*Vitis vinifera* L.)

6.3.1.1.1 In Vivo Studies

Five studies reported *in vivo* experiments assessing the effect of grape seed proanthocyanidin extract (GSPE) or grape seed and skin extract (GSSE) on total cholesterol and LDL-C levels in rats. Administering 375 mg/kg of low molecular weight GSPE to rats on a high-fat diet improved hypertriglyceridemia (Pons et al., 2014). Improvement in cholesterol and triglyceride profiles were seen in rats treated with GSPE. GPSE has also been shown to inhibit lipid buildup in rats. Rats placed on a high-fat diet were provided with 25 mg/kg GSPE for 10 days and subsequently demonstrated decreased levels of hepatic lipid synthesis (Baiges et al., 2010). Lipid profile improvement in rats has also been seen at a dose of 250 mg/kg GSPE (Quesada et al., 2009). Therefore, several studies suggest that treating rats with varying doses of GSPE results in decreased total cholesterol, lipid and LDL-C levels. Similar benefits have been noted with grape skin extract. In obese male Wistar rats on a high-fat diet, GSSE reduced oxidative stress and cholesterol levels. A high-fat diet is associated with the accumulation of triglycerides, total cholesterol and oxidative stress on the hearts and livers of rats. After administering GSSE, the researcher concluded that "GSSE efficiently protected these organs against fat-induced disturbances" in addition to lowering cholesterol and oxidative stress levels in the heart and the lungs (Charradi et al., 2013). Similar effects can be demonstrated in other species of rodents.

Atherosclerotic hamsters given GSPE at 100 mg/kg body weight for 10 weeks showed reduced plasma cholesterol levels (Vinson et al., 2002). When hamsters on a standard or high-fat diet were given GSPE at a dose of 25 mg/kg for 15 days, they showed an improvement in their collective lipid profiles (Caimari et al., 2013). Caimari's study used a lower dose and shorter incubation time in non-atherosclerotic hamsters, but the conclusion supported Vinson's findings. Hamsters are less commonly used in laboratory experiments than are rats, but the data suggest that GSPE reduces total cholesterol levels and lipid buildup in both species of rodents.

6.3.1.1.2 Clinical Studies

Human subjects have exhibited reduced cholesterol and lipid levels after consuming concentrated doses of GSPE. In one experiment, 52 hyperlipidemic subjects administered 200 mg/day red grape seed extract for eight weeks showed reductions in total cholesterol, LDL-C, and Ox-LDL levels (Razavi et al., 2013). Years earlier, 40 hypercholesterolemic subjects receiving 200 mg/day GSPE with niacin-bound chromium demonstrated lower total cholesterol with no significant change in blood pressure (Preuss et al., 2000). In an almost identical study, treating 40 hypercholesterolemic subjects (210–300 mg/dL) with 200 mg/day of GSPE reduced levels of Ox-LDL (Bagchi et al., 2003). To further evaluate the effect of GSPE at varying doses on Ox-LDL levels, 61 subjects with elevated LDL-C levels (100–180 mg/dL) received 200 or 400 mg/dL grape seed extract for 12 weeks and showed signs of reduced Ox-LDL levels (Sano et al., 2007). Ox-LDL levels improved at both doses. Lipid profiles in humans also respond to GSPE. Seventeen hypercholesterolemia subjects receiving 600 mg/day GSPE showed improvements in lipid profiles (Vinson et al., 2001). Interestingly, GSPE may improve the cardiovascular health of chronic smokers. Twenty-eight male smokers given 200 mg GSPE for eight weeks had reduced total cholesterol and LDL-C levels (Weseler et al., 2011), contradicting a much earlier study from 2003. Twenty-four healthy heavy smoker male subjects taking 150 mg/day GSPE for four weeks showed no significant change in total cholesterol, triglycerides, high-density lipoprotein (HDL-C) or LDL-C levels—although thiobarbituric acid reactive substances (TBARS) concentration was reduced (Vigna et al., 2003). It is notable that Weseler's study involved a much higher dosage of GSPE, suggesting that GSPE's cardioprotective effects are dose-related. In summary, GSPE administration reduces total cholesterol, lipid levels and Ox-LDL levels in human subjects, although the effect may be dependent on dosage.

6.3.1.2 Chocolate (*Theobroma cacao* L.)

6.3.1.2.1 In Vitro *Studies*

Cholesterol and triglycerides are also decreased by procyanidins found in chocolate. An *in vitro* experiment tested cocoa extract, pine bark extract and GSPE on hepatic cells. The cocoa proanthocyanidin extract was derived from a cocoa powder containing monomeric (23.7%), dimeric (15.8%), trimeric (18.4%), tetrameric (13.9%) and oligomeric proanthocyanidins (36.2%). HepG2 cells were treated with 25 mg/L of the extracts. Decreases in the levels of free cholesterol, triglycerides and cholesterol ester synthesis in HepG2 cells were noted. The cocoa extract induced a significant decrease in free cholesterol and triglycerides because of higher concentrations of procyanidins (Guerrero et al., 2013).

Cocoa procyanidins have been shown to decrease LDL oxidation. Six healthy subjects between the ages of 23 and 39 ingested 37.5 g of cocoa in 400 mL of water, with each gram of cocoa containing 12.2 mg of monomers, 9.7 mg of dimers and 20.2 mg of trimers through decamers. Centrifuged plasma was cultured with procyanidins to assess the effect of procyanidin chain length on LDL oxidation. Results demonstrated an inverse correlation: as chain length of procyanidins increased, the susceptibility of LDL to oxidation decreased. However, each procyanidin length exhibited a similar level of inhibition when their monomeric units were equivalent. Therefore, procyanidins' antioxidant activity is due to their ring structures and catechol groups rather than chain length or delocalization of charge through polymeric bonds (Steinberg et al., 2002).

Anti-inflammatory cytokine production is accelerated by cocoa procyanidins. Procyanidins show anti-inflammatory effects on the heart by limiting the production of pro-inflammatory cytokines in peripheral blood cells and enhancing the production of anti-inflammatory cytokines (Vestraeten et al., 2005). That study used cocoa procyanidin fractions prepared from cocoa using acetone and water extraction and cultured with peripheral blood cells from healthy volunteers. Each culture was treated with 25 μg/mL of cocoa fractions. Monomers, dimers and trimers produced the most change on cytokines, while tetramers showed minimal effect. Hexamers and heptamers significantly increased IL-1 beta expression by 23% and 20%, respectively (Mao et al., 2000).

Procyanidins inhibit lipoxygenases, as well. Cocoa oligomers at a concentration of 29 μg/mL were tested on rabbit 15- LOX-1 and soybean lipoxygenase, and the size of the oligomers correlated with the lipoxygenase activity. Lipoxygenase inhibition potency decreased from the monomer, was at a minimum at the trimer and tetramers and then increased through the decamers. The researchers determined that procyanidins and other flavanols directly inhibit mammalian 15-lipoxygenase-1 (Schewe and Sies, 2005).

6.3.1.2.2 Clinical Studies

A study of 30 human volunteers indicates that chocolate with a high procyanidin content of 147 mg is more beneficial than chocolate with low procyanidin content of 3.3 mg on leukotriene and prostacyclin levels (Arranz et al., 2013). A second study confirmed these findings. Ten healthy individuals consumed 37 g of low-procyanidin (0.09 mg/g) and 37 g of high-procyanidin chocolate (4.0 mg/g) in the form of Dove Dark Chocolate bars with the two trials separated by one week. Plasma samples were collected before treatment and two and six hours after treatment. The high-procyanidin chocolate caused greater increases in plasma prostacyclin and decreases in plasma leukotrienes. These results suggest a role for procyanidins in decreasing platelet activation (Schramm et al., 2001).

6.3.2 Effects on Blood Pressure, Heart Rate and/or Plasma Oxidative Status

6.3.2.1 Grapes

6.3.2.1.1 In Vivo *Studies*

GSPE lowers blood pressure levels in rats. In 2014, rats on a high-fat diet were administered GSPE at a dose of 375 mg/kg and subsequently demonstrated decreased blood pressure (BP) and reduced oxidative stress levels (Pons et al., 2014). Additionally, ouabain-induced hypertensive rats treated with GSPE at 250 mg/kg for five weeks showed reduced systolic blood pressure levels and improved vascular endothelial function (Liu et al., 2012). That same year, rats given 250 mg/kg GSPE for five weeks had lowered BP levels. The researchers concluded that GSPE improved hypertension in rats (Cui et al., 2012). There is also evidence that GSPE improved BP by influencing hormonal activity. Estrogen-depleted female rats with spontaneous hypertension showed a decrease in arterial hypertension in response to 0.5% GSPE for 10 weeks (Peng et al., 2005).

GSSE has been shown to improve lipotoxicity, inflammation and oxidative stress in rats. Large doses of GSSE administered for 45 days to rats on a high-fat diet demonstrated reduced pancreas lipotoxicity, oxidative stress and inflammation (Aloui et al., 2016). That same year, another study drew a connection between GSSE and oxidative stress. According to this study, GSSE efficiently counteracted almost all bleomycin-induced oxidative stress as well as biochemical and morphological changes in lung tissue (Khazri et al., 2016).

Cardioprotective, renoprotective and neuroprotective properties have been attributed to GSSE through reduced oxidative stress and anti-inflammatory benefits. In a 2016 study, oxidative stress levels were diminished by high-dose GSSE (Oueslati et al., 2016). Additionally, both GSPE and GSSE can actively inhibit inflammation (Xia et al., 2010). A previous study evaluating short-term administration of GSPE and GSSE stated that "a high dosage [of] GSSE protected the brain efficiently against ischemic stroke and should be translated to humans" (Safwen et al., 2015). Finally, lab rats injected with large doses of doxorubicin recovered when the drug was neutralized with an equal dosage of GSSE (Mokni et al., 2012). The experiments as a group support the theory that beneficial oxidative stress and blood pressure levels decreases are seen with GSPE or GSSE administration.

6.3.2.1.2 Clinical Studies

Human subjects also show decreased BP levels with GSPE. When 119 hypertensive adults ingested 150 and 300 mg/day GSPE for 16 weeks, they showed improvements in BP, heart rate (HR) and plasma oxidative and microcirculatory status (Belcaro et al., 2013). Similarly, 70 healthy subjects with a systolic BP between 120 and 159 received 300 mg/d grape seed extract for eight weeks; however, these subjects showed no decrease in ambulatory BP (Ras et al., 2013). Perhaps significantly, Ras's study administers GSPE dosage for half as long as does Belcaro's, again suggesting that GSPE benefits are dose-related or time-related. Findings from an earlier study bolster Belcaro's results; in 2009, 27 adults with metabolic syndrome were given 150 or 300 mg grape seed extract for four weeks and showed a total decrease in systolic and diastolic BP (Sivaprakasapillai et al., 2009). Ras's study notwithstanding, GSPE does seem to lower blood pressure in humans.

6.3.2.2 Cranberry (*Vaciniun macrocarpon* L.)

6.3.2.2.1 In Vitro *Studies*

Cranberry intake is also associated with cardioprotective activity resulting from the presence of proanthocyanidin (PAC) type A (Baranowska and Bartoszek, 2016). Cranberries are unique among fruits in that they are rich in A-type proanthocyanidins (PACs) in contrast to the B-type PACs present in most other fruit. There is encouraging but limited evidence of a cardioprotective effect of cranberries mediated via actions on antioxidant capacity and lipoprotein profiles (Blumberg et al., 2013). In a 2005 study, 20 healthy female volunteers between the ages of 18 and 40 were recruited for an experiment exploring the link between cranberries and plasma antioxidant activity. Subjects consumed 750 mL/day of cranberry juice rich in proanthocyanidin type A for two consecutive weeks. Blood and urine samples were tracked over a period of four weeks, with the total phenolic, proanthocyanidin and catechin content of the supplements and plasma measured. Upon comparing the subjects' lab results with the effects of a placebo, researchers discovered that Vitamin C, total phenolic, total proanthocyanidin and total catechin concentrations were significantly higher within the cranberry juice-subgroup compared with the placebo-receiving group (Duthie et al., 2006).

6.3.3 Effects on Flow-Mediated Dilation (FMD) and Vascular Relaxation

6.3.3.1 Grapes

6.3.3.1.1 In Vitro *Studies*

Using 100 μM/L GSPE to test flow-mediated dilation (FMD) and vascular relaxation in a rabbit model demonstrated a dose-dependent aortic ring dilation (Edirisinghe et al., 2008). This experiment suggests that the cardioprotective properties of GSPE are not limited to only one species.

6.3.3.1.2 Clinical Studies

Similar studies have been attempted on humans. Thirty-five male subjects took 800 mg/day wine, grape and grape seed polyphenols for two weeks and showed no major effect on FMD (van Mierlo et al., 2010). That same year, 50 adults with cardiovascular risk were given 1300 mg/day muscadine grape seed extract for four weeks and subsequently showed no evidence of FMD, although a significant increase in resting brachial diameter was noted (Mellen et al., 2010). A 2004 study corroborated this experiment when 32 adults at high cardiovascular risk (smoking, high cholesterol, hypertension) given 2000 mg/day grape seed extract for four weeks showed improved FMD (Clifton, 2004). All three experiments supported the conclusion that GSPE improves flow-mediated dilation and overall vascular activity.

6.3.3.2 Bilberry (*Vaccinium myrtillus* L.)

6.3.3.2.1 In Vitro *Studies*

A 2016 study was published linking bilberry consumption to cardiovascular disease (CVD) prevention and a reduction in total LDL-C and triglyceride levels. The research involved men (n=11) and women (n=25) who consumed 150 g of frozen stored bilberries three times a week for six weeks. Anthropometric parameters, blood pressure, lipid profile, glucose, liver enzymes, creatinine, albumin, magnesium and anti-radical activity were measured before and after the conclusion of the experiment. According to the data, the consumption of bilberries led to a decrease in the following parameters among women: total cholesterol (P=.017), LDL-C (P=.0347), TG (P=.001), glucose (P=.005), albumin (P=.001), γ-glutamyltransferase (P=.046) and HDL-C. In men, favorable changes were observed in total cholesterol (P=.004), glucose (P=.015), albumin (P=.028), aspartate aminotransferase (P=.012), γ-glutamyltransferase (P=.013) and HDL-C (P=.009). In other words, the regular intake of bilberries can reduce the incidence of cardiovascular disease by decreasing LDL-C/TG and increasing HDL-C levels, due in large part to their high levels of proanthocyanidins (Habanova et al., 2016). Notably, significant decreases were seen in plasma levels of high-sensitivity C-reactive protein (hsCRP), a sensitive biomarker of subclinical inflammation and a predictor of CVD (Karlsen et al., 2010).

6.3.4 Effects of Lipoxygenases on Oxidative Status

6.3.4.1 Chocolate

6.3.4.1.1 In Vitro *Studies*

Procyanidins benefit oxidative status by inhibiting lipoxygenases. Cocoa seed procyanidins ranging in size from monomers to decamers were used at a concentration of 29 μg/mL on rabbit 15- LOX-1 and soybean lipoxygenase. The size of the oligomers of procyanidins were correlated with lipoxygenase activity. Lipoxygenase inhibition potency decreased from the monomer, was at a minimum at the trimer and tetramers, and then increased through the decamers. The researchers determined that procyanidins and other flavanols directly inhibit mammalian 15-lipoxygenase-1 (Schewe and Sies, 2005).

6.3.5 Effects on Oxidative Phosphorylation and Nitric-Oxide Synthase on Vascular Smooth Cells

6.3.5.1 Chocolate

6.3.5.1.1 In Vitro *Studies*

Procyanidin B2 effects have also been tested on rat heart mitochondria. Procyanidin B2 (0.7–17.9 ng/mL) increased State 2 rate of respiration of the mitochondria while inhibiting State 3 respiration. Because it possesses a concentration-dependent uncoupling effect on oxidative phosphorylation, procyanidin B2 may be advantageous in the prevention of cardiovascular disease. Procyanidin B2 may also indirectly affect ROS production (Kopustinskiene et al., 2015).

A study assessing the effect of procyanidins on vascular smooth-muscle cells was completed using New Zealand White rabbit aortic rings and defatted cocoa powder procyanidins. The rings were cultured with 0.0001 mol/L of procyanidins. Monomer through trimer-sized procyanidins did not elicit endothelium-dependent relaxation, while tetramers through decamers did. Polymeric compounds therefore have the ability to cause endothelium-dependent relaxation due to activation of endothelial nitric oxide synthase (Karim et al., 2000).

6.3.5.1.2 Clinical Studies

Flavanol-rich cocoa was tested in 27 healthy human subjects. In the study, 920 mL of cocoa containing 0.682 mg/g procyanidin oligomers was given daily. Eighteen of the participants received 100 μg/kg of nitric oxide synthase inhibitor intravenously, causing an increase in systolic blood pressure and a small increase in diastolic blood pressure. After five days, pulse wave amplitude increased. The results showed that cocoa-induced vasodilation is caused by the activation of nitric oxide. The nitric-oxide synthase inhibitor reversed the vasodilation caused by the cocoa, boosting endothelial function (Fisher et al., 2003).

6.3.5.2 Cranberry

6.3.5.2.1 In Vitro *Studies*

A 2011 study further linked plasma antioxidant capacity with A-type proanthocyanidin—rich cranberries. In a randomized, double-blind, placebo-controlled trial, participants identified with metabolic syndrome (n=15–16/group) were assigned to one of two groups: cranberry juice (480 mL/day) or placebo (480 mL/day) for eight weeks. Anthropometrics, blood pressure measurements, dietary analyses and fasting blood draws were conducted at the introductory screening and the eight weeks of the study. The cranberry juice group demonstrated significantly increased plasma antioxidant capacity and decreased oxidized low-density lipoprotein and malondialdehyde after eight weeks compared to the placebo group. The researchers concluded that two cups/day of concentrated cranberry juice significantly reduces lipid oxidation and increases plasma antioxidant capacity due to the rich concentration of proanthocyanidins (Basu et al., 2011). An additional study also correlated cranberry consumption and overall antioxidant effects; in this study, 10 healthy, nonsmoking men and postmenopausal women age 50–70 years with a body mass index (BMI) of 18.5–29.9 kg/m^2 consumed polyphenol-rich fruits 48 hours before an examination and underwent antioxidant testing afterwards. The researchers concluded that cranberry juice concentrate increased antioxidant capacity in healthy older adults (McKay et al., 2015). In 2000, researchers determined that in nine volunteers consuming 500 mL of cranberry juice, a significant increase in the ability of plasma to reduce potassium nitrosodisulphonate followed. This increase in plasma antioxidant capacity following consumption of cranberry juice was related to the phenolic compounds within cranberries, specifically A-type proanthocyanidins (Pedersen et al., 2000). A 2005 study tested the effects of flavonoid-rich cranberry juice supplementation on plasma lipoprotein levels and LDL oxidation. Twenty-one men were enrolled in a 14-day intervention and instructed to drink cranberry juice

7 mL/kg body weight per day. Physical and metabolic measures, including plasma lipid and oxidized LDL (Ox-LDL) concentrations as well as antioxidant capacity, were performed before and after the intervention. The intervention led to a significant reduction in plasma Ox-LDL levels (−9.9% ± 17.8%, *P*: 178 = .0131) and an increase in antioxidant capacity (+6.5% ± 10.3%, *P*: 180 = .0140). Short-term cranberry juice supplementation is therefore associated with significant increase in plasma antioxidant capacity and reduction in circulating Ox-LDL concentrations (Reul et al., 2005).

6.3.5.3 Bilberry

6.3.5.3.1 In Vivo *Studies*

Plasma triglyceride levels decreased in rats fed with an extract of bilberry leaves (3 g/kg/day) for four days (Cignarella et al., 1996). In a human study, 35 subjects who took 100 g of whole bilberries each day showed an improvement in platelet function, blood pressure and high-density lipoprotein (HDL)-cholesterol levels (Erlund et al., 2008). In another human study, mixed proanthocyanins from bilberry were given as an extract (320 mg/day) for 12 weeks to 60 middle-aged dyslipidemic Chinese subjects; the results showed significant improvements in low-density lipoprotein (LDL)-cholesterol (average decrease of approximately 14%) and HDL-cholesterol (average increase of approximately 14%) (Shao et al., 2009). Bilberry fruit proanthocyanins have also been reported to inhibit smooth-muscle contraction and platelet aggregation, increasing the body's overall vascular health. Proanthocyanins from bilberry were reported to protect against ischemia reperfusion injury in rats that were fed bilberry proanthocyanin extract for 12 days prior to inducing hypertension (Detre et al., 1986). Bilberry proanthocyanidins, therefore, have a variety of antioxidant and anti-inflammatory effects related to their high concentration of proanthocyanidins.

6.3.6 Effects on Plasma Thiobarbituric Acid Reactive Substances (Antioxidant Capacity)

6.3.6.1 Chocolate

6.3.6.1.1 Clinical Studies

Human subjects consumed high- and low-procyanidin-containing chocolate, and investigators measured antioxidant activity by analyzing plasma ThioBarbituric Acid Reactive Substances (TBARs). The procyanidins were found to cause a decrease in plasma TBARs, corresponding to an increase in plasma antioxidant capacity after the consumption of milk chocolate containing 168 mg of flavanols (Arranz et al., 2013).

In another experiment, 20 healthy male and female volunteers were split into four groups and given different amounts of procyanidin-rich chocolate in the form of 0, 27, 53, or 80 g M&M's®. The 27 g sample contained 186 mg of procyanidins. In the control group that did not consume chocolate, there was a decrease in time of inhibition produced by the plasma; however, these decreases were improved in subjects that did consume chocolate. With each increase in amount of chocolate consumed, there was a trend for an increase in the time of inhibition six hours after the consumption. Also in the control group, plasma TBARs measurements increased in the two hour and six hour intervals. This was reversed in those who consumed chocolate. It was concluded that a diet rich in procyanidins can improve the antioxidant potential of plasma (Wang et al., 2000).

6.3.7 Effects of Procyanidins on Platelets with Regard to Oxidative Status

6.3.7.1 Chocolate

6.3.7.1.1 In Vitro Studies

In an *in vitro* study, procyanidin trimers and pentamers from cocoa were cultured with peripheral whole blood cells collected from healthy subjects; platelet agonists were added to some samples. With a concentration of 3 and 10 μg/L, procyanidins increased the fibrinogen binding conformation of the GPllb-llla and hindered platelet activation response to epinephrine (Rein et al., 2000b).

6.3.7.1.2 Clinical Studies

One study showed that the ingestion of a cocoa beverage containing a total of 897 mg of cocoa flavanols including high weights of procyanidins provided for an inhibition of platelet activation and function. The trimers and the pentamers were found to be partly responsible for this. These findings were used to conclude that these compounds have anti-inflammatory effects on the human heart (Rein et al., 2000a).

6.3.8 Effects on Inflammation Markers, Enzymes, Nitrous Oxide Production and Tumor Necrosis Factor (TNF) Secretions

6.3.8.1 Grapes

6.3.8.1.1 In Vivo Studies

An early experiment tested the effect of GSPE on markers of inflammation in mice. Female Swiss-Webster mice were treated with GSPE (25, 50, and 100 mg/kg) alone and in combination with vitamins C, E, and beta-carotenes after induction of peritoneal macrophages using tetra-decanoylphorbol-13-acetate (TPA). GSPE exhibited superior dose-dependent effectiveness in protecting against lipid peroxidation and in preventing DNA fragmentation of the brain (Bagchi et al., 1998). In a later study, GSSE exerted efficient anti-inflammatory effects (Aloui et al., 2016). Another interesting study gauged if GSPE and GSSE protect against arsenic trioxide-induced oxidative stress in rat hearts. In that experiment, rats were infused with arsenic trioxide for 21 days, and then given a dose of GSSE. The GSSE effectively neutralized the arsenic (Sfaxi et al., 2015). As female Swiss-Webster mice were treated with GSPE, they exhibited resistance to lipid peroxidation.

6.3.8.1.2 *Clinical Studies*

Two studies involving humans tested GSPE's effect on inflammatory markers. The first involved 32 diabetic adults at significant cardiovascular risk taking 600 mg/day grape seed extract for four weeks; the human test subjects showed improvement in inflammatory markers and glycemia (Kar et al., 2009). A subsequent study yielded conflicting results. Fifty adults with cardiovascular risk received 1300 mg/day muscadine GSPE for four weeks and subsequently showed no significant change in biomarkers of inflammation, antioxidant capacity or lipid peroxidation, although there was an increase in resting brachial diameter of uncertain clinical significance (Mellen et al., 2010). Mellen concluded that additional study was warranted to assess clinical benefits of grape-derived nutritional supplements.

6.3.9 Effects on Angiotensin-Converting Enzyme (ACE) in Cardiovascular Smooth-Muscle Cells

6.3.9.1 Grapes

6.3.9.1.1 In Vivo *Studies*

Rats on a high-fat diet were given low molecular weight GSPE at a dose of 375 mg/kg and killed after six hours; no alterations in glutathione or plasma ACE were noted (Pons et al., 2014). This is the only available study at present, in either rats or humans, assessing the effect of GSPE on ACE in endovascular smooth-muscle cells.

6.3.10 Effects on Nitrous Oxide (NO) Production in Cardiovascular Cells

6.3.10.1 Grapes

6.3.10.1.1 In Vivo *Studies*

GSPE has a clear effect on nitrous oxide production. Rats given GSPE at a dose of 250 mg/kg body weight for five weeks increased NO production (Cui et al., 2012). An earlier study is confirmatory. GSPE was given to rats at doses up to 1g/L, producing synthesis and release of nitrous oxide in rat isolated aortic rings (Mendes et al., 2003).

6.3.11 Effects on TNF-Alpha Secretion in Inflammation

6.3.11.1 Chocolate

6.3.11.1.1 In Vitro *Studies*

Proanthocyanidin compounds isolated from chocolate were cultured with peripheral blood cells from volunteers at a concentration of 25 μg/mL. Secretion of TNF-alpha was then measured, with monomers and dimers providing the lowest secretory increase. Tetramers through octamers, however, produced a threefold or fourfold increase, while other fractions doubled the baseline amount of secretion. TNF-alpha is a substance essential for infection response; therefore, an increase in its secretion helps decrease inflammation (Mao et al., 2002).

6.3.12 Effect on Foam Cells, Endothelial Cells, Lung Cells, Kidney Cells and Smooth-Muscle Cells

6.3.12.1 Grapes

6.3.12.1.1 Effects on Atherosclerosis/Foam Cells

6.3.12.1.1.1 In Vivo *Studies* GSPE at a dose of 45 µg/mL attenuated foam cell development in the macrophages of mice, implying a benefit in atherosclerosis (Terra et al., 2009).

Hamsters given GSPE at 18.4 mg/kg -70 µg/mL (in bath) for 12 weeks had less plasma cholesterol and atherosclerosis in their aortic rings (Auger et al., 2004). An earlier study treating atherosclerotic hamsters with GSPE at 100 mg/kg body weight for 10 weeks demonstrated reduced foam cells (Vinson et al., 2002). Unfortunately, no comparable study involving humans is available.

6.3.12.1.1.2 In Vitro *Studies* (Protective Effects on Endothelial Cells) Proanthocyanidins applied at 1–20 µM to human umbilical endothelial cells demonstrated a protective effect against peroxynitrile damage, suggesting significant antioxidant effects (Aldini et al., 2003).

6.3.12.1.1.3 In Vivo *Studies* (Effects on Cardiac Cells) GSPE demonstrates beneficial effects on cardiac cells. GSPE applied at 25–100 µM to rat cardiac H9C2 cells increased resistance to cardiac cell apoptosis elicited by reactive oxygen species (Du et al., 2007). According to a 2008 study, both grape flesh and grape skin extracts (when administered orally) improved aortic flow and protected hearts from ischemic reperfusion, and both dietary interventions reduced myocardial infarction size—suggesting both grape flesh and grape skin are equally cardioprotective (Leifert and Abeywardena, 2008). Leifert's study concluded that both grape seeds and grape skin cells are equally effective at cardiovascular protection and have a measurable positive effect on cardiovascular cellular strength.

6.3.12.1.1.4 In Vivo *Studies* (Effects on Lung Cells) Khazri tested putative GSSE protective effects on oxidative stress levels in rat lungs. Twenty-four male Wistar rats were pre-treated for three weeks with vehicle (ethanol 10% control) or GSSE (4 g/kg) prepared from a northern Tunisian cultivar, and then administered a single dose of bleomycin (15 mg/kg) at the seventh day before examining the effects on the lungs. The researchers concluded that GSSE exhibited significant protective properties, and that GSSE exerts potent antioxidant properties that could find potential application in lung protection processes (Khazri et al., 2016).

6.3.12.1.1.5 In Vivo *Studies* (Effects on Kidney Cells) GSSE treatment offers effective protection against diabetes-induced kidney dysfunction in both virgin and pregnant rats (Oueslati et al., 2016). GSSE corrected the surrogate markers of kidney function, as well as the biomarkers of oxidative stress, by acting as a highly potent antioxidant.

6.3.12.2 Chocolate

6.3.12.2.1 In Vitro *Studies (*Effects on Vascular Smooth-Muscle Cells)

Chocolate procyanidins also exert beneficial effects on vascular smooth-muscle cells. Human aortic vascular smooth-muscle cells were cultured and treated with procyanidin B2 (50 μg/mL) or cocoa procyanidin fractions (30 μg/mL). Cocoa procyanidins exhibited inhibitory effects on thrombin-mediated activation and expression of MMP-2 in vascular smooth-muscle cells. Procyanidin B2 could contribute to the inhibition of invasion and migration of vascular smooth-muscle cells treated with red wine polyphenol compounds (RWPC) (Lee et al., 2008).

6.3.13 Antioxidant and Cardioprotective Effects of Pomegranates

6.3.13.1 *In Vitro* Studies

In a 2000 study, researchers concluded that pomegranate juice (PJ) shows antioxidant capacities of up to three times the capacity of red wine (Gil et al., 2000). A subsequent study showed that PJ had the highest antioxidant capacity of all available polyphenol-rich beverages (Seeram et al., 2008). Three years later, a separate study confirmed that Mediterranean pomegranates had the highest antioxidant capacity of the seven selected polyphenol-rich fruits (Shams Ardekani et al., 2011). The polyphenols tested included proanthocyandins, tannins, catechin and epicatechin, with the proanthocyanidins in PJ being more effective than those of similar fruits. A 2011 study specifically tested the antioxidant properties of pomegranates and found a significant correlation between the total antioxidant activity and the proanthocyanidin content. A high correlation ($r^2 = 0.80$) between antioxidant capacity and proanthocyanidin content was found, suggesting that proanthocyanidins are the principal contributor to the antioxidant capacity of pomegranates (El Kar et al., 2011). Additionally, evidence suggests that polyphenolic antioxidants (composed mainly of proanthocyanidins) contained in PJ can cause reduction of oxidative stress and atherogenesis through the activation of redox-sensitive genes and increased eNOS expression (Fuhrman et al., 2010). These studies demonstrate the antioxidant, anti-diabetic and cardioprotective capacity of proanthocyanidins found in pomegranates and pomegranate juice. Additionally, a 2012 study showed that there was a definite correlation between the proanthocyandins on the skin of pomegranates and the overall antioxidant effect on the test subject (Wissam et al., 2012). A pilot study in type 2 diabetic patients with hyperlipidemia found that concentrated PJ decreased cholesterol absorption, had a favorable effect on enzymes concerned in cholesterol metabolism, drastically reduced LDL cholesterol and improved LDL/HDL cholesterol and total/HDL ratios (Esmaillzadeh et al., 2006). In that study, 22 diabetic patients were recruited from the Iranian Diabetes Society and were given a diet high in pomegranate juices with large extracts of proanthocyandins (specifically, 40 g concentrated PJ) for eight weeks. After consumption of concentrated pomegranate juice, significant reductions were seen in total cholesterol ($p < 0.006$), low-density lipoprotein-cholesterol (LDL-C) ($p < 0.006$), LDL-C/

high-density lipoprotein-cholesterol (HDL-C) ($p < 0.001$) and total cholesterol/HDL-C ($p < 0.001$). The presence of proanthocyanidins within pomegranates and PJ produces antioxidant and cardioprotective effects.

6.3.14 Conclusion

Proanthocyanidin extracts from grape seeds, grape skins, cranberries, bilberries, pomegranates and cocoa demonstrate significant benefits on cardiovascular health markers in humans and animals. Beneficial effects include decreases in total cholesterol, total triglyceride and total lipid levels as well as reductions in blood pressure, oxidative stress levels and inflammatory markers. Furthermore, cardiac recovery, endothelial tissue repair and endothelial cellular strength is improved in the presence of these extracts. Additional studies indicate that proanthocyanidin extract prevents cardiac hypertrophy, fibrosis, inflammation, oxidative stress, hypertension and diastolic dysfunction in both rodent and human test subjects. Studies testing the cardioprotective effects of proanthocyanidins invariably indicate a proportional increase in proanthocyanidin intake and cardiovascular health. Although the data regarding proanthocyanidin effects in pomegranates and bilberries is limited, the early returns are encouraging and offer pharmacological potential. Proanthocyanidins show promise in effectively preventing cardiovascular diseases in humans.

6.3.15 Perspective

The preponderance of basic science research on proanthocyanidins suggests beneficial antioxidant effects on multiple organ systems, including the cardiovascular system. Since the leading cause of death globally is cardiovascular disease, the implications for clinical research are clear. If proanthocyanidins are to gain a place in our clinical armamentarium, however, we will need additional research on the relative value of the different proanthocyanidin types, as well as standardization of dosage amounts. We believe that clinical research into the value of proanthocyanidins in the near future is warranted.

REFERENCES

Aldini, G., M. Carini, A. Piccoli, G. Rossoni, and R. M. Facino. "Procyanidins from Grape Seeds Protect Endothelial Cells from Peroxynitrite Damage and Enhance Endothelium-Dependent Relaxation in Human Artery: New Evidences for Cardio-Protection." *Life Sci.* 73, no. 22 (Oct 17, 2003): 2883–98.

Aloui, F., K. Charradi, A. Hichami, S. Subramaniam, N. A. Khan, F. Limam, and E. Aouani. "Grape Seed and Skin Extract Reduces Pancreas Lipotoxicity, Oxidative Stress and Inflammation in High Fat Diet Fed Rats." *Biomed. Pharmacother.* 84 (Dec 2016): 2020–28.

Arranz, S., P. Valderas-Martinez, G. Chiva-Blanch, R. Casas, M. Urpi-Sarda, R. M. Lamuela-Raventos, and R. Estruch. "Cardioprotective Effects of Cocoa: Clinical Evidence from Randomized Clinical Intervention Trials in Humans." *Mol. Nutr. Food Res.* 57, no. 6 (Jun 2013): 936–47.

Auger, C., P. Gerain, F. Laurent-Bichon, K. Portet, A. Bornet, B. Caporiccio, G. Cros, P. L. Teissedre, and J. M. Rouanet. "Phenolics from Commercialized Grape Extracts Prevent Early Atherosclerotic Lesions in Hamsters by Mechanisms Other Than Antioxidant Effect." *J. Agric. Food Chem.* 52, no. 16 (Aug 11, 2004): 5297–302.

Bagchi, D., A. Garg, R. L. Krohn, M. Bagchi, D. J. Bagchi, J. Balmoori, and S. J. Stohs. "Protective Effects of Grape Seed Proanthocyanidins and Selected Antioxidants against Tpa-Induced Hepatic and Brain Lipid Peroxidation and DNA Fragmentation, and Peritoneal Macrophage Activation in Mice." *Gen. Pharmacol.* 30, no. 5 (May 1998): 771–6.

Bagchi, D., C. K. Sen, S. D. Ray, D. K. Das, M. Bagchi, H. G. Preuss, and J. A. Vinson. "Molecular Mechanisms of Cardioprotection by a Novel Grape Seed Proanthocyanidin Extract." *Mutat. Res.* 523–524, no. (Feb–Mar 2003): 87–97.

Baiges, I., J. Palmfeldt, C. Blade, N. Gregersen, and L. Arola. "Lipogenesis Is Decreased by Grape Seed Proanthocyanidins According to Liver Proteomics of Rats Fed a High Fat Diet." *Mol. Cell Proteomics* 9, no. 7 (Jul 2010): 1499–513.

Baranowska, M., and A. Bartoszek. "Antioxidant and Antimicrobial Properties of Bioactive Phytochemicals from Cranberry." *Postepy Hig Med Dosw* (Online) 70, no. 0 (Dec 31, 2016): 1460–68.

Basu, A., N. M. Betts, J. Ortiz, B. Simmons, M. Wu, and T. J. Lyons. "Low-Energy Cranberry Juice Decreases Lipid Oxidation and Increases Plasma Antioxidant Capacity in Women with Metabolic Syndrome." *Nutr Res* 31, no. 3 (Mar 2011): 190–6.

Belcaro, G., A. Ledda, S. Hu, M. R. Cesarone, B. Feragalli, and M. Dugall. "Grape Seed Procyanidins in Pre- and Mild Hypertension: A Registry Study." *Evid. Based Complement. Alternat. Med.* 2013 (2013): 313142.

Blumberg, J. B., T. A. Camesano, A. Cassidy, P. Kris-Etherton, A. Howell, C. Manach, L. M. Ostertag, et al. "Cranberries and Their Bioactive Constituents in Human Health." *Adv Nutr* 4, no. 6 (Nov 2013): 618–32.

Caimari, A., J. M. del Bas, A. Crescenti, L. Arola. "Low Doses of Grape Seed Procyanidins Reduce Adiposity and Improve the Plasma Lipid Profile in Hamsters." *Int. J. Obes. (Lond)* 37, no. 4 (Apr 2013): 576–83.

Charradi, K., S. Elkahoui, F. Limam, and E. Aouani. "High-Fat Diet Induced an Oxidative Stress in White Adipose Tissue and Disturbed Plasma Transition Metals in Rat: Prevention by Grape Seed and Skin Extract." *J. Physiol. Sci.* 63, no. 6 (Nov 2013): 445–55.

Cignarella, A., M. Nastasi, E. Cavalli, and L. Puglisi. "Novel Lipid-Lowering Properties of Vaccinium Myrtillus L. Leaves, a Traditional Antidiabetic Treatment, in Several Models of Rat Dyslipidaemia: A Comparison with Ciprofibrate." *Thromb Res* 84, no. 5 (Dec 1, 1996): 311–22.

Clifton, P. M. "Effect of Grape Seed Extract and Quercetin on Cardiovascular and Endothelial Parameters in High-Risk Subjects." *J. Biomed. Biotechnol.* 2004, no. 5 (2004): 272–78.

Cui, X., X. Liu, H. Feng, S. Zhao, and H. Gao. "Grape Seed Proanthocyanidin Extracts Enhance Endothelial Nitric Oxide Synthase Expression through 5'-Amp Activated Protein Kinase/Surtuin 1-Krupple Like Factor 2 Pathway and Modulate Blood Pressure in Ouabain Induced Hypertensive Rats." *Biol. Pharm. Bull.* 35, no. 12 (2012): 2192–7.

Deaton, C., E. S. Froelicher, L. H. Wu, C. Ho, K. Shishani, and T. Jaarsma. "The Global Burden of Cardiovascular Disease." *J. Cardiovasc. Nurs.* 26, no. 4 Suppl (Jul-Aug 2011): S5–14.

de Pascual-Teresa, S., C. Santos-Buelga, and J. C. Rivas-Gonzalo. "Quantitative Analysis of Flavan-3-Ols in Spanish Foodstuffs and Beverages." *J Agric Food Chem* 48, no. 11 (Nov 2000): 5331–7.

Detre, Z., H. Jellinek, M. Miskulin, and A. M. Robert. "Studies on Vascular Permeability in Hypertension: Action of Anthocyanosides." *Clin Physiol Biochem* 4, no. 2 (1986): 143–9.

Duthie, S. J., A. M. Jenkinson, A. Crozier, W. Mullen, L. Pirie, J. Kyle, L. S. Yap, P. Christen, and G. G. Duthie. "The Effects of Cranberry Juice Consumption on Antioxidant Status and Biomarkers Relating to Heart Disease and Cancer in Healthy Human Volunteers." *Eur J Nutr* 45, no. 2 (Mar 2006): 113–22.

Du, Y., H. Guo, and H. Lou. "Grape Seed Polyphenols Protect Cardiac Cells from Apoptosis via Induction of Endogenous Antioxidant Enzymes." *J. Agric. Food Chem.* 55, no. 5 (Mar 7, 2007): 1695–701.

Edirisinghe, I., B. Burton-Freeman, and C. Tissa Kappagoda. "Mechanism of the Endothelium-Dependent Relaxation Evoked by a Grape Seed Extract." *Clin. Sci. (Lond.)* 114, no. 4 (Feb 2008): 331–7.

El Kar, C., A. Ferchichi, F. Attia, and J. Bouajila. "Pomegranate (Punica granatum) Juices: Chemical Composition, Micronutrient Cations, and Antioxidant Capacity." *J. Food Sci.* 76, no. 6 (Aug 2011): C795–800.

Eren, G., Z. Cukurova, and O. Hergunsel. "Oxidative Stress and the Lung in Diabetes." *Diabetes: Oxidative Stress and Dietary Antioxidants,* London: Academic Press, 2014, pp. 237–245.

Erlund, I., R. Koli, G. Alfthan, J. Marniemi, P. Puukka, P. Mustonen, P. Mattila, and A. Jula. "Favorable Effects of Berry Consumption on Platelet Function, Blood Pressure, and Hdl Cholesterol." *Am J Clin Nutr* 87, no. 2 (Feb 2008): 323–31.

Esmaillzadeh, A., F. Tahbaz, I. Gaieni, H. Alavi-Majd, and L. Azadbakht. "Cholesterol-Lowering Effect of Concentrated Pomegranate Juice Consumption in Type Ii Diabetic Patients with Hyperlipidemia." *Int. J. Vitam. Nutr. Res.* 76, no. 3 (May 2006): 147–51.

Fisher, N. D., M. Hughes, M. Gerhard-Herman, and N. K. Hollenberg. "Flavanol-Rich Cocoa Induces Nitric-Oxide-Dependent Vasodilation in Healthy Humans." *J. Hypertens.* (Dec 2003): 2281–6.

Fuhrman, B., N. Volkova, and M. Aviram. "Pomegranate Juice Polyphenols Increase Recombinant Paraoxonase-1 Binding to High-Density Lipoprotein: Studies In Vitro and in Diabetic Patients." *Nutrition* 26, no. 4 (Apr 2010): 359–66.

Gil, M. I., F. A. Tomas-Barberan, B. Hess-Pierce, D. M. Holcroft, and A. A. Kader. "Antioxidant Activity of Pomegranate Juice and Its Relationship with Phenolic Composition and Processing." *J. Agric. Food Chem.* 48, no. 10 (Oct 2000): 4581–9.

Guerrero, L., M. Margalef, Z. Pons, M. Quinones, L. Arola, A. Arola-Arnal, and B. Muguerza. "Serum Metabolites of Proanthocyanidin-Administered Rats Decrease Lipid Synthesis in Hepg2 Cells." *J. Nutr. Biochem.* 24, no. 12 (Dec 2013): 2092–9.

Habanova, M., J. A. Saraiva, M. Haban, M. Schwarzova, P. Chlebo, L. Predna, J. Gazo, and J. Wyka. "Intake of Bilberries (Vaccinium myrtillus L.) Reduced Risk Factors for Cardiovascular Disease by Inducing Favorable Changes in Lipoprotein Profiles." *Nutr Res* 36, no. 12 (Dec 2016): 1415–22.

Kar, P., D. Laight, H. K. Rooprai, K. M. Shaw, and M. Cummings. "Effects of Grape Seed Extract in Type 2 Diabetic Subjects at High Cardiovascular Risk: A Double Blind Randomized Placebo Controlled Trial Examining Metabolic Markers, Vascular Tone, Inflammation, Oxidative Stress and Insulin Sensitivity." *Diabet. Med.* 26, no. 5 (May 2009): 526–31.

Karim, M., K. McCormick, and C. T. Kappagoda. "Effects of Cocoa Extracts on Endothelium-Dependent Relaxation." *J. Nutr.* 130, no. 8S Suppl (Aug 2000): 2105S–8S.

Karlsen, A., I. Paur, S. K. Bohn, A. K. Sakhi, G. I. Borge, M. Serafini, I. Erlund, et al. "Bilberry Juice Modulates Plasma Concentration of Nf-Kappab Related Inflammatory Markers in Subjects at Increased Risk of Cvd." *Eur J Nutr* 49, no. 6 (Sep 2010): 345–55.

Khazri, O., K. Charradi, F. Limam, M. V. El May, and E. Aouani. "Grape Seed and Skin Extract Protects against Bleomycin-Induced Oxidative Stress in Rat Lung." *Biomed. Pharmacother.* 81 (Jul 2016): 242–9.

Kopustinskiene, D. M., A. Savickas, D. Vetchy, R. Masteikova, A. Kasauskas, and J. Bernatoniene. "Direct Effects Of (-)-Epicatechin and Procyanidin B2 on the Respiration of Rat Heart Mitochondria." *BioMed Res. Int.* 2015 (2015): 232836.

Lee, K. W., N. J. Kang, M. H. Oak, M. K. Hwang, J. H. Kim, V. B. Schini-Kerth, and H. J. Lee. "Cocoa Procyanidins Inhibit Expression and Activation of Mmp-2 in Vascular Smooth Muscle Cells by Direct Inhibition of Mek and Mt1-Mmp Activities." *Cardiovasc. Res.* 79, no. 1 (Jul 1, 2008): 34–41.

Leifert, W. R., and M. Y. Abeywardena. "Cardioprotective Actions of Grape Polyphenols." *Nutr. Res.* 28, no. 11 (Nov 2008): 729–37.

Liu, X., J. Qiu, S. Zhao, B. You, X. Ji, Y. Wang, X. Cui, Q. Wang, and H. Gao. "Grape Seed Proanthocyanidin Extract Alleviates Ouabain-Induced Vascular Remodeling through Regulation of Endothelial Function." *Mol. Med. Rep.* 6, no. 5 (Nov 2012): 949–54.

Mao, T., J. Van De Water, C. L. Keen, H. H. Schmitz, and M. E. Gershwin. "Cocoa Procyanidins and Human Cytokine Transcription and Secretion." *J. Nutr.* 130, no. 8S Suppl (Aug 2000): 2093S–9S.

Mao, T. K., J. van de Water, C. L. Keen, H. H. Schmitz, and M. E. Gershwin. "Modulation of Tnf-Alpha Secretion in Peripheral Blood Mononuclear Cells by Cocoa Flavanols and Procyanidins." *Dev. Immunol.* 9, no. 3 (Sep 2002): 135–41.

McKay, D. L., C. Y. Chen, C. A. Zampariello, and J. B. Blumberg. "Flavonoids and Phenolic Acids from Cranberry Juice Are Bioavailable and Bioactive in Healthy Older Adults." *Food Chem* 168 (Feb 1, 2015): 233–40.

Mellen, P. B., K. R. Daniel, K. B. Brosnihan, K. J. Hansen, and D. M. Herrington. "Effect of Muscadine Grape Seed Supplementation on Vascular Function in Subjects with or at Risk for Cardiovascular Disease: A Randomized Crossover Trial." *J. Am. Coll. Nutr.* 29, no. 5 (Oct 2010): 469–75.

Mendes, A., C. Desgranges, C. Cheze, J. Vercauteren, and J. L. Freslon. "Vasorelaxant Effects of Grape Polyphenols in Rat Isolated Aorta. Possible Involvement of a Purinergic Pathway." *Fundam. Clin. Pharmacol.* 17, no. 6 (Dec 2003): 673–81.

Mensah, G. A., A. E. Moran, G. A. Roth, and J. Narula. "The Global Burden of Cardiovascular Diseases, 1990–2010." *Glob. Heart* 9, no. 1 (Mar 2014): 183–4.

Mokni, M., S. Hamlaoui-Guesmi, M. Amri, L. Marzouki, F. Limam, and E. Aouani. "Grape Seed and Skin Extract Protects against Acute Chemotherapy Toxicity Induced by Doxorubicin in Rat Heart." *Cardiovasc. Toxicol.* 12, no. 2 (Jun 2012): 158–65.

Neto, Calasans, Sylvia Menezes de Athayde, Myriam Fraga, and Museu de Arte da Bahia. *Calasans Neto.* Brazil: Odebrecht, 2007.

Oueslati, N., K. Charradi, T. Bedhiafi, F. Limam, and E. Aouani. "Protective Effect of Grape Seed and Skin Extract against Diabetes-Induced Oxidative Stress and Renal Dysfunction in Virgin and Pregnant Rat." *Biomed. Pharmacother.* 83 (Oct. 2016): 584–92.

Pedersen, C. B., J. Kyle, A. M. Jenkinson, P. T. Gardner, D. B. McPhail, and G. G. Duthie. "Effects of Blueberry and Cranberry Juice Consumption on the Plasma Antioxidant Capacity of Healthy Female Volunteers." *Eur J Clin Nutr* 54, no. 5 (May 2000): 405–8.

Peng, N., J. T. Clark, J. Prasain, H. Kim, C. R. White, and J. M. Wyss. "Antihypertensive and Cognitive Effects of Grape Polyphenols in Estrogen-Depleted, Female, Spontaneously Hypertensive Rats." *Am. J. Physiol. Regul. Integr. Comp. Physiol.* 289, no. 3 (Sep 2005): R771–5.

Pons, Z., L. Guerrero, M. Margalef, L. Arola, A. Arola-Arnal, and B. Muguerza. "Effect of Low Molecular Grape Seed Proanthocyanidins on Blood Pressure and Lipid Homeostasis in Cafeteria Diet-Fed Rats." *J. Physiol. Biochem.* 70, no. 2 (Jun 2014): 629–37.

Preuss, H. G., D. Wallerstedt, N. Talpur, S. O. Tutuncuoglu, B. Echard, A. Myers, M. Bui, and D. Bagchi. "Effects of Niacin-Bound Chromium and Grape Seed Proanthocyanidin Extract on the Lipid Profile of Hypercholesterolemic Subjects: A Pilot Study." *J. Med.* 31, no. 5–6 (2000): 227–46.

Quesada, H., J. M. del Bas, D. Pajuelo, S. Diaz, J. Fernandez-Larrea, M. Pinent, L. Arola, M. J. Salvado, and C. Blade. "Grape Seed Proanthocyanidins Correct Dyslipidemia Associated with a High-Fat Diet in Rats and Repress Genes Controlling Lipogenesis and VLDL Assembling in Liver." *Int. J. Obes. (Lond)* 33, no. 9 (Sep 2009): 1007–12.

Ras, R. T., P. L. Zock, Y. E. Zebregs, N. R. Johnston, D. J. Webb, and R. Draijer. "Effect of Polyphenol-Rich Grape Seed Extract on Ambulatory Blood Pressure in Subjects with Pre- and Stage I Hypertension." *Br. J. Nutr.* 110, no. 12 (Dec 2013): 2234–41.

Razavi, S. M., S. Gholamin, A. Eskandari, N. Mohsenian, A. Ghorbanihaghjo, A. Delazar, N. Rashtchizadeh, M. Keshtkar-Jahromi, and H. Argani. "Red Grape Seed Extract Improves Lipid Profiles and Decreases Oxidized Low-Density Lipoprotein in Patients with Mild Hyperlipidemia." *J. Med. Food* (Mar 2013): 255–8.

Rein, D., T. G. Paglieroni, D. A. Pearson, T. Wun, H. H. Schmitz, R. Gosselin, and C. L. Keen. "Cocoa and Wine Polyphenols Modulate Platelet Activation and Function." *J. Nutr.* 130, no. 8S Suppl (Aug 2000a): 2120S–6S.

Rein, D., T. G. Paglieroni, T. Wun, D. A. Pearson, H. H. Schmitz, R. Gosselin, and C. L. Keen. "Cocoa Inhibits Platelet Activation and Function." *Am. J. Clin. Nutr.* 72, no. 1 (Jul 2000b): 30–5.

Ruel, G., S. Pomerleau, P. Couture, B. Lamarche, and C. Couillard. "Changes in Plasma Antioxidant Capacity and Oxidized Low-Density Lipoprotein Levels in Men after Short-Term Cranberry Juice Consumption." *Metabolism* 54, no. 7 (Jul 2005): 856–61.

Safwen, K., S. Selima, E. Mohamed, L. Ferid, C. Pascal, A. Mohamed, A. Ezzedine, and M. Meherzia. "Protective Effect of Grape Seed and Skin Extract on Cerebral Ischemia in Rat: Implication of Transition Metals." *Int. J. Stroke* 10, no. 3 (Apr 2015): 415–24.

Sano, A., R. Uchida, M. Saito, N. Shioya, Y. Komori, Y. Tho, N. Hashizume. "Beneficial Effects of Grape Seed Extract on Malondialdehyde-Modified Ldl." *J. Nutr. Sci. Vitaminol. (Tokyo)* (Apr 2007): 174–82.

Schewe, T., and H. Sies. "Myeloperoxidase-Induced Lipid Peroxidation of Ldl in the Presence of Nitrite. Protection by Cocoa Flavanols." *BioFactors* 24, no. 1–4 (2005): 49–58.

Schramm, D. D., J. F. Wang, R. R. Holt, J. L. Ensunsa, J. L. Gonsalves, S. A. Lazarus, H. H. Schmitz, J. B. German, and C. L. Keen. "Chocolate Procyanidins Decrease the Leukotriene-Prostacyclin Ratio in Humans and Human Aortic Endothelial Cells." *Am. J. Clin. Nutr.* 73, no. 1 (Jan 2001): 36–40.

Seeram, N. P., M. Aviram, Y. Zhang, S. M. Henning, L. Feng, M. Dreher, and D. Heber. "Comparison of Antioxidant Potency of Commonly Consumed Polyphenol-Rich Beverages in the United States." *J. Agric. Food Chem.* 56, no. 4 (Feb 27, 2008): 1415–22.

Sfaxi, I., K. Charradi, F. Limam, M. V. El May, and E. Aouani. "Grape Seed and Skin Extract Protects against Arsenic Trioxide Induced Oxidative Stress in Rat Heart." *Can. J. Physiol. Pharmacol.* (Jul 29, 2015): 1–9.

Shams Ardekani, M. R., M. Hajimahmoodi, M. R. Oveisi, N. Sadeghi, B. Jannat, A. M. Ranjbar, N. Gholam, and T. Moridi. "Comparative Antioxidant Activity and Total Flavonoid Content of Persian Pomegranate (*Punica granatum* L.) Cultivars." *Iran. J. Pharm. Res.* (Summer 2011): 519–24.

Shao, Z. H., K. R. Wojcik, A. Dossumbekova, C. Hsu, S. R. Mehendale, C. Q. Li, Y. Qin, et al. "Grape Seed Proanthocyanidins Protect Cardiomyocytes from Ischemia and Reperfusion Injury Via Akt-Nos Signaling." *J Cell Biochem* 107, no. 4 (Jul 1, 2009): 697–705.

Sivaprakasapillai, B., I. Edirisinghe, J. Randolph, F. Steinberg, and T. Kappagoda. "Effect of Grape Seed Extract on Blood Pressure in Subjects with the Metabolic Syndrome." *Metabolism* 58, no. 12 (Dec 2009): 1743–6.

Steinberg, F. M., R. R. Holt, H. H. Schmitz, and C. L. Keen. "Cocoa Procyanidin Chain Length Does Not Determine Ability to Protect Ldl from Oxidation When Monomer Units Are Controlled." *J. Nutr. Biochem.* 13, no. 11 (Nov 2002): 645–52.

Terra, X., J. Fernandez-Larrea, G. Pujadas, A. Ardevol, C. Blade, J. Salvado, L. Arola, and M. Blay. "Inhibitory Effects of Grape Seed Procyanidins on Foam Cell Formation In Vitro." *J. Agric. Food Chem.* 57, no. 6 (Mar 25, 2009): 2588–94.

van Mierlo, L. A., P. L. Zock, H. C. van der Knaap, and R. Draijer. "Grape Polyphenols Do Not Affect Vascular Function in Healthy Men." *J. Nutr.* 140, no. 10 (Oct 2010): 1769–73.

Verstraeten, S. V., J. F. Hammerstone, C. L. Keen, C. G. Fraga, and P. I. Oteiza. "Antioxidant and Membrane Effects of Procyanidin Dimers and Trimers Isolated from Peanut and Cocoa." *J. Agric. Food Chem.* 53, no. 12 (Jun 15, 2005): 5041–8.

Vigna, G. B., F. Costantini, G. Aldini, M. Carini, A. Catapano, F. Schena, A. Tangerini, *et al.* "Effect of a Standardized Grape Seed Extract on Low-Density Lipoprotein Susceptibility to Oxidation in Heavy Smokers." *Metabolism* 52, no. 10 (Oct 2003): 1250–7.

Vinson, J. A., M. A. Mandarano, D. L. Shuta, M. Bagchi, and D. Bagchi. "Beneficial Effects of a Novel Ih636 Grape Seed Proanthocyanidin Extract and a Niacin-Bound Chromium in a Hamster Atherosclerosis Model." *Mol. Cell. Biochem.* 240, no. 1–2 (Nov 2002): 99–103.

Wang, J. F., D. D. Schramm, R. R. Holt, J. L. Ensunsa, C. G. Fraga, H. H. Schmitz, and C. L. Keen. "A Dose-Response Effect from Chocolate Consumption on Plasma Epicatechin and Oxidative Damage." *J. Nutr.* 130, no. 8S Suppl (Aug 2000): 2115S–9S.

Weseler, A. R., E. J. Ruijters, M. J. Drittij-Reijnders, K. D. Reesink, G. R. Haenen, and A. Bast. "Pleiotropic Benefit of Monomeric and Oligomeric Flavanols on Vascular Health—A Randomized Controlled Clinical Pilot Study." *PLOS ONE* 6, no. 12 (2011).

Wissam, Z., B. Ghada, A. Wassim, and K. Warid. "Effective Extraction of Polyphenols and Proanthocyanidins from Pomegranate's Peel." *Int. J. Pharm. Pharm. Sci.* 4 Suppl 3 (May 9, 2012). Accessed August 10, 2017. http://www.ijppsjournal.com/Vol4Suppl3/4003.pdf.

Xia, E. Q., G. F. Deng, Y. J. Guo, and H. B. Li. "Biological Activities of Polyphenols from Grapes." *Int. J. Mol. Sci.* 11, no. 2 (Feb 4, 2010): 622–46.

7 Extra-Virgin Olive Oil and Blood Pressure

A New Approach to the Treatment of Cardiovascular Disease

Ana Belén Segarra, Magdalena Martínez-Cañamero, Germán Domínguez-Vías, Marina Hidalgo, Manuel Ramírez-Sánchez and Isabel Prieto

CONTENTS

7.1 INTRODUCTION

Hypertension is recognized at present as a major health problem and one of the main risk factors for stroke in industrialized countries (Lackland and Weber, 2015). Although developed countries present the fastest increment in blood pressure, current evidence shows it is a global problem in all regions of the world (Pearson, 1999). The accelerated rate of this global burden needs a global response for hypertension that includes the identification of the risk population, design of prevention strategies and the implementation of cost-effective treatment. In this sense, the use of natural products and diet could be an effective management tool to use in preventing the development of hypertension and could be associated with a reduction in hypertension related outcomes, including metabolic syndrome and stroke. Indeed, the reduction in blood pressure levels is mainly the result of changes in lifestyle and other nonpharmacologic interventions (Lackland et al., 2014).

The high risk of adverse cardiovascular outcomes associated with high blood pressure is the result of individuals unaware of uncontrolled hypertension for prolonged periods of time (Lackland and Weber, 2015; Tibazarwa and Damasceno, 2014). This concept has led to the implementation of public programs and health strategies to control blood pressure in the population and reduce cardiovascular and renal disease, including the control of body weight, dietary sodium intake, level of physical activity and other dietary factors as fruit and vegetable intake, alcohol consumption and the amount of fat in the diet. For example, the reduction in mean blood pressure through a decrease in salt intake represents a cost-effective approach to prevent hypertension and cardiovascular diseases (Beaglehole et al., 2012). These strategies are implemented to reduce blood pressure values in individuals in the prehypertension category (systolic blood pressure 120–129 mmHg or diastolic blood pressure 80–89 mmHg).

On the basis of the results of the Oxford Vascular Study, Rothwell et al. postulated in 2004 that strokes and ischaemic attacks were related to episodic hypertension and blood pressure variability rather than chronically high mean blood pressure (Poulter et al., 2015; Rothwell et al., 2004). Several trials suggested that various types of drugs exerted differential effects on blood pressure variability, with calcium-channel blockers seemingly the most effective and the β-blockers the least (Webb et al., 2010). Episodic hypertension with very low levels of mean systolic blood pressure is associated with a higher risk of a cardiovascular event than a constant hypertension without blood pressure variability. Long-term variability in blood pressure could be associated with the use of drugs with short duration of action (Rothwell et al., 2010; Webb et al., 2010).

Conventional antihypertensives are usually associated with many side effects. However, for primary health care, natural products are better accepted by the human body and have fewer side effects. During recent decades, a number of concerted efforts have been made to search for plants with hypotensive and antihypertensive therapeutic values. Dietary and lifestyle changes, as well as drugs, can improve blood pressure control and decrease the risk of cardiovascular disease (Tabassum and Ahmad, 2011).

7.2 DIETS AND HYPERTENSION

The Dietary Approach to Stop Hypertension (DASH diet) has been recommended by the American Heart Association for the non-pharmacological management of hypertension since the last decade of the 20th century (Appel et al., 2006; Moore et al, 2001; Sacks et al., 1995; Siervo et al., 2015). This dietary pattern promotes the intake of vegetable foods, including fruits, whole grains and nuts; low-fat dairy products, poultry and fish; and attempts to reduce the consumption of red meat, sweets, high-sugar beverages and high saturated fat and cholesterol. Thus, the DASH diet promotes high intake of protective nutrients such as fiber, vegetable protein, K, Ca and Mg, and a lower intake of saturated fatty acids, cholesterol, refined carbohydrates and sodium. The DASH diet has been demonstrated to significantly reduce systolic and diastolic blood pressure in hypertensive subjects when compared with a standard American diet (Appel et al., 1997). At the same time, it improved cardiovascular risk factors and also appeared to have beneficial effects in individuals with high cardio-metabolic risk on insulin sensitivity (Shirani et al., 2013), inflammation (Azadbakht et al., 2011), oxidative stress (Asemi et al., 2013a), fasting glucose (Azadbakht et al., 2005) and total plasmatic cholesterol (Chen et al., 2010). However, these effects on cardio-metabolic biomarkers are unclear, and other studies have observed non-significant effects on blood pressure (Asemi et al., 2013b; Nowson et al., 2004, 2005), fasting glucose (Appel et al., 2003; Lopes et al., 2003) and total cholesterol (Blumenthal et al., 2010).

Another dietary pattern useful in stopping hypertension is the Mediterranean Diet (MD), which coincides with the more relevant nutrient levels for hypertension prevention. The MD was identified as a healthy diet when Ancel Keys et al. (1986) established the relationship between the diet and different coronary risk factors in the Seven Countries Study. The results of this study demonstrated the protective effects of the traditional Mediterranean diet on mortality and cardiovascular disease. The MD was ascribed to the list of Intangible Cultural Heritage of UNESCO in November 2010 (decision 5.COM 6.41). However, from 1960 to the present, the prevalence of cardiovascular and metabolic diseases has increased globally, as well as in the north of Europe and in the Mediterranean countries, probably associated to a loss of the healthy lifestyle, including mainly the traditional dietary pattern (Beaglehole, 1999). The main risk factor for developing cardiovascular disease is high blood pressure. Different studies have demonstrated that when subjects adopt a MD diet during at least one year, both systolic and diastolic levels were reduced in individuals with normal blood pressure or mild hypertension; the effect was higher for systolic blood pressure (Nissensohn et al., 2016). The results from the PREDIMED study (Primary Prevention of Cardiovascular Disease with a Mediterranean Diet) showed that the difference in the reduction of blood pressure between each of two MD groups (one supplemented with olive oil and the other with nuts) and a low-fat diet group was similar after two years of study (Estruch et al., 2013). It s interesting to take account of the fact that subjects used olive oil as the main added fat even in the fat-reduced diet, which is in keeping with the high consumption of olive oil in Spain (around 4 L of extra-virgin olive oil per capita in 2013). On the other hand, hypertension is a heterogeneous disorder, and several factors may contribute to its development: endothelial dysfunction, insulin resistance, sodium intake, stress and genetics (Chobanian et al., 2003). Finally, the MD health benefits cannot be attributed to a single food, and several components of this diet have been associated with an effect on blood pressure. The nutrients most relevant for hypertension are fats, such as olive oil and fish oil, as well as the phytochemicals and the antioxidants (Ortega, 2006; Schwingshackl et al., 2011; Widmer et al., 2015).

7.3 EXTRA-VIRGIN OLIVE OIL COMPOSITION

Extra-virgin olive oil is composed primarily of triacylglycerols (99%), free fatty acids, mono- and diacylglycerols and others compounds, such as hydrocarbons, sterols, aliphatic alcohols, tocopherols and pigments. However, the compounds that contribute to the unique character of this oil are a set of phenolic and volatile compounds (Figure 7.1).

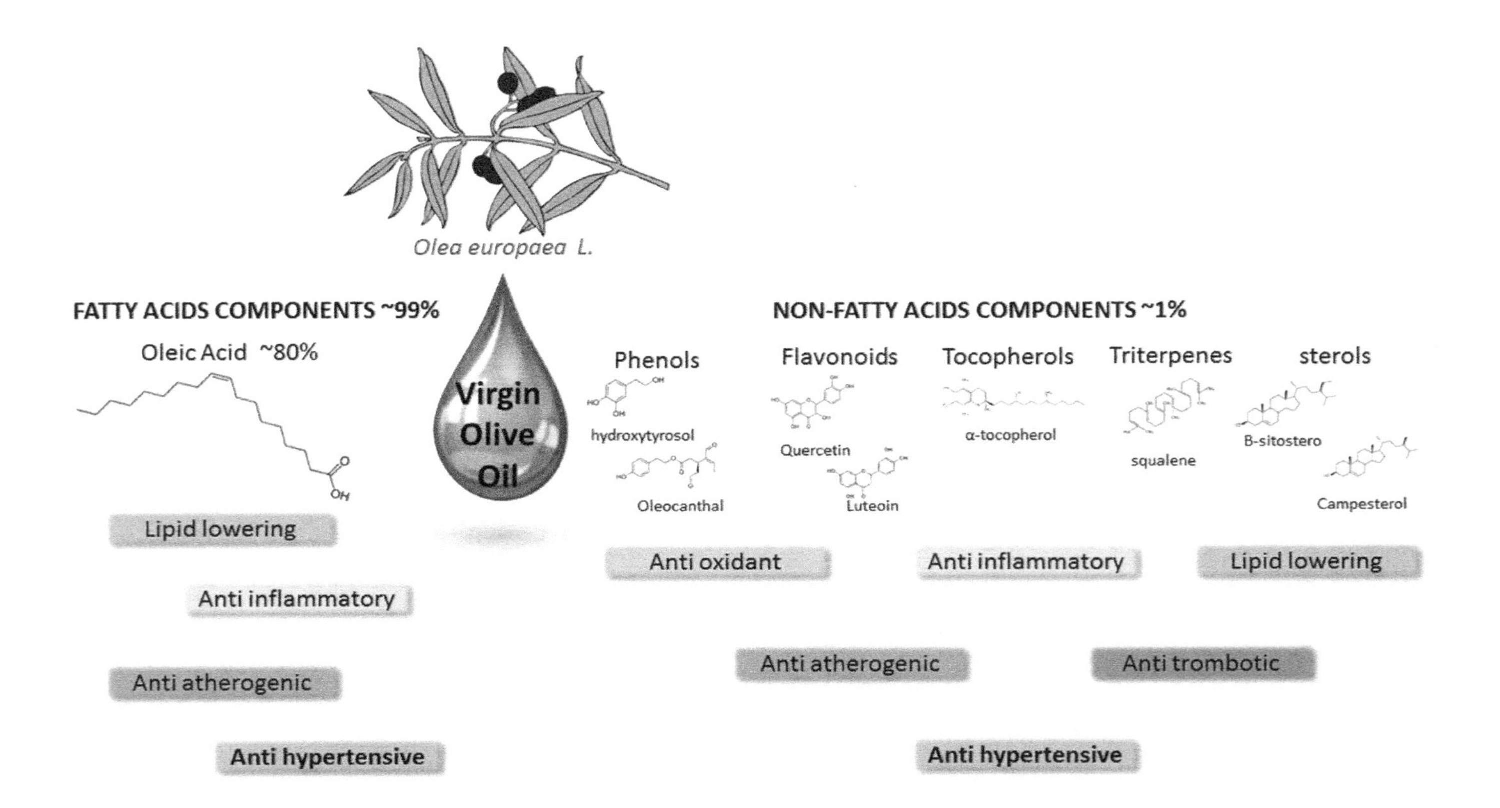

FIGURE 7.1 The most relevant components in extra-virgin olive oil and their main physiological effects related to blood pressure control and the development of hypertension.

The main fatty acids present in olive oil are oleic (C18:1), linoleic (C18:2), linolenic (C18:3), stearic (C18:0), palmitic (C16:0), palmitoleic (C16:1) and myristic (C14:0) acids. Other fatty acids are found in trace amounts (heptadecanoic and eicosanoic) (Scano et al., 1999).

The fatty acid composition of different extra-virgin olive oil may differ depending on the latitude, the climate, the variety and the stage of maturity of the fruit. Oleic acid is the first one to be synthetized in the fruit, and there is an antagonistic relationship between oleic and palmitic, palmitoleic and linoleic acids (Ninni, 1999). Spanish extra-virgin olive oils have a high percentage of oleic acid but are low in linoleic and palmitic.

The presence of partial glycerides (diacylglycerols and monoacylglycerols) in these oils is due either to incomplete triacylglycerol synthesis or hydrolytic reactions. Storage conditions affect the distribution of fatty acids, and their ratio depends on oil acidity (Pérez-Camino et al., 2002; Spyros et al., 2004).

Two hydrocarbons are present in considerable amounts in extra-virgin olive oil (squalene and β-carotene). Squalene is the major constituent of the unsaponifiable matter partially responsible for the beneficial health effects of olive oil and its chemopreventive action against certain cancers (Rao et al., 1998). The main carotenoids present in olive oil are lutein and β-carotene, and the other pigment present is chlorophyll.

The E-vitamer α-tocopherol is found in the free form. The levels reported indicate a wide range of milligrams of α-tocopherol per kg oil that depends on the cultivar and technological factors, and refining or hydrogenation causes loss of tocopherols (Belitz et al., 2004).

The most important aliphatic and aromatic alcohols are fatty alcohols and diterpene alcohols. Alkanols and alkenols with less than ten carbon atoms in their molecules, which are present in free and esterified form, and some aromatic alcohols (benzyl alcohol and 2-phenylethanol) are constituents of the olive oil volatile fraction.

Four classes of sterols can be found in olive oil: common sterols (4-desmethylsterols), 4α-methylsterols, triterpene alcohols (4, 4-dimethylsterols) and triterpene dialcohols. The maincomponents of this sterol fraction are β-sitosterol and campesterol.

The phenolic substances are secondary plant metabolites, chemically characterized by the presence of one or more aromatic rings with one or more hydroxyl substituents (Stefani and Rigacci, 2014). The olive tree (*Olea europaea* L., family *Oleaceae*) produces its own battery of polyphenols that includes flavonols, lignans and glycosides (Rigacci and Stefani, 2016).

These polyphenols are found in the lipid and water fractions of extra-virgin olive oil and include the phenolic alcohols, hydroxytyrosol (3,4-dihydroxyphenylethanol, 3,4-DHPEA) and tyrosol (p-hydroxyphenylethanol, p-HPEA) and their secoiridoid precursors, including oleuropein, the main responsible for the bitter taste of olive leaves and drupes, and oleocanthal, the main responsible for the burning sensation when consuming extra-virgin olive oil (Dinda et al., 2011). Phenolic concentration in extra-virgin olive oil depends on several variables, such as the olive cultivar and the ripening stage of fruit (Ranalli et al., 2009), environmental factors, extraction conditions, systems and storage conditions, and time (Servili and Montedoro, 2002). All these compounds have demonstrated antioxidant and anti-inflammatory activities, which play an important role in the health properties of extra-virgin olive oil.

7.4 EFFECTS OF EXTRA-VIRGIN OLIVE OIL ON BLOOD PRESSURE

Different observational epidemiological and experimental studies support the evidence that olive oil, as the main fat source of the MD, is associated with a reduced risk of cardiovascular disease and a lower prevalence of hypertension (Lopez et al., 2016). The population–based International Study of Macro/Micronutrients and Blood Pressure (INTERMAP), a cross-sectional epidemiological study of 4,680 men and women aged 40–59 years from Japan, the People's Republic of China, the United Kingdom and the United States, has demonstrated that oleic acid intake from vegetable (mainly olive oil) but not animal sources is inversely related to diastolic blood pressure, and this effect was more intense in subjects without pharmacological treatment (Miura et al., 2013).

This observation is in agreement with previous results that indicated positive effects of extra-virgin olive oil intake on hypertension. In a randomized double-blind crossover trial, a daily intake of extra-virgin olive oil (30 g/day in women and 40 g/day in men) was able to significantly reduce systolic and diastolic blood pressure, and reduced by 48% the dosage of pharmacological treatment (Ferrara et al., 2000). This effect of extra-virgin olive oil was also found in a cohort of non-diabetic hypertensive elderly people who reduced their systolic blood pressure after a extra-virgin olive oil intake of 60 g/day during four weeks, getting values comparable to those in normotensive subjects (Perona et al., 2004). The effects of olive oil on blood pressure were also analyzed in a sub-study of the PREDIMED trial. After administration of a MD supplemented with 51 g/day of olive oil during a one-year period, the composition of the plasmatic membrane of erythrocytes from hypertensive participants had undergone changes, with increased levels of phosphatidylethanolamine and decreased levels of lysophosphatidylcholine and sphingomyelin, affecting membrane molecular morphology and packaging (Barceló et al., 2009). Although the evidence derived from human studies also supports the concept that extra-virgin olive oil may exert an antihypertensive activity, these reports are limited.

7.5 EXTRA-VIRGIN OLIVE OIL AND MEMBRANE FUNCTION

The first study which demonstrated the effect of olive oil on hypertension in an animal model was carried out on Dahl salt-sensitive rats (Ganguli et al., 1986). This effect has also been demonstrated in spontaneously hypertensive rats (SHR) after feeding six weeks on a diet with 10% of extra-virgin olive oil. Aortic rings of SHR animals had attenuated phenylephrine-induced contractions and a greater relaxation in response to acetylcholine, which suggested an improvement of endothelial function consistent with a change in membrane composition (Herrera et al., 2001).

The changes in lipid composition have also been established in adipose tissue, where a diet enriched in extra-virgin olive oil and high oleic sunflower normalized the altered composition of triglyceride molecular species and phospholipid fatty acids in SHR compared to animals fed a chow diet (Perona and Ruiz-Gutierrez, 2004).

When Spontaneously Hypertensive Rats (SHR) were fed with high-fat diets containing 15% of refined olive oil, pomace olive oil or pomace olive oil supplemented in oleanolic acid during 12 weeks, no significant differences were found in blood pressure. However, diets rich in pomace olive oil improved endothelial dysfunction in SHR aorta by mechanisms associated with enhanced eNOS expression; these effects were most likely associated with some minor components from pomace olive oil (Rodriguez-Rodriguez et al., 2007). The hypotensive effects of extra-virgin olive oil, triolein and oleic acid have also been studied in SHR rats. Both oleic acid and triolein mimicked the hypotensive effects of extra-virgin olive oil and showed a high impact on biophysical and functional membrane properties of aortas (Terés et al., 2008).

7.6 EXTRA-VIRGIN OLIVE OIL AND THE RENIN-ANGIOTENSIN SYSTEM

Several pathophysiologic factors are involved in the relationship between hypertension and the other components of cardiovascular disease, including inappropriate activation of the renin-angiotensin system (RAS). There is growing evidence that enhanced activation of this hormonal system is a key factor in the development of endothelial dysfunction and hypertension. Systemic and local renin-angiotensin systems interact in the control of blood pressure. The role of intra-renal angiotensin in the regulation of sodium and water balance is well known (Prieto et al., 2003), and brain angiotensin plays a major role in blood pressure control (Banegas et al., 2006). On the other hand, the aminopeptidases that participate in the enzymatic cascade of the RAS also play an important role in blood pressure control, and their study offers new perspectives for the understanding of central blood pressure control and the treatment of hypertension (Banegas et al., 2006).

A high-fat diet enhances the expression of RAS components (Schüler et al., 2017), induces an increase of Angiotensin Converting Enzyme (ACE) serum concentrations (Li et al., 2015) and enhances the hypertensive response to angiotensin II administration. This response is mediated, at least in part, by increased activity of the brain renin-angiotensin system and proinflammatory cytokines (Xue et al., 2016), establishing a link between dietary fat, the RAS, hypertension and cardiovascular disease. However, these effects depend on the kind of fat source. The intake of extra-virgin olive oil, a fat that has been demonstrated to exert beneficial effects on hypertension and cardiovascular disease, also affects several components of plasmatic and local RAS, but these changes are not related to the development of hypertension (Figure 7.2).

When mice fed diets with different degrees of dietary fatty acid saturation, serum total cholesterol levels are higher in mice fed diets containing saturated fat (lard and coconut) than in those fed diets with unsaturated fat (sunflower, fish and olive oil). Serum Aspartyl and Glutamyl Aminopeptidase activities are increased progressively also with the degree of saturation of the dietary fatty acids. Glutamyl and Aspartyl Aminopeptidases play a major role in the regulation of RAS. Glutamyl Aminopeptidase (Angiotensinase A) metabolizes Angiotensin II in Angiotensin III, and Aspartyl Aminopeptidase converts Angiotensin I in Angiotensin 2–10, which is further converted in Angiotensin III by the ACE enzyme. Therefore, the increase of serum Glutamyl and Aspartyl Aminopeptidase activities suggest a heightened metabolism of Ang I and II, which leads to an increase in Ang III levels (Arechaga et al., 2001).

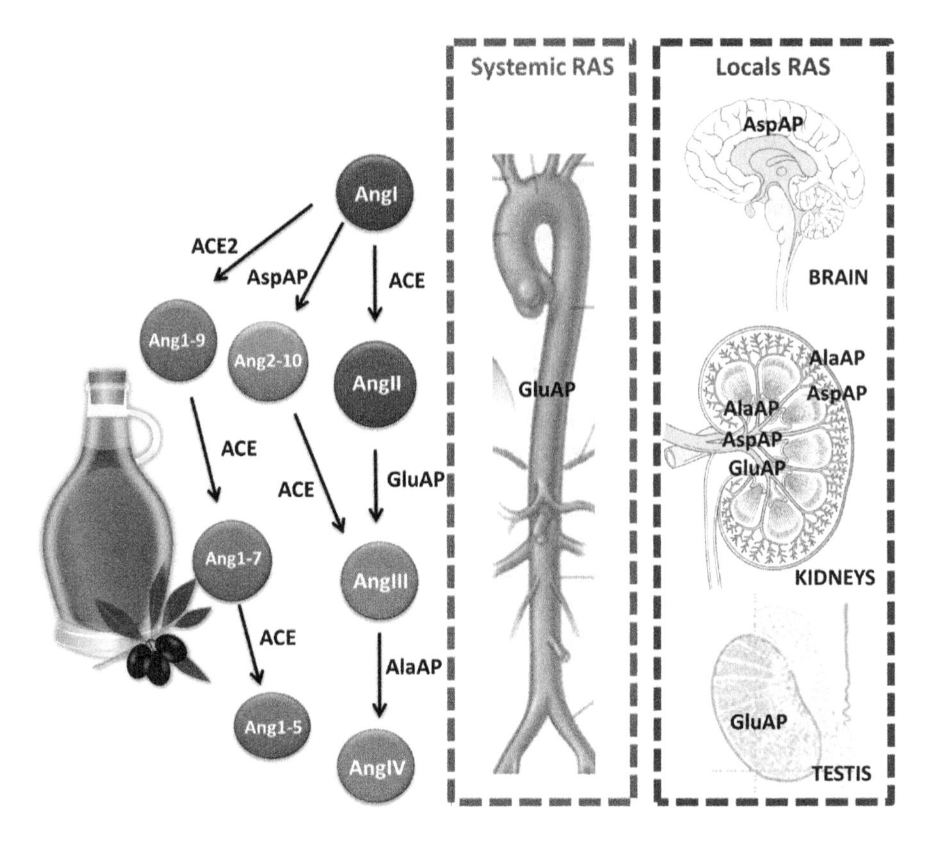

FIGURE 7.2 The Renin-Angiotensin pathway, with the main peptides and enzymes, and the most important effects of extra-virgin olive oil intake on both systemic and locals systems (brain, kidneys and testis). ACE=Angiotensin Converting Enzyme; ACE2=Angiotensin Converting Enzyme 2; GluAP=Glutamyl Aminopeptidase Activity; AlaAP=Alanyl Aminopeptidase Activity; AspAP=Aspartyl Aminopeptidase Activity.

Significant differences were also found in mice fed with a diet enriched in olive oil (20% by weight), where serum Glutamyl Aminopeptidase activity decreased more than 50% compared with animals fed with a standard diet. These changes also affected Glutamyl and Aspartyl Aminopeptidase Activities in different tissues, including lung, adrenal, atrium and testis, indicating functional changes in systemic and locals RAS (Ramírez-Expósito et al., 2001).

It has been proposed that dietary fat composition influences physiologic and metabolic functions modifying the cholesterol-phospholipid composition of cellular membranes. Modifications in the composition and physical properties of the membranes may lead to alterations in the activities of membrane-bound enzymes. These effects have been demonstrated, for example, in testis, where the type of fat used in the diet influences the local activities of Glutamyl and Aspartyl Aminopeptidase and, therefore, may influence the autocrine/paracrine functions of locally synthesized angiotensin (Arechaga et al., 2002). More recently, Domínguez-Vías et al. (2017) have demonstrated in male Wistar rats that a diet enriched (20%) in butter plus cholesterol increased membrane-bound Glutamyl Aminopeptidase activity in testis after 24 weeks, and significant and positive correlations were established between plasma triglycerides and total cholesterol. In contrast, a diet enriched in extra-virgin olive oil increased soluble DPPIV (T-cell activation antigen CD36), an enzymatic activity that influences testis development/spermatogenesis by regulating the immune states, and is considered as co-stimulator of T-cells to participate in immunoreaction.

The kind of fat in the diet also modifies the profile of fatty acids in the brain and affects aminopeptidase activities in this tissue, and a relationship has been described between specific brain fatty acid changes and angiotensin metabolism in adult male rats (Segarra et al., 2011). After an experimental period with different fat sources (fish, olive or coconut oil), the distribution of fatty acids in the frontal cortex reflects the composition of the diets. Although there were no significant differences in Aminopeptidase Activities between the three diets, highly significant correlations between fatty acids in the cortex and soluble and membrane Aminopeptidase Activities were found. Fish and coconut oils particularly affected the correlation of fatty acids with membrane-bound Glutamyl Aminopeptidase Activity, whereas olive oil influenced Aspartyl Aminopeptidase Activity.

Nevertheless, the main local RAS implicated in the control of blood pressure is the intra-renal RAS. Villarejo et al. demonstrated in 2015 that the intake of a diet enriched in extra-virgin olive oil (20%) during 12 weeks delayed the progressive increase in systolic blood pressure in male SHR rats, compared with a standard chow diet. This effect was related to changes in local renal aminopeptidases, suggesting an opposite response between the renal cortex and medulla: while in the renal cortex the AngII effects may be counterbalanced, in the renal medulla it may be potentiated. The lower levels in systolic blood pressure in the extra-virgin olive oil were parallel with lower amounts of urine nitric oxide and 8-isoprostanes, which may suggest a decrease in the levels of oxidative stress in these animals, probably associated with the effect of different minor components of extra-virgin olive oil, such as hidroxytyrosol and oleouropeine.

The role of the gut microbiota in the development of hypertension has also been related to the intake of high-fat diets. However, the fat source seems to be more important than the amount. Swiss Webster mice fed with standard chow or two high-fat diets enriched with extra-virgin olive oil or butter, respectively, showed a distinctive pattern on the intestinal microbiome as indicated by next-generation sequencing. The butter diet elicited the highest values of systolic blood pressure, correlating positively with the percentage of *Desulfovibrio* sequences in feces, which in turn showed significantly higher mean values in the butter diet than in the extra-virgin olive oil diet (Prieto et al., 2018). Additionally, the beneficial effect of extra-virgin olive oil on the development of hypertension in SHR rats is associated to specific changes in gut microbiota, with significant differences between a standard chow diet and a diet enriched in extra-virgin olive oil, when selected bacterial groups were quantified by RT-PCR using *Lactobacillus, clostridia XIV* and universal primers. Interestingly, the results suggested an increase in the microbial diversity of the feces of the SHR rats fed the extra-virgin olive oil diet, and a significant inverse correlation between the abundance of *Clostridia XIV* and the systolic blood pressure values (Hidalgo et al., 2018). These results indicate a link between the diet and the prevalence of some taxa in the gut microbiome, and support the possibility that, at least in part, the effect of extra-virgin olive oil on blood pressure may be mediated by the modulation of intestinal microbiota.

REFERENCES

Appel LJ, Moore TJ, Obarzanek E, Vollmer WM, Svetkey LP, Sacks FM, Bray GA, et al. A clinical trial of the effects of dietary patterns on blood pressure. DASH Collaborative Research Group. *N. Engl. J. Med.* 1997; 336(16):1117–24.

Appel LJ, Champagne CM, Harsha DW, Cooper LS, Obarzanek E, Elmer PJ, Stevens VJ, et al. Effects of comprehensive lifestyle modification on blood pressure control: Main results of the PREMIER clinical trial. *JAMA* 2003; 289(16):2083–93.

Appel LJ, Brands MW, Daniels SR, Karanja N, Elmer PJ, Sacks FM, American Heart Association. Dietary approaches to prevent and treat hypertension: A scientific statement from the American Heart Association. *Hypertension* 2006; 47(2):296–308.

Arechaga G, Martínez JM, Prieto I, Ramírez MJ, Sánchez MJ, Alba F, De Gasparo M, Ramírez M. Serum aminopeptidase A activity of mice is related to dietary fat saturation. *J. Nutr.* 2001; 131(4):1177–9.

Arechaga G, Prieto I, Segarra AB, Alba F, Ruiz-Larrea MB, Ruiz-Sanz JI, de Gasparo M, Ramirez M. Dietary fatty acid composition affects aminopeptidase activities in the testes of mice. *Int. J. Androl.* 2002; 25(2):113–8.

Asemi Z, Samimi M, Tabassi Z, Sabihi SS, Esmaillzadeh A. A randomized controlled clinical trial investigating the effect of DASH diet on insulin resistance, inflammation, and oxidative stress in gestational diabetes. *Nutrition* 2013a; 29(4):619–24.

Asemi Z, Tabassi Z, Samimi M, Fahiminejad T, Esmaillzadeh A. Favourable effects of the Dietary Approaches to Stop Hypertension diet on glucose tolerance and lipid profiles in gestational diabetes: A randomised clinical trial. *Br. J. Nutr.* 2013b; 109(11):2024–30.

Azadbakht L, Mirmiran P, Esmaillzadeh A, Azizi T, Azizi F. Beneficial effects of a Dietary Approaches to Stop Hypertension eating plan on features of the metabolic syndrome. *Diabetes Care* 2005; 28(12):2823–31.

Azadbakht L, Surkan PJ, Esmaillzadeh A, Willett WC. The Dietary Approaches to Stop Hypertension eating plan affects C-reactive protein, coagulation abnormalities, and hepatic function tests among type 2 diabetic patients. *J. Nutr.* 2011; 141(6):1083–8.

Banegas I, Prieto I, Vives F, Alba F, de Gasparo M, Segarra AB, Hermoso F, Durán R, Ramírez M. Brain aminopeptidases and hypertension. *J. Renin Angiotensin Aldosterone Syst.* 2006; 7(3):129–34.

Barceló F, Perona JS, Prades J, Funari SS, Gomez-Gracia E, Conde M, Estruch R, Ruiz-Gutiérrez V. Mediterranean-style diet effect on the structural properties of the erythrocyte cell membrane of hypertensive patients: The Prevencion con Dieta Mediterranea Study. *Hypertension* 2009; 54(5):1143–50.

Beaglehole R. International trends in coronary heart disease mortality and incidence rates. *J. Cardiovasc. Risk* 1999; 6(2):63–8.

Beaglehole R, Bonita R, Horton R, Ezzati M, Bhala N, Amuyunzu-Nyamongo M, Mwatsama M, Reddy KS. Measuring progress on NCDs: One goal and five targets. *Lancet* 2012; 380(9850):1283–5.

Belitz H-D, Grosch W, Schieberle P (eds), *Food Chemistry* (third edition). Berlin: Springer Verlag. 2004.

Blumenthal JA, Babyak MA, Sherwood A, Craighead L, Lin PH, Johnson J, Watkins LL, et al. Effects of the dietary approaches to stop hypertension diet alone and in combination with exercise and caloric restriction on insulin sensitivity and lipids. *Hypertension* 2010; 55(5):1199–205.

Chen ST, Maruthur NM, Appel LJ. The effect of dietary patterns on estimated coronary heart disease risk: Results from the Dietary Approaches to Stop Hypertension (DASH) trial. *Circ. Cardiovasc. Qual. Outcomes* 2010; 3(5):484–9.

Chobanian AV, Bakris GL, Black HR, Cushman WC, Green LA, Izzo JL Jr, Jones DW, et al. Seventh report of the Joint National Committee on Prevention, Detection, Evaluation, and Treatment of High blood pressure. *Hypertension* 2003; 42(6):1206–52.

Dinda B, Debnath S, Banik R. Naturally occurring iridoids and secoiridoids. An updated review, part 4. *Chem. Pharm. Bull.* 2011; 59(7):803–33.

Domínguez-Vías G, Segarra AB, Martínez-Cañamero M, Ramírez-Sánchez M, Prieto I. Influence of a virgin olive oil versus butter plus cholesterol-enriched diet on testicular enzymatic activities in adult male rats. *Int. J. Mol. Sci.* 2017; 18(8). pii: E1701.

Estruch R, Ros E, Martínez-González MA. Mediterranean diet for primary prevention of cardiovascular disease. *N. Engl. J. Med.* 2013; 369(7):676–7.

Ferrara LA, Raimondi AS, d'Episcopo L, Guida L, Dello Russo A, Marotta T. Olive oil and reduced need for antihypertensive medications. *Arch. Intern. Med.* 2000; 160(6):837–42.

Ganguli MC, Tobian L, Iwai J. Reduction of blood pressure in salt-fed Dahl salt-sensitive rats with diets rich in olive oil, safflower oil or calcium biphosphate but not with calcium carbonate. *J. Hypertens. Suppl.* 1986; 4(5):S168–9.

Herrera MD, Pérez-Guerrero C, Marhuenda E, Ruiz-Gutiérrez V. Effects of dietary oleic-rich oils (virgin olive and high-oleic-acid sunflower) on vascular reactivity in Wistar-Kyoto and spontaneously hypertensive rats. *Br. J. Nutr.* 2001; 86(3):349–57.

Hidalgo M, Prieto I, Abriouel H, Villarejo AB, Ramírez-Sánchez M, Cobo A, Benomar N, Gálvez A, Martínez-Cañamero M. Changes in gut microbiota linked to a reduction in systolic blood pressure in spontaneously hypertensive rats fed an extra virgin olive oil-enriched diet. *Plant Foods Hum. Nutr.* 2018. Doi:10.1007/s11130-017-0650-1.

Keys A, Menotti A, Karvonen MJ, Aravanis C, Blackburn H, Buzina R, Djordjevic BS, et al. The diet and 15-year death rate in the seven countries study. *Am. J. Epidemiol.* 1986; 124(6):903–15.

Lackland DT, Roccella EJ, Deutsch AF, Fornage M, George MG, Howard G, Kissela BM, et al. Factors influencing the decline in stroke mortality: A statement from the American Heart Association/American Stroke Association. *Stroke* 2014; 45(1):315–53.

Lackland DT, Weber MA. Global burden of cardiovascular disease and stroke: Hypertension at the core. *Can. J. Cardiol.* 2015; 31(5):569–71.

Li C, Culver SA, Quadri S, Ledford KL, Al-Share QY, Ghadieh HE, Najjar SM, Siragy HM. High-fat diet amplifies renal renin angiotensin system expression, blood pressure elevation, and renal dysfunction caused by Ceacam1 null deletion. *Am. J. Physiol. Endocrinol. Metab.* 2015; 309(9):E802–10.

Lopes HF, Martin KL, Nashar K, Morrow JD, Goodfriend TL, Egan BM. DASH diet lowers blood pressure and lipid-induced oxidative stress in obesity. *Hypertension* 2003; 41(3):422–30.

Lopez S, Bermudez B, Montserrat-de la Paz S, Jaramillo S, Abia R, Muriana FJ. Virgin olive oil and hypertension. *Curr. Vasc. Pharmacol.* 2016; 14(4):323–9.

Miura K, Stamler J, Brown IJ, Ueshima H, Nakagawa H, Sakurai M, Chan Q, et al. Relationship of dietary monounsaturated fatty acids to blood pressure: The International Study of Macro/Micronutrients and blood pressure. *J. Hypertens.* 2013; 31(6):1144–50.

Moore TJ, Svetkey L, Appel L, George B, Vollmer W. 2001. *The DASH Diet for Hypertension.* NY, New York: Simon & Schuster.

Ninni V. A statistical approach to the biosynthetic route of the fatty acids in olive oil: Crosssectional and time series analyses. *J.Sci.Food Agric.* 1999; 79:2113–21.

Nissensohn M, Román-Viñas B, Sánchez-Villegas A, Piscopo S, Serra-Majem L. The effect of the Mediterranean diet on hypertension: A systematic review and meta-analysis. *J. Nutr. Educ. Behav.* 2016; 48(1):42–53.e1.

Nowson CA, Worsley A, Margerison C, Jorna MK, Frame AG, Torres SJ, Godfrey SJ. Blood pressure response to dietary modifications in free-living individuals. *J. Nutr.* 2004; 134(9):2322–9.

Nowson CA, Worsley A, Margerison C, Jorna MK, Godfrey SJ, Booth A. Blood pressure change with weight loss is affected by diet type in men. *Am. J. Clin. Nutr.* 2005; 81(5):983–9.

Ortega R. Importance of functional foods in the Mediterranean diet. *Public Health Nutr.* 2006; 9(8A):1136–40.

Pearson TA. Cardiovascular disease in developing countries: Myths, realities, and opportunities. *Cardiovasc. Drugs Ther.* 1999; 13(2):95–104.

Pérez-Camino MC, Moreda W, Mateos R, Cert A. Determination of esters of fatty acids with low molecular weight alcohols in olive oils. *J. Agric. Food Chem.* 2002; 50(16):4721–5.

Perona JS, Ruiz-Gutierrez V. Virgin olive oil normalizes the altered triacylglycerol molecular species composition of adipose tissue in spontaneously hypertensive rats. *J. Agric. Food Chem.* 2004; 52(13):4227–33.

Perona JS, Cañizares J, Montero E, Sánchez-Domínguez JM, Catalá A, Ruiz-Gutiérrez V. Virgin olive oil reduces blood pressure in hypertensive elderly subjects. *Clin. Nutr.* 2004; 23(5):1113–21.

Poulter NR, Prabhakaran D, Caulfield M. Hypertension. *Lancet* 2015; 386(9995):801–12.

Prieto I, Hermoso F, de Gasparo M, Vargas F, Alba F, Segarra AB, Banegas I, Ramírez M. Aminopeptidase activity in renovascular hypertension. *Med. Sci. Monit.* 2003; 9(1):BR31-6.

Prieto I, Hidalgo M, Segarra AB, Martínez-Rodríguez AM, Cobo A, Ramírez M, Abriouel H, Gálvez A, Martínez-Cañamero M. Influence of a diet enriched with virgin olive oil or butter on mouse gut microbiota and its correlation to physiological and biochemical parameters related to metabolic syndrome. PLoS One. 2018 Jan 2;13(1):e0190368. doi: 10.1371/journal.pone.0190368. eCollection 2018.

Ramírez-Expósito MJ, Martínez-Martos JM, Prieto I, Alba F, Ramírez M. Angiotensinase activity in mice fed an olive oil-supplemented diet. *Peptides* 2001; 22(6):945–52.

Ranalli A, Marchegiani D, Contento S, Girardi F, Nicolosi MP, Brullo MD. Variations of the iridoid oleuropein in Italian olive varieties during growth and maturation. *Eur. J. Lipid Sci. Technol.* 2009; 111:678–87.

Rao CV, Newmark HL, Reddy BS. Chemopreventive effect of squalene on colon cancer. *Carcinogenesis* 1998; 19(2):287–90.

Rigacci S, Stefani M. Nutraceutical properties of olive oil polyphenols. An itinerary from cultured cells through animal models to humans. *Int. J. Mol. Sci.* 2016; 17(6). pii:E843.

Rodriguez-Rodriguez R, Herrera MD, de Sotomayor MA, Ruiz-Gutierrez V. Pomace olive oil improves endothelial function in spontaneously hypertensive rats by increasing endothelial nitric oxide synthase expression. *Am. J. Hypertens.* 2007; 20(7):728–34.

Rothwell PM, Coull AJ, Giles MF, Howard SC, Silver LE, Bull LM, Gutnikov SA, et al. Change in stroke incidence, mortality, case-fatality, severity, and risk factors in Oxfordshire, UK from 1981 to 2004 (Oxford Vascular Study). *Lancet* 2004; 363(9425):1925–33.

Rothwell PM, Howard SC, Dolan E, O'Brien E, Dobson JE, Dahlöf B, Sever PS, Poulter NR. Prognostic signifi cance of visit-to-visit variability, maximum systolic blood pressure, and episodic hypertension. *Lancet* 2010; 375(9718):895–905.

Sacks FM, Obarzanek E, Windhauser MM, Svetkey LP, Vollmer WM, McCullough M, Karanja N, et al. Rationale and design of the Dietary Approaches to Stop Hypertension trial (DASH). A multicenter controlled-feeding study of dietary patterns to lower blood pressure. *Ann. Epidemiol.* 1995; 5(2):108–18.

Scano P, Casu M, Lai A, Saba G, Dessi MA, Deiana M, Corongiu FP, Bandino G. Recognition and quantitation of cis-vaccenic and EicosenoicFatty acids in olive oils by C-13 nuclear magnetic resonance spectroscopy. *Lipids* 1999; 34(7):757–9.

Schüler R, Osterhoff MA, Frahnow T, Seltmann AC, Busjahn A, Kabisch S, Xu L, et al. High-saturated-fat diet increases circulating angiotensin-converting enzyme, which is enhanced by the rs4343 polymorphism defining persons at risk of nutrient-dependent increases of blood pressure. *J. Am. Heart Assoc.* 2017; 6(1). pii: e004465.

Schwingshackl L, Strasser B, Hoffmann G. Effects of monounsaturated fatty acids on cardiovascular risk factors: A systematic review and meta-analysis. *Ann. Nutr. Metab.* 2011; 59(2–4):176–86.

Segarra AB, Ruiz-Sanz JI, Ruiz-Larrea MB, Ramírez-Sánchez M, de Gasparo M, Banegas I, Martínez-Cañamero M, Vives F, Prieto I. The profile of fatty acids in frontal cortex of rats depends on the type of fat used in the diet and correlates with neuropeptidase activities. *Horm. Metab. Res.* 2011; 43(2):86–91.

Servili M, Montedoro GF. Contribution of phenolic compounds in virgin olive oil quality. *Eur. J. Lipid Sci. Technol.* 2002; 104(9–10):602–13.

Shirani F, Salehi-Abargouei A, Azadbakht L. Effects of Dietary Approaches to Stop Hypertension (DASH) diet on some risk for developing type 2 diabetes: A systematic review and meta-analysis on controlled clinical trials. *Nutrition* 2013; 29(7–8):939–47.

Siervo M, Lara J, Chowdhury S, Ashor A, Oggioni C, Mathers JC. Effects of the Dietary Approach to Stop hypertension (DASH) diet on cardiovascular risk factors: A systematic review and meta-analysis. *Br. J. Nutr.* 2015; 113(1):1–15.

Spyros A, Philippidis A, Dais P. Kinetics of diglyceride formation and isomerization in virgin olive oils by employing 31P NMR spectroscopy. Formulation of a quantitative measure to assess olive oil storage history. *J. Agric. Food Chem.* 2004; 52(2):157–64.

Stefani M, Rigacci S. Beneficial properties of natural phenols: Highlight on protection against pathological conditions associated with amyloid aggregation. *BioFactors* 2014; 40(5):482–93.

Tabassum N, Ahmad F. Role of natural herbs in the treatment of hypertension. *Pharmacogn. Rev.* 2011; 5(9):30–40.

Terés S, Barceló-Coblijn G, Benet M, Alvarez R, Bressani R, Halver JE, Escribá PV. Oleic acid content is responsible for the reduction in blood pressure induced by olive oil. *Proc. Natl. Acad. Sci. U. S. A.* 2008; 105(37):13811–6.

Tibazarwa KB, Damasceno AA. Hypertension in developing countries. *Can. J. Cardiol.* 2014; 30(5):527–33.

Villarejo AB, Ramírez-Sánchez M, Segarra AB, Martínez-Cañamero M, Prieto I. Influence of extra virgin olive oil on blood pressure and kidney angiotensinase activities in spontaneously hypertensive rats. *Planta Med.* 2015; 81(8):664–9.

Webb AJS, Fischer U, Mehta Z, Rothwell PM. Effects of antihypertensive-drug class on interindividual variation in blood pressure and risk of stroke: A systematic review and meta-analysis. *Lancet* 2010; 375(9718):906–15.

Widmer RJ, Flammer AJ, Lerman LO, Lerman A. The Mediterranean diet, its components, and cardiovascular disease. *Am. J. Med.* 2015; 128(3):229–38.

Xue B, Yu Y, Zhang Z, Guo F, Beltz TG, Thunhorst RL, Felder RB, Johnson AK. Leptin mediates high-fat diet sensitization of angiotensin II-elicited hypertension by upregulating the brain renin-angiotensin system and inflammation. *Hypertension* 2016; 67(5):970–6.

8 Role of Herbs and Spices in Cardiovascular Health

Haroon Khan, Ada Popolo, Marya and Luca Rastrelli

CONTENTS

8.1 INTRODUCTION

Cardiovascular diseases (CVDs) are considered the leading cause of morbidity and mortality in developed and developing countries. According to the World Health Organization (WHO), every year about 17.1 million people die from cardiovascular problems like heart attack, stroke and heart failure, and it is estimated that the occurrence of deaths may increase to 23.6 million by 2030 (WHO, 2009). CVDs are mainly associated with various factors such as elevated serum total cholesterol, raised LDL and an increase in LDL oxidation, high platelet aggregation, hypertension and smoking (Rahman and Lowe, 2006). Augmented oxidative stress also plays a crucial role in the pathophysiology of CVDs such as atherosclerosis, heart failure, hypertension, cardiac hypertrophy and ischemia-reperfusion (Cave et al., 2006). A number of bioactive compounds obtained from terrestrial plants may reduce the risk of cardiovascular diseases due to their anti-inflammatory, anti-ischemic, anti-angiogenic, antioxidative, anti-hypercholesterolemic and antiplatelet activities (Pastor-Villaescusa et al., 2015).

There are several synthetic therapeutic agents currently available for the treatment of different cardiovascular disorders, but most of them have undesirable side effects, are inefficient and have pharmacokinetic problems. Moreover, the interaction of cardiovascular medication with other drugs limits the use of certain medications. The identification of alternative therapies with several clinical applications, efficacy and a reasonable safety profile is the target of several research groups around the world (Rangel-Huerta et al., 2015).

About 30% of known drugs are based on natural products (Xie et al., 2015). Research on natural products continues to explore a variety of lead structures that may be used as templates by pharmaceutical industries for the development of new drugs (de Fátima et al., 2014). It is important to mention here that some success has been achieved with natural products for the treatment of a variety of CVDs. Therefore, numerous research groups are focusing their attention on exploring natural products for that purpose.

8.2 ROLE OF FRUITS AND VEGETABLES IN CVDS

Sufficient consumption of fruits and vegetables efficiently decreases the risk of cardiovascular diseases, as fruits and vegetables are a rich sources of flavonoids which have significant ability in the protection of coronary heart diseases (Wang et al., 2011). For centuries, soybeans have been used extensively as food and medicine in Asia. Epidemiologists have long noted that Asian populations who consume soy foods as a dietary staple have a lower incidence of CVDs than those who consume a typical Western diet (D'Adamo and Sahin, 2014). Soy contains various classes of bioactive compounds, isoflavones being particularly important. The three major isoflavones found in soybeans are genistein, daidzein and glycitein. These isoflavones, which are weak estrogens, may act as antiestrogens by competing with the most potent naturally occurring endogenous estrogens, 17β-estradiol, for binding to the estrogen receptor and reduced risk of cardiovascular diseases like atherosclerosis (González Cañete and Durán Agüero, 2014). Researchers analyzed the functional and anatomopathological effects of soybean extract and isoflavones on post-myocardial ischemia (MI) and observed a protective effect 30 days after the MI in the soybean extract group (Miguez et al., 2012) (Table 8.1).

Isoflavones have been shown to mitigate hypertension by targeting mechanisms involving vasodilation; in particular, interaction with the estrogen-response element of genes related to endothelial nitric oxide (NO) synthase increases endogenous NO production, which improves brachial artery flow (Jackson et al., 2011). However, data of epidemiological studies are controversial and have not clarified the correlation between soy intake and prevention of CVD mortality In fact, it appears that different soy foods may vary in their biological efficacy and protective effects. For example, soy isoflavones for their estrogenic activity may protect against the sharp rise in CVD incidence after menopause, when endogenous estrogen concentrations are depleted (Vitale et al., 2013). It has been shown that in postmenopausal women, six months of isoflavone supplementation improves endothelial vasodilation and results in a significant reduction in cellular adhesion molecules such as Intercellular Adhesion Molecule 1, Vascular Cell Adhesion Protein 1 and E-selectin (Colacurci et al., 2005).

Furthermore, there is emerging evidence on the diverging biological effects among individuals capable of metabolizing the soy isoflavone daidzein to the more bioactive metabolite equol. For example, in exploring the effect of soy on CVD risk factors, it was found that blood pressure, plasma triglycerides (TGs) and C-reactive protein (CRP) concentrations were significantly lowered among equol- and *O*-desmethylangolensin (ODMA)-producing pre-hypertensive, postmenopausal

TABLE 8.1
List of Fruits and Vegetables that Are Used for the Treatment of CVD with Possible Underlying Mechanisms

Botanical Origin	Common Name	Effects	Mechanism(s)	References
Glycine ussuriensis (Fabaceae)	Soybean	Atherosclerosis	Prevent atherosclerosis by binding to the estrogen receptor	Nagata (2000)
Solanum lycopersicum (Solanaceae)	Tomatoes	Hypertension	Reduced systolic and diastolic blood pressure	Paran et al. (2009)
Theobroma cacao (Malvaceae)	Cocoa	Antihyperlipidemic, anti-inflammatory and antiplatelet activity	Increase HDL-cholesterol level, decrease glucose and IL-1β level	Sarriá et al. (2015)
Vitis vinifera (Vitaceae)	Grapes	Atherosclerosis	Inhibit LDL oxidation and platelet aggregation, improve endothelial function, reduce blood pressure, reduce inflammation	Castilla et al. (2006)
Punica granatum (Punicaceae)	Pomegranate	Hypertension	Reduce systolic and diastolic blood pressure	Sahebkar et al. (2017)
Morusnigra (Moraceae)	Blackberry	Prevent endothelial cells damage	Protect against peroxynitrite-induced oxidative stress	Serraino et al. (2003)
Morus alba L. (Moraceae)	Mulberry	Antiatherogenic antioxidative	Inhibit foam cell formation and LDL oxidation	Liu et al. (2008)
Vaccinium corymbosum and *Vaccinium macrocarpon* (Ericaceae)	Blueberry and cranberry	Anti-inflammatory	Reduced TNF-α induced upregulation of inflammatory mediators in human microvascular endothelial cells	Youdim et al. (2002)
Vaccinium corymbosum (Ericaceae)	Blueberry	Anti-inflammatory	Inhibit iNOS expression Inhibit foam cell formation	Pergola et al. (2006)
Prunus virginiana (Rosaceae)	Chokeberry	Anti-hyperlipidemic	Enhance lipid metabolism	Valcheva-Kuzmanova et al. (2007)
Vaccinium corymbosum (Ericaceae) and *Fragaria × ananassa* (Rosaceae)	Blueberry and strawberry	Prevent dyslipidemia human umbilical vein endothelial cells damage	Recruitment of TNF receptor associated factors, inhibition of pro-inflammatory signaling	Prior et al. (2009) Xia et al. (2007)

Chinese women, compared to non-producers. In addition, racial differences in the ability to convert daidzein into equol have been reported (Liu et al., 2014).

Another possible explanation for the hypotensive effects could be the fact that soy foods and legumes in general are rich in arginine, which is a precursor of NO in the l-arginine-nitric oxide pathway (Moncada and Higgs, 1993). It is thought that arginine aids in regulation of blood pressure through increased production and improved NO bioavailability in the vascular endothelium (Dong et al., 2011). Two meta-analyses have concluded that supplementation with l-arginine significantly improves blood pressure and endothelial function in adults (Dong et al., 2011; Bai et al., 2009). Hypotensive activity of soy-containing foods could also have been related to the increased arginine content relative to lysine. In fact, both amino acids compete for the same transporter in the intestinal lumen, so increased lysine relative to arginine could limit uptake of the latter and thus affect its bioconversion and consequent downstream hypotensive effects (Vasdev et al., 2008). Although several studies have come to conflicting conclusions, it is clear that the hypotensive effect of isoflavones is best achieved in persons with established hypertension (Ramdath et al., 2017). Much of the focus on soy has been directed toward the hypocholesterolemic properties of bioactive peptides in soy protein, which exert their effects primarily through mechanisms involving the LDL-C receptor (LDLR), cholecystokinin secretion and consequent bile acid regulation (Torres et al., 2006; Maki et al., 2010).

Since it was recognized that isoflavones could be the bioactive component attributed to soy protein, isoflavones lacking soy bean was studied in seven trials and compared with casein or milk protein, or various animal proteins. Two of the studies showed small significant decreases in LDL cholesterol (Gil-Izquierdo et al., 2012). These studies were very carefully controlled feeding studies, all meals being formulated according to strict nutritional specifications, and complete meals were provided to the participants. They were specifically designed to sort out the effects of the protein from the isoflavones and showed an effect of protein but not of isoflavones on LDL cholesterol (Sacks et al., 2006).

Nevertheless, there exists some controversy about the hypocholesterolemic effect of soy protein.

Although the lipid effects of soy protein are much more modest than initially reported, they are still relevant clinically. At population levels, as epidemiologic and intervention data suggest, for every 1% reduction in LDL-cholesterol, there is a corresponding 1%–2% reduction in cardiovascular events (CVEs), and for every 2%–3% increase in HDL-cholesterol, there is a reduction in CVEs by 2%–4% that is independent of LDL-cholesterol (Ramdath et al., 2017).

Several meta-analyses, all reported by Ramdath and co-workers, support the beneficial effects of soy in cholesterol lowering and have culminated in the U.S. Food and Drug Administration's approval of a health claim that "25 g of soy protein a day, as a part of a diet low in saturated fat and cholesterol, may reduce the risk of cardiovascular disease" (Ramdath et al., 2017). Red fruits and vegetables like tomatoes, watermelon, apricot, pink guava and grapefruits are rich sources of lycopene, which is a dietary carotenoid. It is actually the most potent antioxidant related with decreased risk of chronic diseases like CVDs (Nguyen and Schwartz, 1999). A clinical trial was conducted by adding tomato extracts to the treatment regimen of hypertensive patients. The findings revealed that the combination therapy of tomato extracts with low-dose ACE inhibitors, inhibitors of calcium channels or with low-dose diuretics showed significant decrease in systolic and diastolic blood pressure by more than 10 mmHg and 5 mmHg, respectively. Besides this, the compliance level of patients was high and no side effects were recorded (Paran et al., 2009).

Epidemiological studies provide indisputable evidence supporting the direct role of lycopene in prevention of CVDs. In fact, the Framingham Heart Offspring Study (Jacques et al., 2013), indicates that there is a strong inverse association between lycopene intake and incidence of myocardial infarction, angina pectoris and coronary insufficiency. Furthermore, low plasma lycopene levels were reported in hypertension, myocardial infarction, stroke and atherosclerosis. For this reason, decreased α-carotene and lycopene concentrations in the blood have recently been proposed as possible criteria for prognosis of public health in the United States. According to the results of meta-analysis, a plasma carotenoid level <1 μM translates into a very high risk of health consequences. Moderate health risk is proclaimed to be associated with carotenoid concentrations in the range 1.5–2.5 μM. Values for carotenoid concentrations from 2.5 to 4 μM suggest a moderate risk, whereas carotenoid concentrations over 4 μM are proposed to have the lowest risk of health consequences. According to the same report, over 95% of the US population falls into the moderate or high risk category of the carotenoid health index (Petyaev, 2016).

In another study, the cocoa products that are rich in dietary products significantly improved cardiovascular health by increasing HDL-cholesterol level and by decreasing glucose and IL-1β level (Sarriá et al., 2015). Cocoa and chocolate have also been found to be rich sources of antioxidant flavonoids, mainly (–)-*epi*catechin and other flavan-3-ols (Figure 8.1).

Thus, cocoa products play a crucial role in cardiovascular protection by decreasing inflammation, inhibiting platelet activity and reducing blood pressure and vasodilation (Engler and Engler, 2006).

In a review of 28 human intervention studies conducted between 2000 and 2007 on the effect of cocoa consumption (Cooper et al., 2008), the main outcomes were improved endothelial function, decreased susceptibility of low-density lipoprotein (LDL) to oxidation, inhibition of platelet aggregation and activation and decreased levels of F2-isoprostanes. In many studies, regular consumption of cocoa, or a flavanol extract of cocoa, reduced blood pressure, blood cholesterol, F2-isoprostanes and susceptibility of LDL to oxidation. At the molecular level, many of the effects are mediated through interactions with nitric oxide metabolism in the blood vessel endothelium (Kerimi and Williamson, 2015), improving endothelial dysfunction, increasing vasodilation and lowering blood pressure. Repeated administration of high-flavanol cocoa produces a longer-term effect that is characterized by an increase in the baseline level of brachial artery flow-mediated dilation (FMD). FMD value largely reflects NO-mediated arterial function (including that of coronary arteries), and low levels are associated with cardiovascular events and elevated risk factors for CVDs. This effect can be mediated by changes in gene expression and protein synthesis (endothelial nitric oxide synthase, eNOS) (Schewe et al., 2008). Besides the improvement of NO bioavailability and bioactivity, the positive endothelial effects of cocoa can be correlated with other mechanisms. Modulation of PGI_2 and

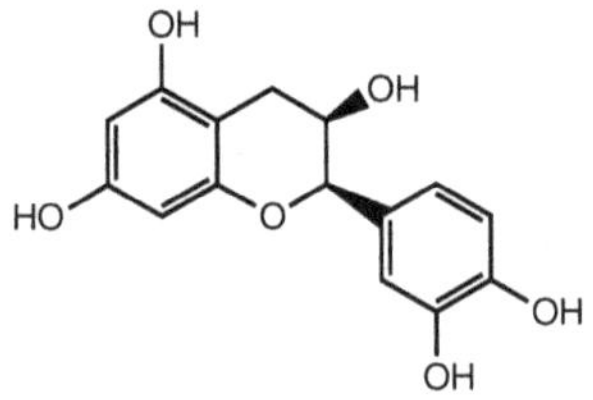

(-) Epicatechin 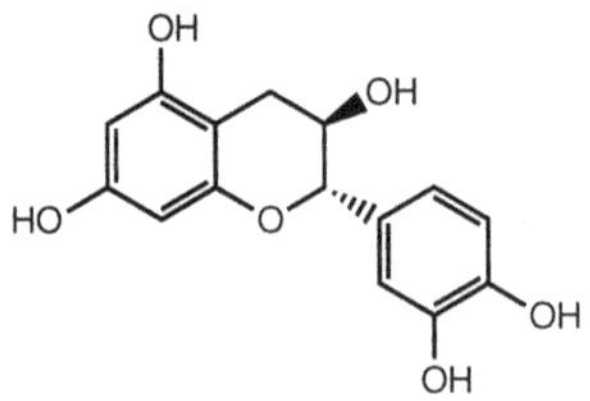 (+) Epicatechin

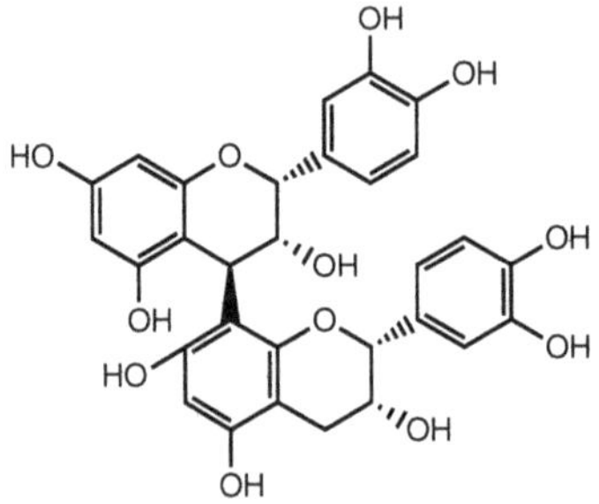

Procyanidin B2 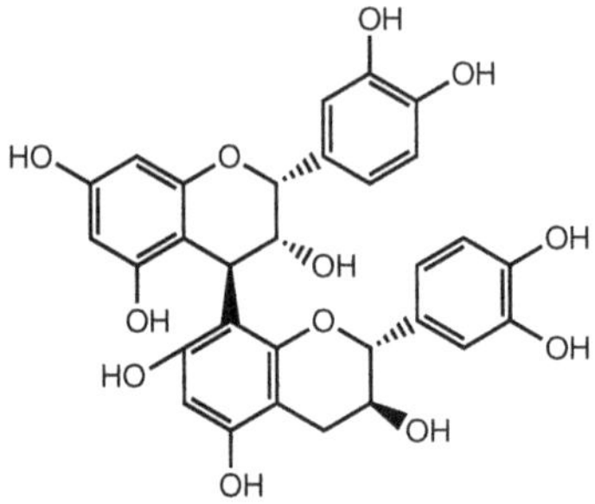 Procyanidin B1

FIGURE 8.1 Main flavonoids with cardiovascular health protective activity present in cocoa.

leukotrienes, reduction of xanthine oxidase and myeloperoxidase activities, suppression of the pro-inflammatory cytokines IL-1β, IL-2 and IL-8 production and inhibition of ET-1 release (Kerimi and Williamson, 2015) seem to be involved. Furthermore, decrease of the biomarkers associated with vascular damage like monocyte CD62L expression and the formation of elevated endothelial microparticles (McFarlin et al., 2015) and mobilization of functionally unaltered circulating angiogenic cells (EPCs) (Heiss et al., 2010) have been reported. These data indicate that the most significant changes remain as endothelial function, blood pressure and cholesterol levels, and, together with additional unpublished intervention studies, have enabled a claim on cocoa flavanols to be accepted by the European Food Safety Authority (EFSA NDA Panel, 2014) on endothelium-dependent vasodilation (Williamson, 2017).

Recent randomized and placebo-controlled trials showed that utilization of pomegranate juice efficiently reduced the systolic and diastolic blood pressure independently of dose and duration of consumption (Sahebkar et al., 2017). The hypotensive effect seems to be attributable to the pomegranate's ability to act as an ACE inhibitor. In fact, ACE inhibition induces blood vessels relaxation and dilatation, thus lowering blood pressure and allowing more blood and oxygen to get to the heart (Aviram and Rosenblat, 2012). Consumption of natural fruits and vegetables, such as grapes, decreased the risk of cardiovascular diseases, as it contains a wide variety of polyphenols, including flavonoids, phenolic acids and resveratrol. Several studies suggest that grape-derived poliphenols could reduce the risk of atherosclerosis through various mechanisms, such as inhibition of LDL oxidation and platelet aggregation, reduction of inflammation, and activation of sirtuin-1 a protein that prevent senescence (Castilla et al., 2006). Grapes, wine and resveratrol play a vital role in reducing morbidity and mortality risk due to cardiovascular complications. Wines obtained from grapes possess variety of antioxidants like resveratrol and proanthocyanidins, which play an important role in cardiac protection activities. Proanthocyanidins are mainly found in grape seeds, while resveratrol is found mainly in grape skin (Bertelli and Das, 2009). Various types of berry fruits, such as blueberry, cranberry, chokeberry and strawberry are rich in natural antioxidants (vitamin C and E), micronutrients (folic acids, calcium) and phytochemicals (polyphenols) (Bhagwat et al., 2011) and have a positive and profound impact on human health and diseases (Seeram, 2008). Vitamin E reduces cardiovascular diseases such as MI, stroke or cardiovascular death in individuals with DM and haptoglobin (Blum et al., 2010). Besides this, high glycemia, lipids and lipids oxidation are associated with coronary artery diseases (CAD). Therefore, human intervention studies proved that berry fruits have the ability to significantly decrease chances of CVD and maintain normal vascular function and blood pressure by improving LDL oxidation, lipid peroxidation, total plasma antioxidant activity, glucose metabolism and dyslipidemia (Krentz, 2003). In a clinical trial, the researchers suggested that consumption of anthocyanin-rich strawberries for one month improved lipid profile and platelet function in healthy volunteers (Alvarez-Suarez et al., 2014).

Olive oil is considered as the main source of fat in Mediterranean Diet, which is mainly associated with low morbidity and mortality rate for CVDs (Knoops et al., 2004). Olive oil, particularly, the virgin olive oil-rich diets, decreases pro-thrombotic environment, modifies platelet aggregation, coagulation and fibrinolysis (Covas et al., 2006).

8.3 ROLE OF HERBS IN CVDS

Generally, the botanical term herb refers to plants with a non-woody stem that produce seeds and which die at the end of the growing season, but in herbal medicine, this term is used not only to refer to herbaceous plants but also to refer to various parts of plants like bark, fruits, flowers, leaves and seeds of trees, shrubs and woody vines and their extracts (Eisenberg et al., 1993). In cardiovascular patients, the most commonly used herbs are echinacea, garlic, ginseng, ginkgo biloba and glucosamine, which are used to prevent hyperlipidemia, impaired blood flow and other cardiovascular problems (Yeh et al., 2006).

For centuries, garlic has been used as food and medicine. The active constituent of garlic is allicin, which shows a variety of pharmacological activities. Other organosulfur compounds stem from the decomposition of allicin produced by chopping garlic and include diallyl trisulfide, diallyl disulfide, diallyl sulfide, ajoene and dithins I (Figure 8.2). Several studies suggest that regular use of garlic prevents various cardiovascular disorders. In fact, garlic effectively reduces the risk of stroke and heart attack by lowering total and LDL-cholesterol and triacylglycerol (Warshafsky et al., 1993), and exhibits cardioprotective properties against

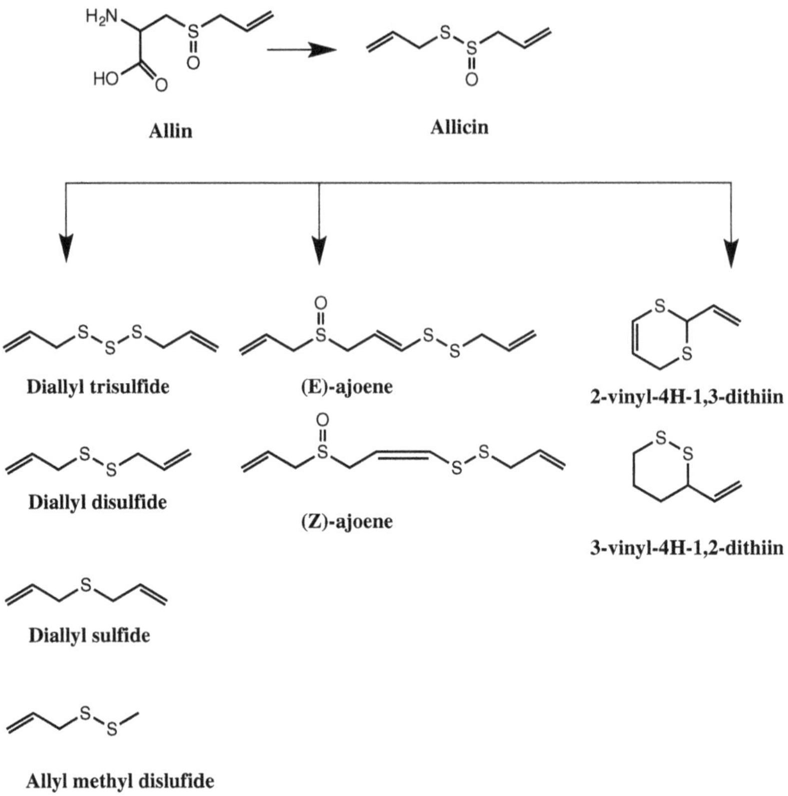

Allin

Allicin

Diallyl trisulfide

(E)-ajoene

2-vinyl-4H-1,3-dithiin

Diallyl disulfide

(Z)-ajoene

3-vinyl-4H-1,2-dithiin

Diallyl sulfide

Allyl methyl dislufide

FIGURE 8.2 Bioactive organosulfur compounds from garlic.

cardiotoxicity, arrhythmia, hypertrophy, ischemia-reperfusion injury, cardiac and mitochondrial dysfunction and myocardial infarction (Khatua et al., 2013; Supakul et al., 2014). Furthermore, garlic contains ajoenes, allyl methyl trisulfide and vinyldithins that increase fibrinolytic activity (Kendler, 1987), reduce serum levels of thromboxane B_2 (Ali and Thomson, 1995), inhibit platelet aggregation (Bordia et al., 1977). Allicin from garlic may exhibit an antimyocardial fibrosis effect by downregulation of the TGF β/Smads signal transduction, which ultimately leads to production of adhesion glycoproteins (Li et al., 2016). S-propargyl-cysteine has shown cardioprotection in ischemic heart disease (Wen and Zhu, 2015), while garlic-derived polysulfides may be useful in the treatment of myocardial ischemic disease (Lavu et al., 2011). In fact, Benavides et al. (2007) showed that garlic-derived polysulfides such as diallyl trisulphide (DATS) and diallyl disulfide (DADS) are H_2S donors in the presence of thiols and thiol-containing compounds (i.e., glutathione), independent of the H_2S-forming enzymes cystathionine γ-lyase (CSE), cystathionine β-synthase (CBS) and 3-mercaptopyruvate sulfurtransferase (3-MST). H_2S, much like NO, is an endogenously produced gaseous signaling molecule that plays a critical role in many physiologic processes and has been shown to exert cytoprotective actions in various models of CVD and cardiovascular injury (Kondo et al., 2013).

Onions contain α-sulfinyl disulfide group, which showed significant fibrinolytic activity by suppressing platelet aggregation (Kendler, 1987). The flavonoids and quercetin in onion could be recommended for preventing and treating various cardiovascular diseases by controlling cholesterol level and enhancing antioxidant ability (Lu et al., 2015; Majewska-Wierzbicka and Czeczot, 2012). It has been demonstrated that *S*-methylcysteine and flavonoids of onion can decrease the levels of blood glucose, serum lipids, oxidative stress and lipid peroxidation while increasing insulin secretion and antioxidant enzyme activity (Akash et al., 2014). Besides this, onion also efficiently inhibits arachidonic acid cascade in platelets (Kawakishi and Morimitsu, 1994). The flour derived from flaxseeds that are used in making bread and bakery products possesses low saturated and high polyunsaturated fat content. It also contains phytosterol and mucilage that significantly reduce total and LDL-cholesterol concentration (Cunnane et al., 1993; Bierenbaum et al., 1993). The use of lemongrass (*Cymbopogon citratus*) effectively reduces cholesterol concentration in hypercholesterolemic patients, as they are rich in geraniol and citral (Elson et al., 1989). The powdered fenugreek seeds also significantly reduced LDL-cholesterol and triacylglycerol concentration (Sharma and Raghuram, 1990). The non-saponin fractions present in ginseng roots remarkably inhibit platelet aggregation by inhibiting thromboxane A_2 production (Park et al., 1995). The evening primrose oil also exhibited significant activity due to presence of α- and γ- linoleic acids. The evening primrose oil has the ability to increase blood clotting time, reduce platelet aggregation and blood lipid concentration (Renaud et al., 1982). Besides this, other herbal medicines that are used to treat or prevent cardiovascular problems are mentioned in Table 8.2.

8.4 ROLE OF SPICES IN CVDS

Various epidemiological studies revealed that certain diets rich in fruits, herbs and spices are related to low risk of cardiovascular diseases and play an important role in the prevention and treatment of heart diseases. Garlic (*Allium sativum*) is a common spice belonging to the family Alliaceae, which originated in Central Asia. It is mostly used as flavoring agent, in traditional medicine and as a functional food (Stavric, 1994). Indian physician Charaka, the father of Ayurvedic medicine born in 300 BC, claimed that garlic has the ability to strengthen the heart and maintain the fluidity of blood and, therefore, acts as heart tonic (Fenwick et al., 1985). It has also been proved that garlic has the ability to inhibit enzymes that are involved in lipid aggregation, decrease platelet aggregation, prevent lipid peroxidation of oxidized erythrocytes and LDL, increase antioxidant status and inhibit ACE inhibitors (Rahman, 2001). Allicin, the active constituent of garlic, efficiently lowers the fatty streaks formation in aortic sinus (Abramovitz et al., 1999). Another study revealed that it also increased blood fibrinolytic activity (Kleijnen et al., 1989). Hence, it is clear that garlic has multiple properties useful in the prevention of cardiovascular diseases and therefore can be used as dietary supplement for prevention of cardiac

TABLE 8.2
List of Herbs that Are Used for the Treatment of CVD's with Possible Underlying Mechanism(s)

Botanical Origin	Common Name	Effects	Mechanism(s)	References
Allium sativum (Amaryllidaceae)	Garlic	Anti-hypercholesterolemic	Decrease total and LDL-cholesterol and triacylglycerol Increase fibrinolysis Inhibit platelet aggregation	Warshafsky et al. (1993)
Allium cepa (Amaryllidaceae)	Onion	Antiplatelet, fibrinolytic activity	Inhibit arachidonic acid cascade in platelets	Kawakishi and Morimitsu (1994)
Linumusitatissimum (Linaceae)	Flaxseed	Anti-hypercholesterolemic	Decrease total and LDL- cholesterol and triacylglycerol	Bierenbaum et al. (1993)
Plantago psyllium (Plantaginaceae)	Psyllium	Anti-hypercholesterolemic		Sprecher et al. (1993)
Trigonella foenum-graecum (Fabaceae)	Fenugreek	Anti-hypercholesterolemic	Decrease total and LDL- cholesterol and triacylglycerol concentration	Sharma and Raghuram (1990)
Panax ginseng (Araliaceae)	Asian ginseng	Antiplatelet	Inhibit thromboxane A_2 production	Park et al. (1995)
Oenothera biennis (Onagraceae)	Evening primrose oil	Antiplatelet	Increase blood clotting time Reduce platelet aggregation and blood lipid concentration	Renaud et al. (1982)
Ginkgo biloba (Ginkgoaceae)	Ginkgo	Improved blood flow and vasodilation	Inhibit platelet aggregation factor	Kleijnen and Knipschild (1992)
Crataegus oxyacantha (Rosaceae)	Hawthorn	Angina	Dilation of smooth muscles of coronary vessels and inhibition of biosynthesis of thromboxane A_2	Vibes et al. (1994)

diseases. Curcumin (diferuloylmethane) is the chief component of turmeric and is derived from the rhizome of the East Indian plant *Curcuma longa*. Turmeric, a spice used in traditional medicinal system, is used as an anti-inflammatory agent, in stomach and liver problems and as a cosmetic (Aggarwal et al., 2007). It has been evidenced that curcumin has the ability to reduce thromboxane and platelet aggregation (Keihanian et al., 2018).

Curcumin also has the ability to decrease triglycerides, serum cholesterol and cardiomyocytic apoptosis. Furthermore, curcumin increases the level of HDL and pro-inflammatory cytokines while downregulating NF-κB COX (Soni and Kuttan, 1992). Regular consumption of curcumin is probably an alternative way of modifying cholesterol-related parameters, as evidenced by a study that measured the effect of curcumin extract on weight, glucose and lipid profiles in patients with metabolic syndrome. After 12 weeks of curcumin extract intake, there was an elevation in the high-density lipoprotein cholesterol level, whereas the level of low-density lipoprotein cholesterol was decreased significantly (Yang et al., 2014). (Figure 8.3) Ginger (*Zingiber officinale*) is a medicinal plant widely used almost all over the world since ancient times for various ailments such as rheumatism, muscular aches, cramps, sore throat, hypertension, fever, infectious diseases, vomiting and indigestion (Ali et al., 2008). Gingerol is the active constituent present in ginger has the ability to relax blood vessels, stimulate blood flow and relieve pain. A study showed that gingerol obtained from *Zingiber* restrains the function of platelets by inhibiting thromboxane formation (Fuhrman et al., 2000). Black pepper is an extensively used spice and

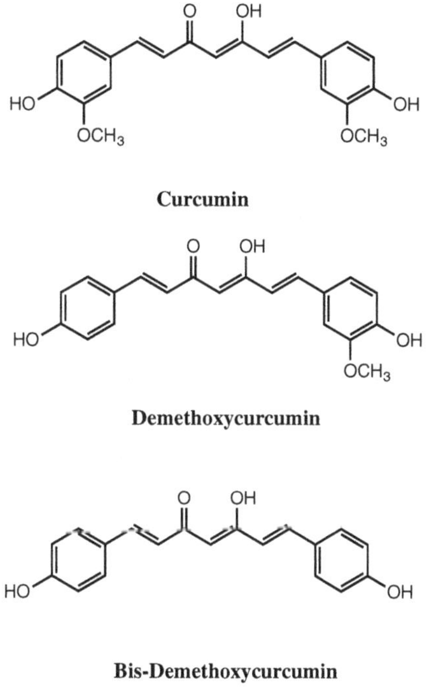

O OH
HO OH
OCH_3 OCH_3

Curcumin

O OH
HO OH
OCH_3

Demethoxycurcumin

O OH
HO OH

Bis-Demethoxycurcumin

FIGURE 8.3 Structure of curcuminoids.

is considered as the king of spices. It has been evidenced that its active constituent piperine, has the ability to protect against *in vitro* oxidative damage by inhibiting free radicals and ROS. It also has the ability to reduce lipid peroxidation *in vivo* (Srinivasan, 2007). Vanadium, mainly found in black pepper, is responsible for enhancing recovery of cardiac function in myocardial infarction (MI) and pressure overload-induced hypertrophy by activating Akt signaling through inhibition of protein tyrosine phosphatases (Bhuiyan and Fukunaga, 2009). The bark and leaves of cinnamon are extensively used as spice and may show antioxidant and antimicrobial activity (Table 8.3).

The Indian Materia Medica (Nadkarni, 1996) and *Indian Medicinal Plants – A Compendium of 500 species* (Vaidyaratnam, 1994) considered cinnamon as an herbal drug that has cardiovascular effects. The extracts of various species of *Cinnamomum* suppressed raised serum total cholesterol and triglyceride level. Furthermore, their chloroform fractions significantly inhibited HMG-CoA reductase in liver, an enzyme which is responsible for cholesterol biosynthesis (Amin and El-Twab, 2009). Traditionally, coriander is well documented in the treatment for diabetes and cholesterol patients (Burdock and Carabin, 2009). Its seeds have significant hypolipidemic action, which can reduce the level of triglycerides and cholesterol in the tissues of animals (Chithra and Leelamma, 1997). An *in vivo* study revealed that the coriander seed oil exhibited significant decrease in TC, TGs, TAG and LDL-c levels in plasma, while increasing HDL-c (Ramadan et al., 2008). Besides this, the leaf spice extracts of coriander are rich with natural antioxidants. Therefore, coriander can significantly inhibit human platelet aggregation in order to prevent thrombosis, which is an important event in cardiovascular diseases (Suneetha and Krishnakantha, 2005).

8.5 MICROALGAE

For thousands of years, algae have been part of the human diet. In addition to their nutritional value, algae are increasingly marketed as "functional foods" or "nutraceuticals"; these terms describe foods that contain bioactive compounds, or phytochemical products that, beyond the basic food role, may represent a health benefit. Macro- and microalgal biomass are an interesting natural source of a great variety of biologically active compounds, such as carotenoids, phycobilins, fatty acids, vitamins, polysaccharides and sterols. Carotenoids consumption is leading to a potential reduction of cardiovascular risk. Carotenoids are associated with blood pressure, reduction of pro-inflammatory cytokines and improvement of insulin sensitivity in muscle, liver and adipose tissues. Fucoxanthin is a carotenoid present in edible brown seaweed (Figure 8.4), and its beneficial effects on cardiovascular diseases were recently reviewed (Gammone et al., 2015). Animal studies have shown that long-term intake of fucoxanthin causes weight loss. For example, in a study conducted by Maeda et al. (2005), it has been shown that fucoxanthin or dried powder of seaweed (*Undaria pinnatifida*) significantly reduced the white adipose tissue in in Male Wistar rats and female KK-Ay mice after four weeks of treatment. The mechanisms may be related to an upregulation of the mitochondrial uncoupling proteins (UCP) that

TABLE 8.3
List of Spices that Are Used for the Treatment of CVDs with Possible Underlying Mechanism(s)

Botanical Origin	Common Name	Active Constituents	Mechanism/Uses	References
Allium sativum (Alliaceae)	Garlic	Allicine	Heart tonic by maintaining fluidity of blood and strengthen the heart	Fenwick et al. (1985)
Curcuma longa (Zingiberaceae)	Turmeric	Curcumin	Downregulate NF-κB transcriptional factor by inhibiting IKK, improve pro-inflammatory cytokines during cardiopulmonary bypass and decrease myocytic apoptosis after cardiac ischemia	Soni and Kuttan (1992)
Zingiber officinale (Zingebraceae)	Ginger	Gingerol	Prevent platelet aggregation by inhibiting thromboxane A2 formation	Fuhrman et al. (2000)
Piper longum and *piper nigrum* (Piperaceae)	Black pepper	Piperine and vanadium	Promote cardiac functions by activating Akt signaling through inhibition of protein tyrosine phosphatases	Bhuiyan and Fukunaga (2009)
Cinnamomum cassia (Lauraceae)	Cinnamon	Hypolipidemic action	Suppressed raised serum total cholesterol and triglyceride level by inhibiting HMG-CoA reductase in liver.	Amin and El-Twab (2009)
Coriandrum sativum (Apiaceae)	Coriander	Hypolipidemic action	Reduce the level of triglycerides and cholesterol in the tissues	Chithra and Leelamma (1997)
Murraya koenigii (Rutaceae)	Curry	Antiplatelet	Arachidonic acid inhibition	Suneetha and Krishnakantha (2005)

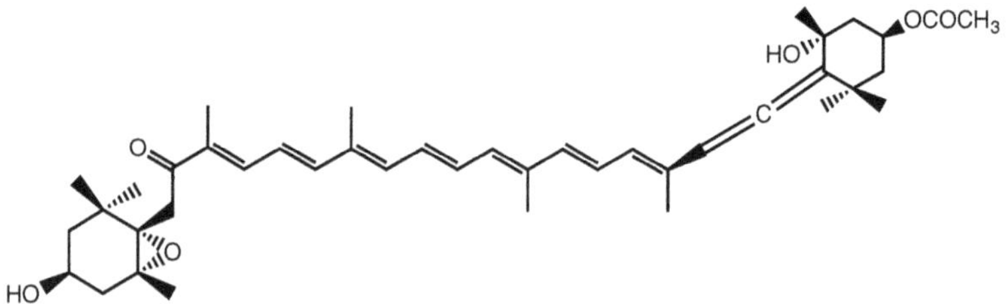

Fucoxanthin

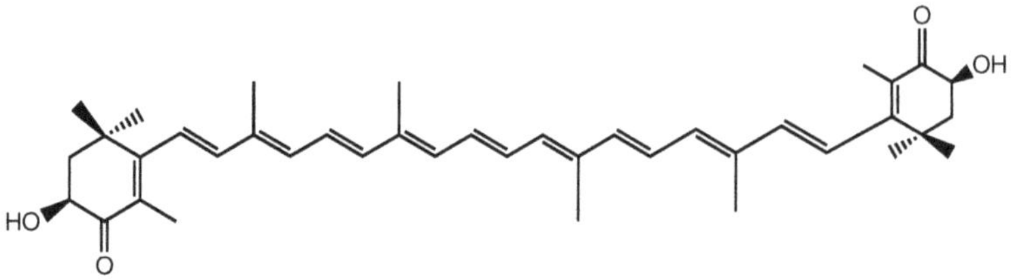

Astaxanthin

FIGURE 8.4 Carotenoids fucoxanthin and astaxanthin from seaweed.

separate oxidative phosphorylation from ATP synthesis with energy dissipated as heat (Maeda et al., 2005) (Table 8.4).

Other mechanisms include suppression of the differentiation of the adipocytes and of lipid accumulation through the inhibition of glycerol-3-phosphate dehydrogenase or the inhibition of the PPAR-γ receptor related to side effects such as edema and body weight. PPARs are expressed in places in the cardiovascular system such as endothelial cells, vascular smooth muscle cells and monocytes/macrophages. It has been shown that they play an important role in the modulation of inflammatory, fibrotic and hypertrophic responses (Das and Chakrabarti, 2006).

Astaxanthin is a red color pigment belonging to the family xanthophylls. It is mainly found in marine sources such as in microalgae (*Haematococcus pluvialis*), krill, plankton, fish and other sea foods. Various astaxanthin isomers have been characterized on the basis of the configuration of the two hydroxyl groups on the molecule.

Humans cannot synthesize it and depend on food in order to obtain it (Schweigertm, 1998). Astaxanthins are responsible for decreasing LDL-C and TG, increasing HDL-C and decreasing markers of lipid peroxidation, inflammation and thrombosis (Fassett and Coombes, 2012). Phlorotannins can also play a role in seaweeds' antihypertensive effects. Phlorotannins are phenolic compounds exclusive to brown algae that consist of dehydro-oligomers or dehydro-polymers of phloroglucinol in these derivatives molecular arrangements of phloroglucinol units and linkages between aryl-aryl or diaryl ether bonds and the number of hydroxyl groups give a wide assortment of structures. Several of these compounds, such as phloroglucinol, eckol and dieckol, have shown antioxidant, anti-tumor and anti-cancer activities (Montero et al., 2016). (Figure 8.5).

TABLE 8.4
List of Natural Beverages that Are Used for the Treatment of CVDs with Possible Underlying Mechanism(s)

Natural Products	Common Name	Active Constituents	Mechanisms	References
Camellia sinensis (Theaceae)	Green and black tea	(−)-epigallocatechin-3-gallate	Reduce surrogate markers of atherosclerosis and lipid peroxidation	Stangl et al. (2006); Basu and Lucas (2007)
Theobroma cocoa (Malvaceae)	Cocoa	Catechin, *epi*catechin	Decrease inflammatory cytokines production, enhance anti-inflammatory cytokines level	Mao (2000)
Cabernet sauvignon (Vitaceae)	Red wine	Resveratrol	Inhibit LDL oxidation, decrease systemic BP	Rifici et al. (2002)
Pinus pinaster (Pinaceae)	Pine bark	Pycnogenol	Protect endothelial cells enhance endothelial derived vasorelaxation, strengthen capillaries, preserve vascular integrity, inhibit platelet aggregation by inhibiting thromboxane A2 formation	Fitzpatrick (1998); Wei (1997)
Hibiscus sabdariffa (Malvaceae)	Roselle	Hypertension, hypolipidemic and antiatherosclerotic	Decrease serum cholesterol, triglyceride and LDL-C	Chen et al. (2003)
Coffea arabica, Coffea robusta	Coffee	Caffeine	Improve endothelial-dependent vasodilatation by enhancing NO production	Umemura et al. (2006)

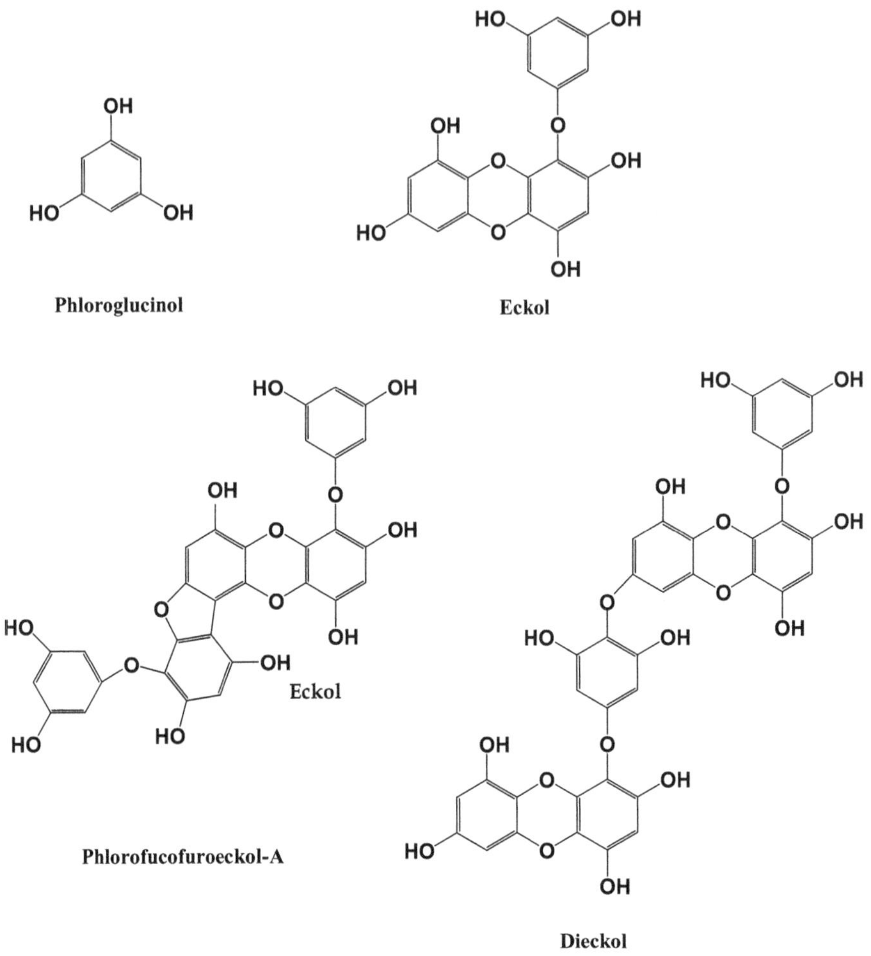

Phloroglucinol

Eckol

Phlorofucofuroeckol-A

Dieckol

FIGURE 8.5 Chemical structures of some phlorotannins from seaweed.

8.6 MISCELLANEOUS

Besides water, tea is the most consumed beverage globally. Depending on the level of fermentation, tea can be divided into green tea which is unfermented, black tea which is fully fermented and oolong which is partially fermented (Cheng, 2006). Green and black tea are considered to be the most promising tool for cardiovascular protection (Stangl et al., 2006). Generally, consumption of green tea is found to be greater than black tea and thus plays a vital role in health benefits. The (–)-*epi*gallocatechin-3-gallate (major catechin), mainly found in green tea, significantly improved the potential human health effects by improving cardiovascular health. Green tea also enhances weight loss and antioxidant effects. Besides tea, other food sources of catechins, such as chocolate, apple, pear, grapes and red wine, are also very popular and highly consumed (Basu and Lucas, 2007). A local soft drink material and medicinal

TABLE 8.5
List of Most Frequent Cardiac Interactions between Herbal Products and Drugs

Herb	Common Name	Uses	Adverse Cardiac Interactions	References
Citrus aurantium (Rutaceae)	Bitter orange	Weight loss	Enhance stimulatory effects of caffeine e.g. tachycardia, cardiac arrest, ventricular fibrillation, transient collapse and blackout	Jordan et al. (2004)
Hypericum perforatum (Hypericaceae)	St. John's wort	Depression, anxiety, sleep disorders, common cold, herpes and in HIV	Reduce the efficacy of immunosuppressants in cardiac transplant patients and cause transplant rejection, increase HR and BP with iMAO, decrease concentration of digoxin and statin	Sugimoto et al. (2001); Johne et al. (1999)
Leonurus cardiac (Lamiaceae)	Motherwort	Cardiac debility, tachycardia, anxiety, insomnia, amnorrhea, hypotensive and diuretic	Potentiate antithrombotic antiplatelet effect thus increase risk of bleeding	Zou (1989)
Panax ginseng (Araliaceae)	Ginseng	MI, CHF, angina pectoris and anti-diabetic	Hypertension, behavioral changes, hypoglycemia and decrease effects of warfarin	Siegel (1979); Chun-Sum et al. (2004)
Ginkgo biloba (Ginkgoaceae)	Ginkgo	Poor circulation, cognitive disorders	Increase risk of bleeding with antiplatelet, anticoagulant and antithrombotic such as warfarin, aspirin, COX-2 inhibitors	Barnes et al. (2004)
Allium sativum (Amaryllidaceae)	Garlic	High cholesterol, hypertension, heart disease	Increase risk of bleeding with antiplatelet or anticoagulant such as warfarin	German et al. (1995); Rose et al. (1990)
Crataegus oxyacantha (Rosaceae)	Hawthorn	CHF, angina, bradyarrhythmia and cerebral insufficiency	Potentiate effects of digoxin, CCB's and nitrates, increase risk of bleeding with antiplatelet or anticoagulant	Guo et al. (2008); Vibes et al. (1994))
Citrus paradise (Rutaceae)	Grapefruit juice	Promote cardiovascular health and weight loss	Potentiate effects of statins, CCB's and cyclosporins, cause severe hypotension, myopathy or liver toxicity	Bailey et al. (1998a,b)

(Continued)

TABLE 8.5 (CONTINUED)
List of Most Frequent Cardiac Interactions between Herbal Products and Drugs

Herb	Common Name	Uses	Adverse Cardiac Interactions	References
Serenoa repens (Arecaceae)	Saw palmetto	Benign prostatic hypertrophy, diuretic, urinary antiseptic	Increase risk of bleeding with antiplatelet or anticoagulant agents cause cholestatic hepatitis, acute pancreatitis	Goepel et al. (1999); Bressler (2005)
Aconitum napellus (Ranunculaceae)	Aconite	Mild diaphoretic, decrease PR, atrial and ventricular fibrillation	Cause bradycardia and hypotension to fatal ventricular arrhythmia	(Lowe et al. (2005), Smith et al. (2005)
Pausinystalia yohimba (Rubiaceae)	Yohimbine	Sexual disorder, exhaustion, α_2-adrenergic receptor antagonist	Increase risk of HR and hypertensive crises with iMAO, decrease ACE inhibitors and β-blockers effects	Tam et al. (2001)
Glycyrrhiza glabra (Fabaceae)	Licorice	Ulcer, cirrhosis, infections	Decrease effect of antihypertensive drugs thus cause hypertension, hypokalemia and edema	Mansoor (2001)
Stephania tetrandra (Menispermaceae)	Tetrandine	Hypertension and angina	Lowers plasma glucose and cause hepato and renal toxicity	Seeff (2007)
Echinacea purpurea (Asteraceae)	Echinacea	Stimulate immune system and prevent infections	Increase QT interval and hepatotoxic effects such as with amiodarone, statins, fibrates and niacin	Freeman and Spelman (2008)
Salvia miltiorrhiza (Lamiaceae)	Danshen	Coronary artery disease and menstrual abnormalities	Increase risk of bleeding with antiplatelet or anticoagulant, increase side effects of digoxin	Izzat et al. (1998)

herb, *Hibiscus sabdariffa* significantly exhibited antiatherosclerotic and hypolipidemic effects by reducing total cholesterol and triglycerides in rabbits (Chen et al., 2003). Furthermore, coffee is an important source of caffeine and is also considered as the most widely used beverage globally. Various studies supported that moderate use of coffee confers cardiovascular benefit, while heavy consumption of coffee is toxic to the heart (Umemura et al., 2006).

8.7 ADVERSE CARDIAC INTERACTIONS OF HERBAL PRODUCTS

Various problems arise with the consumption of herbal products. Most herbal products are not scientifically evaluated, and very limited information is available regarding their safety and efficacy. The European Medicines Agency (EMA) pays great attention to herbal substances, preparations and combinations with a focus on safety and efficacy. The preparation of monographs is entrusted to the Committee on Herbal Medicinal Products (HMPC), which compiles and assesses scientific data on herbal substances. Furthermore, particular attention is paid to interactions between herbal products and drugs. A list of the most frequent interactions is reported in Table 8.5.

8.8 CONCLUSIONS

It can be concluded with confidence that the natural products (NPs) are diverse in nature and have outstanding potential to effectively address the underlying CVDs. NPs are freely available and comparatively affordable to a majority of the communities of the world. Moreover, natural organic compounds are generally considered as harmless, but safety confirmation needs to be evaluated.

REFERENCES

Abramovitz, D., Gavri, S., Harats, D. et al. 1999. Allicin-induced decrease in formation of fatty streaks (atherosclerosis) in mice fed a cholesterol-rich diet. *Coronary Artery Disease* 10(7): 515–520.

Aggarwal, B.B., Sundaram, C., Malani, N., Ichikawa, H. 2007. Curcumin: the Indian solid gold. In: *The Molecular Targets and Therapeutic Uses of Curcumin in Health and Disease*. Springer, 1–75.

Akash, M.S.H., Rehman, K., Chen, S. 2014. Spice plant *Allium cepa*: dietary supplement for treatment of type 2 diabetes mellitus. *Nutrition* 30(10): 1128–1137.

Ali, M., Thomson, M. 1995. Consumption of a garlic clove a day could be beneficial in preventing thrombosis. *Prostaglandins, Leukotrienes and Essential Fatty Acids* 53(3): 211–212.

Ali, B.H., Blunden, G., Tanira, M.O., Nemmar, A. 2008. Some phytochemical, pharmacological and toxicological properties of ginger (Zingiber officinale Roscoe): a review of recent research. *Food and Chemical Toxicology* 46(2): 409–420.

Alvarez-Suarez, J.M., Giampieri, F., Tulipani, S., et al. 2014. One-month strawberry-rich anthocyanin supplementation ameliorates cardiovascular risk, oxidative stress markers and platelet activation in humans. *Journal of Nutritional Biochemistry* 25(3): 289–294.

Amin, K.A., El-Twab, T.M.A. 2009. Oxidative markers, nitric oxide and homocysteine alteration in hypercholesterolimic rats: role of atorvastatine and cinnamon. *International Journal of Clinical and Experimental Medicine* 2(3): 254.

Aviram, M., Rosenblat, M. 2012. Pomegranate protection against cardiovascular diseases. *Evidence-Based Complementary and Alternative Medicine* 2012: 382763.

Bai, Y., Sun, L., Yang, T., Sun, K., Chen, J., Hui, R. 2009. Increase in fasting vascular endothelial function after short-term oral L-arginine is effective when baseline flow-mediated dilation is low: meta-analysis of randomized controlled trials. *American Journal of Clinical Nutrition* 89: 77–84.

Bailey, D.G., Kreeft, J.H., Munoz, C., Freeman, D.J., Bend, J.R. 1998a. Grapefruit juice—felodipine interaction: effect of naringin and 6′, 7′-dihydroxybergamottin in humans. *Clinical Pharmacology & Therapeutics* 64(3): 248–256.

Bailey, D.G., Malcolm, J., Arnold, O., David Spence, J. 1998b. Grapefruit juice–drug interactions. *British Journal of Clinical Pharmacology* 46(2): 101–110.

Barnes, P.M., Powell-Griner, E., McFann, K., Nahin, R.L. 2004. Complementary and alternative medicine use among adults: United States, 2002. *Advance Data* 2 (343): 1–19.

Basu, A., Lucas, E.A. 2007. Mechanisms and effects of green tea on cardiovascular health. *Nutrition Reviews* 65(8): 361–375.

Benavides, G.A., Squadrito, G.L., Mills, R.W., et al. 2007. Hydrogen sulfide mediates the vasoactivity of garlic. *Proceedings of the National Academy of Sciences USA* 104: 17977–17982.

Bertelli, A.A., Das, D.K. 2009. Grapes, wines, resveratrol, and heart health. *Journal of cardiovascular pharmacology* 54(6): 468–476.

Bhagwat, S., Haytowitz, D.B., Holden, J.M. 2011. USDA database for the flavonoid content of selected foods, Release 3.1. Beltsville: US Department of Agriculture: 03-01.

Bhuiyan, M.S., Fukunaga, K. 2009. Cardioprotection by vanadium compounds targeting Akt-mediated signaling. *Journal of Pharmacological Sciences* 110(1): 1–13.

Bierenbaum, M.L., Reichstein, R., Watkins, T.R. 1993. Reducing atherogenic risk in hyperlipemic humans with flax seed supplementation: a preliminary report. *Journal of the American College of Nutrition* 12(5): 501–504.

Blum, S., Vardi, M., Brown, J.B., et al. 2010. Vitamin E reduces cardiovascular disease in individuals with diabetes mellitus and the haptoglobin 2-2 genotype. *Pharmacogenomics* 11(5): 675–684.

Bordia, A.K., Joshi, H., Sanadhya, Y., Bhu, N. 1977. Effect of essential oil of garlic on serum fibrinolytic activity in patients with coronary artery disease. *Atherosclerosis* 28(2): 155–159.

Bressler, R. 2005. Herb-drug interactions. Interactions between saw palmetto and prescription medications. *Geriatrics* 60(11): 32–34.

Burdock, G.A., Carabin, I.G. 2009. Safety assessment of coriander (Coriandrum sativum L.) essential oil as a food ingredient. *Food and Chemical Toxicology* 47(1): 22–34.

Castilla, P., Echarri, R., Davalos, A., et al. 2006. Concentrated red grape juice exerts antioxidant, hypolipidemic, and antiinflammatory effects in both hemodialysis patients and healthy subjects. *The Journal of Nutrition* 84: 252–262.

Cave, A.C., Brewer, A.C., Narayanapanicker, A., et al. 2006. NADPH oxidases in cardiovascular health and disease. *Antioxidants & Redox Signaling* 8(5–6): 691–728.

Chen, C.C., Hsu, J.D., Wang, S.F., et al. 2003. Hibiscus sabdariffa extract inhibits the development of atherosclerosis in cholesterol-fed rabbits. *Journal of Agricultural and Food Chemistry* 51(18): 5472–5477.

Cheng, T.O. 2006. All teas are not created equal: the Chinese green tea and cardiovascular health. *International Journal of Cardiology* 108(3): 301–308.

Chithra, V., Leelamma, S. 1997. Hypolipidemic effect of coriander seeds (Coriandrum sativum): mechanism of action. *Plant Foods for Human Nutrition* 51(2): 167–172.

Chun-Sum, Y., Wei, G., Dey, L., Karrison, T. 2004. Brief Communication: American Ginseng *Reduces* Warfarin's Effect in Healthy Patients. *Annals of Internal Medicine* 141(1): 23.

Colacurci, N., Chiantera, A., Fornaro, F., et al. 2005. Effects of soy isoflavones on endothelial function in healthy postmenopausal women. *Menopause* 12: 299–307.

Cooper, K.A., Donovan, J.L., Waterhouse, A.L., et al. 2008. Cocoa and health: a decade of research. *British Journal of Nutrition* 99: 1–11.

Covas, M.I., Nyyssonen, K., Poulsen, H.E., Kaikkonen, J., Zunft, H.J., Kiesewetter, H. 2006. The effect of polyphenols in olive oil on heart disease risk factors. *Annals of Internal Medicine* 145: 333–341.

Cunnane, S.C., Ganguli, S., Menard, C., et al. 1993. High α-linolenic acid flaxseed (Linum usitatissimum): some nutritional properties in humans. *British Journal of Nutrition* 69(2): 443–453.

D'Adamo, C.R., Sahin, A. 2014. Soy foods and supplementation: a review of commonly perceived health benefits and risks. *Alternative Therapies in Health and Medicine* 20 Suppl 1: 39–51.

Das, Saibal K., Chakrabarti, R. 2006. Role of PPAR in cardiovascular diseases. *Recent Patents on Cardiovascular Drug Discovery* 1.2: 193–209.

de Fátima, A., Terra, B.S., da Silva, C.M., da Silva, D.L., Araujo, D.P., da Silva Neto, L., Nascimento de Aquino, R.A. 2014. From nature to market: examples of natural products that became drugs. *Recent Patents Biotechnology* 8 (1): 76–88.

Dong, J.Y., Qin, L.Q., Zhang, Z., et al. 2011. Effect of oral L-arginine supplementation on blood pressure: a meta-analysis of randomized, double-blind, placebo-controlled trials. *American Heart Journal* 162: 959–965.

EFSA NDA Panel (European Food Safety Authority Panel on Dietetic Products, Nutrition and Allergies) 2014. Scientific Opinion on the modification of the authorisation of a health claim related to cocoa flavanols and maintenance of normal endothelium-dependent vasodilation pursuant to Article 13(5) of Regulation (EC) No 1924/20061 following a request in accordance with Article 19 of Regulation (EC) No 1924/2006. *EFSA Journal* 12: 3654.

Elson, C., Underbakke, G., Hanson, P., Shrago, E., Wainberg, R., Qureshi, A. 1989. Impact of lemongrass oil, an essential oil, on serum cholesterol. *Lipids* 24(8): 677–679.

Engler, M.B., Engler, M.M. 2006. The emerging role of flavonoid-rich cocoa and chocolate in cardiovascular health and disease. *Nutrition Reviews* 64(3): 109–118.

Eisenberg, D.M., Kessler, R.C., Foster, C., Norlock, F.E., Calkins, D.R., Delbanco, T.L. 1993. Unconventional medicine in the United States–prevalence, costs, and patterns of use. *New England Journal of Medicine* 328(4): 246–252.

Fassett, R., Coombes, J. 2012. Astaxanthin in cardiovascular health and disease. *Molecules* 17: 2030–2048.

Fenwick, G.R., Hanley, A.B., Whitaker, J.R. 1985. The genus Allium—part 1. *Critical Reviews in Food Science & Nutrition* 22(3): 199–271.

Fitzpatrick, K. 1998. Endothelium-dependent vascular effects of Pycnogenol. *Journal of Cardiovascular Pharmacology* 32: 509–515.

Freeman, C., Spelman, K. 2008. A critical evaluation of drug interactions with Echinacea spp. *Molecular Nutrition & Food Research* 52(7): 789–798.

Fuhrman, B., Rosenblat, M., Hayek, T., Coleman, R., Aviram, M. 2000. Ginger extract consumption reduces plasma cholesterol, inhibits LDL oxidation and attenuates development of atherosclerosis in atherosclerotic, apolipoprotein E-deficient mice. *The Journal of Nutrition* 130(5): 1124–1131.

Gammone, M.A., Riccioni, G., D'Orazio, N. 2015. Carotenoids: potential allies of cardiovascular health? *Food Nutrition Research* 59: 26762.

German, K., Kumar, U., Blackford, H. 1995. Garlic and the risk of TURP bleeding. *BJU International* 76(4): 518–518.

Gil-Izquierdo, A., Penalvo, J.L., Gil, J.I., Medina, S., Horcajada, M.N., Lafay, S., Silberberg, M., Llorach, R., Zafrilla, P., Garcia-Mora, P., Ferreres, F. 2012. Soy isoflavones and cardiovascular disease epidemiological, clinical and -omics perspectives. *Current Pharmaceutical Biotechnologies* 13(5): 624–631.

Goepel, M., Hecker, U., Krege, S., Rübben, H., Michel, M.C. 1999. Saw palmetto extracts potently and noncompetitively inhibit human α1-adrenoceptors in vitro. *The Prostate* 38(3): 208–215.

González Cañete, N., Durán Agüero, S. 2014. Soya isoflavones and evidences on cardiovascular protection. *Nutrition Hospitalaira* 29(6): 1271–1282.

Guo, R., Pittler, M.H., Ernst, E. 2008. Hawthorn extract for treating chronic heart failure. *Cochrane Database Systematic Review* (1): CD005312.

Heiss, C., Jahn, S., Taylor, M., et al. 2010. Improvement of endothelial function with dietary flavanols is associated with mobilization of circulating angiogenic cells in patients with coronary artery diseases. *Journal of the American College of Cardiology* 56: 218–224.

Izzat, M.B., Yim, A.P., El-Zufari, M.H. 1998. A taste of Chinese medicine! *The Annals of Thoracic Surgery* 66(3): 941–942.

Jackson, R.L., Greiwe, J.S., Schwen, R.J. 2011. Emerging evidence of the health benefits of S-equol, an estrogen receptor β-agonist. *Nutrition Reviews* 69: 432–448.

Jacques, P. A., Lyass, A., Massaro, J. M., Vasan, R. S., D'Agostino, R. B. Sr. 2013. Relation of lycopene intake and consumption of tomato products to incident cardiovascular disease. *British Journal of Nutrition* 110(3): 545–551.

Johne, A., Brockmöller, J., Bauer, S., Maurer, A., Langheinrich, M., Roots, I. 1999. Pharmacokinetic interaction of digoxin with an herbal extract from St John's wort (Hypericum perforatum). *Clinical Pharmacology & Therapeutics* 66 (4): 338–345.

Jordan, S., Murty, M., Pilon, K. 2004. Products containing bitter orange or synephrine: suspected cardiovascular adverse reactions. *Canadian Medical Association Journal* 171(8): 993.

Kawakishi, S., Morimitsu, Y. 1994. Sulfur chemistry of onions and inhibitory factors of the arachidonic acid cascade. In: *Food Phytochemicals for Cancer Prevention I.* Chapter 8, pp. 120–127.

Keihanian, F., Saeidinia, A., Bagheri, R.K., Johnston, T.P., Sahebkar, A. 2018. Curcumin, hemostasis, thrombosis, and coagulation. *Journal of Cellular Physiology* 233: 4497–4511.

Kendler, B.S. 1987. Garlic (Allium sativum) and onion (Allium cepa): a review of their relationship to cardiovascular disease. *Preventive Medicine* 16(5): 670–685. ACS Publications.

Kerimi, A., Williamson, G. 2015. The cardiovascular benefits of dark chocolate. *Vascular Pharmacology* 71: 11–15.

Khatua, T.N., Adela, R., Banerjee, S.K. 2013. Garlic and cardioprotection: insights into the molecular mechanisms. *Canadian Journal of Physiology and Pharmacology* 91(6): 448–458.

Kleijnen, J., Knipschild, P. 1992. Ginkgo biloba for cerebral insufficiency. *British Journal of Clinical Pharmacology* 34(4): 352–358.

Kleijnen, J., Knipschild, P., Riet, G.T. 1989. Garlic, onions and cardiovascular risk factors. A review of the evidence from human experiments with emphasis on commercially available preparations. *British Journal of Clinical Pharmacology* 28(5): 535–544.

Knoops, K.T., de Groot, L.C., Krombout, D., Perrin, A.E., Moreiras-Varela, O., Menotti, A. 2004. Mediterranean diet, lifestyle factors, and 10-year mortality in elderly European men and women. The HALE Project. *Journal of American Medical Association* 292: 1433–1439.

Kondo, K., Bhushan, S., King, A.L., et al. 2013. H(2)S protects against pressure overload-induced heart failure via upregulation of endothelial nitric oxide synthase. *Circulation* 127:1116–1127.

Krentz, A.J. 2003. Lipoprotein abnormalities and their consequences for patients with type 2 diabetes. *Diabetes, Obesity and Metabolism* 5(s1): S19–27.

Lavu, M., Bhushan, S., Lefer, D.J. 2011. Hydrogen sulfide-mediated cardioprotection: mechanisms and therapeutic potential. *Clinical Science* 120(6): 219–229.

Li, S.C., Ma, L.N., Chen, J., Li, Y.K. 2016. Effect of allicin on myocardial fibrosis after myocardial infarction in rats and its relationship with TGFβ/Smads signal transduction. *China Journal of Chinese Materia Medica* 41(13): 2517–2521.

Liu, L.K., Lee, H.J., Shih, Y.W., Chyau, C.C., Wang, C.J. 2008. Mulberry anthocyanin extracts inhibit LDL oxidation and macrophage-derived foam cell formation induced by oxidative LDL. *Journal of Food Science* 73(6): H113–21.

Liu, Z.M., Ho, S.C., Chen, Y.M., Liu, J., Woo, J. 2014. Cardiovascular risks in relation to daidzein metabolizing phenotypes among Chinese postmenopausal women. *PLoS ONE* 9: e87861.

Lowe, L., Matteucci, M.J., Schneir, A.B. 2005. Herbal aconite tea and refractory ventricular tachycardia. *New England Journal of Medicine* 353(14): 1532–1532.

Lu, T.M., Chiu, H.F., Shen, Y.C., Chung C.-C., Venkatakrishnan, K., Wang, C.K. 2015. Hypocholesterolemic efficacy of quercetin rich onion juice in healthy mild hypercholesterolemic adults: a pilot study. *Plant Foods for Human Nutrition* 70(4): 395–400.

Maeda, H., et al. 2005. Fucoxanthin from edible seaweed, Undaria pinnatifida, shows anti-obesity effect through UCP1 expression in white adipose tissues. *Biochemical and Biophysical Research Communications* 332.2 (2005): 392–397.

Majewska-Wierzbicka, M., Czeczot, H. 2012. Flavonoids in the prevention and treatment of cardiovascular diseases. *Polski Merkuriusz Lekarski* 32(187): 50–54.

Maki, K.C., Butteiger, D.N., Rains, T.M. 2010. Effects of soy protein on lipoprotein lipids and fecal bile acid excretion in men and women with moderate hypercholesterolemia. *Journal of Clinical Lipidology* 4: 531–542.

Mansoor, G.A. 2001. Herbs and alternative therapies in the hypertension clinic. *American Journal of Hypertension* 14 (9): 971–975.

Mao, T. 2000. Effect of cocoa procyanidins on the secretion of IL-4 in peripheral blood mononuclear cells. *Journal of Medicinal Food* 3: 107–114.

McFarlin, B.K., Venable, A.S., Henning A.L., et al. 2015. Hill D.W. Natural cocoa consumption: potential to reduce atherogenic factors? *Journal of Nutritional Biochemistry* 26: 626–632.

Miguez, A.C., Francisco, J.C., Barberato S.H., et al. 2012. The functional effect of soybean extract and isolated isoflavone on myocardial infarction and ventricular dysfunction: the soybean extract on myocardial infarction. *Journal of Nutritional Biochemistry* 23: 1740–1748.

Moncada, S., Higgs, A. 1993. The L-arginine-nitric oxide pathway. *New England Journal of Medicine* 329: 2002–2012.

Montero, L., Sánchez-Camargo, A. P., García-Cañas, V., Tanniou, A., Stiger-Pouvreau, V., Russo, M., Rastrelli, L., Cifuentes, A., Herrero, M., Ibáñez, E. 2016. Anti-proliferative activity and chemical characterization by comprehensive two-dimensional liquid chromatography coupled to mass spectrometry of phlorotannins from the brown macroalga *Sargassum muticum* collected on North-Atlantic coasts. *Journal of Chromatography A* 1428: 115–125.

Nadkarni, K.M. 1996. [Indian materia medica]; Dr. KM Nadkarni's Indian materia medica: with Ayurvedic, Unani-Tibbi, Siddha, allopathic, homeopathic, naturopathic & home remedies, appendices & indexes. 1: Popular Prakashan.

Nagata, C. 2000. Ecological study of the association between soy product intake and mortality from cancer and heart disease in Japan. *International Journal of Epidemiology* 29: 832–836.

Nguyen, M.L., Schwartz, S.J. 1999. Lycopene: chemical and biological properties. *Food Technology (USA)* 53(2): 38–45.

Paran, E., Novack, V., Engelhard, Y.N., Hazan-Halevy, I. 2009. The effects of natural antioxidants from tomato extract in treated but uncontrolled hypertensive patients. *Cardiovascular Drugs and Therapy* 23(2): 145–151.

Park, H.J., Rhee, M.H., Park, K.M., Nam, K.Y., Park, K.H. 1995. Effect of non-saponin fraction from Panax ginseng on cGMP and thromboxane A2 in human platelet aggregation. *Journal of Ethnopharmacology* 49(3): 157–162.

Pastor-Villaescusa, B., Rangel-Huerta, O.D., Aguilera, C.M., Gil, A. 2015. A systematic review of the efficacy of bioactive compounds in cardiovascular disease: carbohydrates, active lipids and nitrogen compounds. *Annals of Nutrition and Metabolism* 66(2–3): 168–181.

Pergola, C., Rossi, A., Dugo, P., Cuzzocrea, S., Sautebin, L. 2006. Inhibition of nitric oxide biosynthesis by anthocyanin fraction of blackberry extract. *Nitric Oxide* 15(1): 30–39.

Petyaev, I.M. 2016. Lycopene deficiency in ageing and cardiovascular disease. *Oxidative Medicine and Cellular Longevity*: 2016: article ID 3218605, 6 pages.

Prior, R.L., Wu, X., Gu, L., et al. 2009. Purified berry anthocyanins but not whole berries normalize lipid parameters in mice fed an obesogenic high fat diet. *Molecular Nutrition & Food Research* 53(11): 1406–1418.

Rangel-Huerta, O.D., Pastor-Villaescusa, B., Aguilera, C.M., Gil, A. 2015. A systematic review of the efficacy of bioactive compounds in cardiovascular disease: phenolic compounds. *Nutrients* 7(7): 5177–5216.

Ramadan, M.F., Amer, M.M.A., Awad, A.E.-S. 2008. Coriander (Coriandrum sativum L.) seed oil improves plasma lipid profile in rats fed a diet containing cholesterol. *European Food Research and Technology* 227(4): 1173–1182.

Rahman, K. 2001. Historical perspective on garlic and cardiovascular disease. *The Journal of Nutrition* 131(3): 977S–979S.

Rahman, K., Lowe, G.M. 2006. Garlic and cardiovascular disease: a critical review. *The Journal of nutrition* 136(3): 736S–740S.

Ramdath, D.D., Padhi, E.M., Sarfaraz, S., Renwick, S., Duncan, A.M. 2017. Beyond the cholesterol-lowering effect of soy protein: a review of the effects of dietary soy and its constituents on risk factors for cardiovascular disease. *Nutrients* 9(4): 324.

Renaud, S., McGregor, L., Morazain, R., et al. 1982. Comparative beneficial effects on platelet functions and atherosclerosis of dietary linoleic and γ-linolenic acids in the rabbit. *Atherosclerosis* 45(1): 43–51.

Rifici, V., Schneider, S., Khachadurian, A.K. 2002. Lipoprotein oxidation mediated by J774 murine macrophages is inhibited by individual red wine polyphenols but not by ethanol. *Journal of Nutrition* 132: 2532–2537.

Rose, K.D., Croissant, P.D., Parliament, C.F., Levin, M.B. 1990. Spontaneous spinal epidural hematoma with associated platelet dysfunction from excessive garlic ingestion: a case report. *Neurosurgery* 26(5): 880–882.

Sacks, F.M., Lichtenstein. A., Van Horn, L., Harris, W., Kris-Etherton, P., Winston, M.; American Heart Association Nutrition Committee. 2006. Soy protein, isoflavones, and cardiovascular health: an American Heart Association Science Advisory for professionals from the Nutrition Committee. *Circulation* 113(7): 1034–1044.

Sahebkar, A., Ferri, C., Giorgini, P., Bo, S., Nachtigal, P., Grassi, D. 2017. Effects of pomegranate juice on blood pressure: a systematic review and meta-analysis of randomized controlled trials. *Pharmacological Research* 115: 149–161.

Sarriá, B., Martínez-López, S., Sierra-Cinos, J.L., et al. 2015. Effects of bioactive constituents in functional cocoa products on cardiovascular health in humans. *Food Chemistry* 174: 214–218.

Schewe, T., Steffen, Y., Sies, H. 2008. How do dietary flavanols improve vascular function? A position paper. *Archives of Biochemistry and Biophysics* 476: 102–106.

Schweigertm, F. 1998. *Metabolism of Carotenoids in Mammals.* Basel, Switzerland: Birkhauser Verlag.

Seeram, N.P. 2008. Berry fruits: compositional elements, biochemical activities, and the impact of their intake on human health, performance, and disease. *Journal of Agricoltural and Food Chemistry*. 56 (3): 627–629.

Serraino, I., Dugo, L., Dugo, P. et al. 2003. Protective effects of cyanidin-3-O-glucoside from blackberry extract against peroxynitrite-induced endothelial dysfunction and vascular failure. *Life Sciences* 73(9): 1097–1114.

Sharma, R., Raghuram, T. 1990. Hypoglycaemic effect of fenugreek seeds in non-insulin dependent diabetic subjects. *Nutrition Research* 10(7): 731–739.

Siegel, R.K. 1979. Ginseng abuse syndrome: problems with the panacea. *JAMA* 241(15): 1614–1615.

Seeff, L.B. 2007. Herbal hepatotoxicity. *Clinics in Liver Disease* 11(3): 577–596.

Smith, S.W., Shah, R.R., Hunt, J.L., Herzog, C.A. 2005. Bidirectional ventricular tachycardia resulting from herbal aconite poisoning. *Annals of Emergency Medicine* 45(1): 100–101.

Soni, K., Kuttan, R. 1992. Effect of oral curcumin administration on serum peroxides and cholesterol levels in human volunteers. *Indian Journal of Physiology and Pharmacology* 36: 273–273.

Sprecher, D.L., Harris, B.V., Goldberg, A.C., et al. 1993. Efficacy of psyllium in reducing serum cholesterol levels in hypercholesterolemic patients on high-or low-fat diets. *Annals of Internal Medicine* 119(7 Part 1): 545–554.

Srinivasan, K. 2007. Black pepper and its pungent principle-piperine: a review of diverse physiological effects. *Critical Reviews in Food Science and Nutrition* 47(8): 735–748.

Stangl, V., Lorenz, M., Stangl, K. 2006. The role of tea and tea flavonoids in cardiovascular health. *Molecular Nutrition & Food Research* 50(2): 218–228.

Stavric, B. 1994. Role of chemopreventers in human diet. *Clinical Biochemistry* 27(5): 319–332.

Sugimoto, K., Ohmori, M., Tsuruoka, S., et al. 2001. Different effects of St John's wort on the pharmacokinetics of simvastatin and pravastatin. *Clinical Pharmacology & Therapeutics* 70(6): 518–524.

Suneetha, W.J., Krishnakantha, T. 2005. Antiplatelet activity of coriander and curry leaf spices. *Pharmaceutical Biology* 43(3): 230–233.

Supakul, L., Pintana, H., Apaijai, N., Chattipakorn, S., Shinlapawittayatorn, K., Chattipakorn, N. 2014. Protective effects of garlic extract on cardiac function, heart rate variability, and cardiac mitochondria in obese insulin-resistant rats. *European Journal of Nutrition* 53(3): 919–928.

Torres, N., Torre-Villalvazo, I., Tovar, A.R. 2006. Regulation of lipid metabolism by soy protein and its implication in diseases mediated by lipid disorders. *Journal of Nutritional Biochemistry* 17: 365–373.

Tam, S.W., Worcel, M., Wyllie, M. 2001. Yohimbine: a clinical review. *Pharmacology & Therapeutics* 91(3): 215–243.

Umemura, T., Ueda, K., Nishioka, K., Hidaka, T., Takemoto, H., Nakamura, S. 2006. Effects of acute administration of caffeine on vascular function. *American Journal of Cardiology* 98: 1538–1541.

Vaidyaratnam, P.Vs. 1994. Indian medicinal plants: a compendium of 500 species. *Orient Longman Ltd, Madras* 4: 59–64.

Valcheva-Kuzmanova, S., Kuzmanov, K., Mihova, V., Krasnaliev, I., Borisova, P., Belchevam, A. 2007. Antihyperlipidemic effect of Aronia melanocarpa fruit juice in rats fed a high-cholesterol diet. *Plant Foods for Human Nutrition* 62(1): 19–24.

Vasdev, S., Gill, V. 2008. The antihypertensive effect of arginine. *International Journal of Angiology Spring* 17 (1): 7–22.

Vibes, J., Lasserre, B., Gleye, J., Declume, C. 1994. Inhibition of thromboxane A2 biosynthesis in vitro by the main components of Crataegus oxyacantha (Hawthorn) flower heads. *Prostaglandins, Leukotrienes and Essential Fatty Acids (PLEFA)* 50(4): 173–175.

Vitale, D.C., Piazza, C., Melilli, B., Drago, F., Salomone, S. 2013. Isoflavones: estrogenic activity, biological effect and bioavailability. *European Journal of Drug Metabolism and Pharmacokinetics* 38(1): 15–25.

Wang, C.Z., Mehendale, S.R., Calway, T., Yuan, C.S. 2011. Botanical flavonoids on coronary heart disease. *American Journal of Chinese Medicine* 39 (4): 661–671.

Warshafsky, S., Kamer, R.S., Sivak, S.L. 1993. Effect of garlic on total serum cholesterol a meta-analysis. *Annals of Internal Medicine* 119(7 Part 1): 599–605.

Wei, Z. 1997. Pycnogenol enhances endothelial cell antioxidant defenses. *Redox Report* 3: 219–224.

Wen, Y.D., Zhu, Y.Z. 2015. The pharmacological effects of S-propargyl-cysteine, a novel endogenous H2S- producing compound. *Handbook of Experimental Pharmacology* 230: 325–336.

WHO. 2009. World Health Organization cardiovascular diseases (CVDs) fact sheet N° 317. September 2009. Accessed April 12, 2010. Available at: http://www.who.int/mediacentre/factsheets/fs317/en/index.html.

Williamson, G. 2017. The role of polyphenols in modern nutrition. *Nutrition Bulletin* 42(3): 226–235.

Xia, M., Ling, W., Zhu, H., et al. 2007. Anthocyanin prevents CD40-activated proinflammatory signaling in endothelial cells by regulating cholesterol distribution. *Arteriosclerosis, Thrombosis, and Vascular Biology* 27(3): 519–524.

Xie, T., Song, S., Li, S., Ouyang, L., Xia, L., Huang, J. 2015. Review of natural product databases. *Cell Proliferation* 48(4): 398–404.

Yang, Y.S., Su, Y.F., Yang, H.W., Lee, Y.H., Chou, J.I., Ueng, K.C. 2014. Lipid-lowering effects of curcumin in patients with metabolic syndrome: a randomized, double-blind, placebo-controlled trial. *Phytotherapy Research* 28: 1770–1777.

Yeh, G.Y., Davis, R.B., Phillips, R.S. 2006. Use of complementary therapies in patients with cardiovascular disease. *The American Journal of Cardiology* 98(5): 673–680.

Youdim, K.A., McDonald, J., Kalt, W., Joseph, J.A. 2002. Potential role of dietary flavonoids in reducing microvascular endothelium vulnerability to oxidative and inflammatory insults. *The Journal of Nutritional Biochemistry* 13(5): 282–288.

Zou, Q.J.B. 1989. Effect of mother wort on blood hyper viscosity. *American Journal of Chinese Medicine* 17(12): 6570.

9 Serum Albumin Binding of Natural Substances and Its Influence on the Biological Activity of Endogenous and Synthetic Ligands for G-Protein-Coupled Receptors

Sarah Engelbeen and Patrick M.L. Vanderheyden

CONTENTS

9.1 SERUM ALBUMIN AS A BINDER OF MULTIPLE NATURAL AND SYNTHETIC MOLECULES

Serum albumin belongs to a family of soluble blood plasma proteins in mammals. It is produced in the liver and represents the most abundant protein in the blood. Its major biological function is to maintain the oncotic or colloid osmotic pressure, which is essential for the appropriate distribution of body fluids between the blood vessels and the tissues (Nicholson et al., 2000). The primary function of albumin is to act as a transporter protein for endogenous and apolar substances such as thyroid hormones, steroids and fatty acids (Kragh-Hansen, 1981). The general structure of albumin encompasses several alpha helices and contains eleven distinct binding sites for apolar compounds. It possesses one heme group and can bind up to nine fatty acids in one serum albumin molecule (Fanali et al., 2012). Serum albumins are present in all mammals, and the bovine form, that is, bovine serum albumin (BSA) is used in a wide array of cellular and biochemical assays and procedures. They include immunodetection (such as Enzyme-Linked ImmunoSorbent Assay or ELISA), mammalian cell culture media, protein chemistry and screening for enzyme and receptor interacting substances (Mather, 1998; Vazquez et al., 2011; Xiao and Isaacs, 2012). These latter assays have enormously expanded our knowledge on the molecular mechanisms of endogenous molecules such as hormones, neurotransmitters, enzyme substrates and synthetic analogues. These methodologies allow us to isolate and characterize active components of natural origin such as plants, invertebrates, fungi and bacteria. Furthermore, such identified molecules can serve as leads to synthesize even more active and specific molecules. Traditional medicine, which is mainly based on phytotherapy, can benefit from knowledge of the molecular mechanism(s) and an accurate determination of the potency of these substances from certain plants and specific parts/extracts of them. In relation to this, serum albumin has the capability and flexibility to bind a very broad spectrum of natural substances, some of which are used in the treatment of hypertension, hypercholesterolemia or as antioxidant. In relation to this, we will focus in this paragraph on the ability of plant-derived polyphenols/tannins to interact with serum albumin. Several processed food products and beverages such as black tea, matured red wine, coffee and cocoa contain these so-called polyphenolic compounds (Quideau et al., 2011). There have been numerous human intervention trials to investigate or provide evidence for the beneficial effects of these polyphenol-rich foods on cardiovascular diseases, neurodegeneration and cancer (Del Rio et al., 2013). In this context, most of the mechanistic properties of this large group of compounds have been studied in cell lines and/or in other *in vitro* models. A particular group of polyphenols, the green tea catechins, represent an example in which epidemiological, clinical as well as experimental studies established the multiple effects of these compounds. They include cardioprotective and antioxidant effects, reduction of plasma lipids, improvement of endothelial functions and anti-inflammatory, anti-proliferative and anti-thrombotic effects (Babu and Liu, 2008). As any biological effect of these ingested compounds is mediated by their presence in the plasma, it is of great importance to be aware of their interaction with serum albumin and whether this might affect their biological activity. Several studies indeed revealed the serum binding capacity of tannins/polyphenols. A few of these studies are cited below.

Tannins, which are a group of polyphenolic compounds, include three major categories: gallic acid esters, phloroglucinol derivatives and flavone-derived substances (Hemingway and Karchesy, 1989). Initially, the BSA binding capacity of tannins was determined by immobilizing certain polyphenolics on chromatography paper followed by the quantification of the amount of BSA bound (Dawra et al., 1988). Ellagic acid and quercetin were found to bind BSA with a capacity of 297 and 78 μg BSA/mg, respectively (Dawra et al., 1988). In a similar approach, the polyphenols extracted from coffee pulp, that is, tannic acid, chlorogenic acid and catechin, were found to bind to BSA (Vélez et al., 1985). Furthermore, sodium dodecyl sulfate (SDS) and native gel electrophoresis revealed the formation of BSA complexes with the water-soluble polyphenols oenotehin B, corilagin, (+)catechin, procyanidin B3 and gallic acid derivatives (Kusuda et al., 2006). Polyphenolic compounds isolated from *Opuntia ficus indica* were found to elicit calcium response in T-cell lines, and this effect was reduced by including fatty acid free BSA, suggesting their interaction with serum albumin (Aires et al., 2004). By using the technique of quartz crystal microbalance, it was possible to monitor the binding of molecules to the surface of proteins, in which it was found that thearubigin, which is one of the major polyphenols of black tea, can form complexes with BSA (Chitpan et al., 2007).

Catechins, which are the major polyphenolic compounds in green tea, are found to bind to human and bovine serum albumin (Zinellu et al., 2014). The flavonoids display higher affinity for human serum albumin when compared to BSA, and the galloyl moiety increases the binding affinity. The authors in this study suggest that the ability of serum albumin binding can modulate the plasma concentration and thus the biological activity of these compounds. In this context, the BSA binding of the polyphenols (+)-catechin, (–)-epicatexhin, (–)-epicatexhin-gallate, malvidin-3-glucoside, tannic acid, procyanidin B4, procyanidin B2 gallate and procyanidin oligomers were compared by measuring the quenching of the protein intrinsic fluorescence (Soares et al., 2007), in which it was found that binding affinity increased with the size of the molecules and with the presence of galloyl groups. Fluorescence quenching and circular dichroism were used to characterize the BSA binding of the polyphenols resveratrol, genistein and curcumin (Bourassa et al., 2010). This study indicated that these polyphenols elicited a change of the BSA conformation.

Another systematic study compared the BSA binding affinity of tea tannins, including tea catechins, grape seed proanthocyanidins, mimosa 5-deoxy proanthocyanidins and sorghum procyanidins (Frazier et al., 2010). These BSA binding interactions were found to be exothermic and involved different binding sites. In this context, thirteen isoflavones isolated from *Puerariae lobate* flowers were found to bind to BSA (Liu et al., 2012). The binding affinity depended on the methoxylation and hydroxylation of these compounds and was applied to isolate active compounds from complex mixtures. Tannins isolated from grape seeds were also found to interact with BSA (Ferrer-Gallego et al., 2012). Their affinity increased with their molecular mass and according to their hydrophobicity. In the same way substitution of the C-3 of catechins has an important impact on their binding affinity to serum albumin (Ikeda et al., 2017). Moreover, the presence of a gallate moiety on catechin increases its affinity for BSA (Pal et al., 2012). Interestingly, it was found that while these catechins have moderate antioxidant activity, in the presence of BSA this activity was substantially enhanced (Almajano et al., 2007; Bonoli-Carbognin et al., 2008). Obviously it is not possible within the scope of this chapter to provide a complete overview of the serum binding properties of natural substances that might be beneficial in the treatment of chronic diseases. However, the studies cited above stress the importance of evaluating the ability of serum albumin to modulate the biological activity of such components. In relation to this, in the next part we will discuss the ability of BSA to influence the interaction of natural and endogenous compounds with the cellular/molecular targets and focus on G-protein-coupled receptors, in particular.

9.2 INFLUENCE OF SERUM ALBUMIN ON CELLULAR AND BIOCHEMICAL READ-OUT SYSTEMS

In vitro biochemical assays to screen natural products/extracts include assays such as radioligand binding to receptors, functional read-out and/or viability assays on cells and enzyme measurements. In order to be able to appreciate the potency of these compounds, these *in vitro* determinations should approach the *in vivo* circumstances (Vergote et al., 2009; Vazquez et al., 2011). This chapter focusses on the influence of proteins and particularly serum albumin, on the binding of endogenous, synthetic and natural substances to their molecular target. Conceptually, two basic mechanisms can be discerned by which serum albumin can be able to influence the target binding of these molecules. As all the biochemical assays are carried out in water-based buffers at the physiological pH of 7.4, the solubilization of mainly hydrophobic substances may represent a serious hurdle in the screening of apolar extracts/compounds. Therefore, inclusion of serum albumin (mostly BSA) might be helpful during the solubilization step as well as during the incubation of the cells/membranes/proteins containing the receptor or enzyme of interest (Mather, 1998). Moreover, complexation with albumin might also enhance (or not) the stability of the compounds and eventually favor the active conformation of larger peptides or proteins. Another benefit could come from the reduction of non-specific compound interactions such as with non-target membrane constituents, plastic or glass binding (Goebel-Stengel et al., 2011). On the other hand, serum albumin can also have an adverse effect on the enzyme/receptor binding by reducing the free concentration of the compounds (Kragh-Hansen, 1981; Fanali et al., 2012).

9.3 INFLUENCE OF BSA ON LIGAND BINDING TO G-PROTEIN-COUPLED RECEPTORS

In the next part of this chapter we will discuss a few concrete examples in which the binding and effects of ligands on G-protein-coupled receptors can be influenced by serum albumins such as BSA. BSA can act at different levels of the ligand-receptor interaction. For instance, Vergote et al. used the 125I-vasoactive intestinal peptide ([^{125}I]-VIP) binding assay in rat lung membranes to systematically evaluate the effects of BSA on ligand adsorption as well as its total and non-specific binding (Vergote et al., 2009). These authors draw our attention to the importance of the experimental conditions and in particular the absence or presence of BSA on the screening of peptide receptor ligands. Below we discuss the current information regarding the ability of BSA to modulate the binding and potency of three different GPCR ligands, that is, angiotensin II (AT_1 receptors), cholecystokinin-8 (CCK receptors) and free fatty acids (FFA receptors).

9.3.1 Influence on the Binding of Ligands of the Renin-Angiotensin System

First, we will discuss the consequences of the interaction between BSA and angiotensins. The octapeptide angiotensin II is generated by the renin-angiotensin system (RAS) and is a potent vasoactive hormone with a wide range of (patho-)physiological effects that include vasoconstriction, increase of the renal sodium re-absorption, increase of the production of the mineralocorticoid hormone aldosterone, enhance growth of smooth-muscle cells and cardiomyocytes and increase inflammation (De Gasparo et al., 2000). Most of these detrimental effects are mediated by its interaction with angiotensin type 1 receptors (AT1R), a typical member of the large family of G-protein-coupled receptors. The first evidence that BSA influences ligand-AT1R interaction came from functional and binding studies in which angiotensin II receptor activation was inhibited by non-peptide AT1R antagonists. It was found that the Ki values of biphenyl-tetrazoles containing two acidic groups such as BMS-18560 and EXP3174 were substantially higher in the presence of 0.22% compared to 0.01% BSA (23 times for BMS-18560 and eight times for EXP3174 (Dickinson et al., 1994). Similarly, the binding affinity of two other diacid AT1R blockers, that is, DUP 532 and GR 117,289, were also decreased in the presence of BSA (Chiu et al., 1991; Robertson et al., 1992). Based on these results, these authors concluded that (at least partially) the presence of the diacid function results in substantial BSA binding of these drugs and therefore reduces their free concentration to bind to the AT1R. For instance, the group of Andrew T. Chiu demonstrated that up to 99% of DUP 532 is bound to BSA, while this was 'only' 60% for losartan (Chiu et al., 1991). In a later, more systematic study, Maillard, and colleagues studied the ability of different AT1R antagonists to compete with 125I-angiotensin II- AT1R binding in the absence or presence of human serum albumin. Interestingly, the fold decrease in affinity by serum albumin varied between only 1.2 for losartan (a non-diacidic molecule) up to >5000 for DUP-532 and inversely correlated with their Kd for human serum albumin as determined by surface plasmon resonance measurements (Maillard et al., 2001). In other words, the stronger these antagonists bind to serum albumin, the more pronounced the receptor binding affinity is influenced. Such pronounced binding to plasma proteins (such as BSA) is hypothesized to increase the plasma residence of diacid AT1R antagonists and therefore prolong their receptor blockade. In order to gain insight into the molecular interactions between AT1R antagonists and human serum albumin (HSA), a docking and molecular dynamic simulation was undertaken by Jinyu Li and colleagues (Li et al., 2010). These analyses revealed that all non-peptide AT1R antagonists bind to two distinct binding sites on HSA with high- and low-affinity, respectively, corresponding to site II (subdomain IIIA of HSA) and site I. Furthermore, the relative contribution of electrostatic and hydrophobic interactions for valsartan, which possesses a typical the biphenyl-tetrazole moiety, was higher than that of telmisartan, which has a distinct bis-benzimidazole structure (Li et al., 2010). In this context, Dickinson et al. suggested that all di-acidic AT1R antagonists bind at two binding sites of BSA, respectively: the bilirubin site 3 and the common drug binding site 6, according to the classification of Kragh-Hansen (Kragh-Hansen, 1981; Dickinson et al., 1994).

Interestingly, the binding of angiotensin II to BSA (or human serum albumin) has not been investigated systematically. However, the *in vivo* peripheral hemodynamic and central behavioral effects are investigated in rats after intraperitoneal administration of free angiotensin II and when it is covalently coupled to BSA or the calcium-binding neurospecific protein S100b, using the carbodiimide coupling method (Tolpygo et al., 2014, 2015). It was found that protein-angiotensin II complexes induced a similar increase in goal-seeking drinking behavior as compared to the free peptide. These protein-angiotensin II complexes also caused a robust increase in systolic blood pressure, though it was slightly less compared to free angiotensin II (Tolpygo et al., 2014, 2015). Interestingly, these effects on blood pressure were inhibited by the angiotensin-(1–7) fragment, which is proposed to behave as a functional antagonist of angiotensin II via interaction with a distinct GPCR (Mas-receptor). Moreover, these inhibitory effects were preserved by administration of angiotensin-(1–7)-BSA complex (Tolpygo et al., 2012).

Nevertheless, it is not known whether BSA modulates the ability of angiotensin II to activate AT1R. In order to systematically investigate this, we have measured angiotensin II-mediated calcium responses in the absence and presence of 0.1% BSA. As read-out system, we used CHO-K1 cells that are stably transfected with the bioluminescent protein aequorin and that upon calcium binding emit a strong light signal that was measured in a 96-well reader (denoted as CHO-AEQ cells). Transfecting these cells with the human AT1R enabled us to measure characteristic transient calcium responses by angiotensin II. The response was determined by calculating the area under the curve of transient rise in luminescence after angiotensin II exposure (Bahem

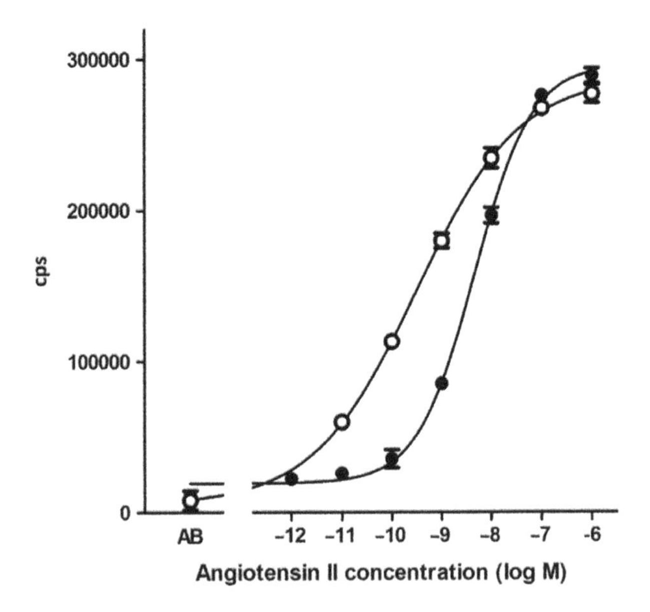

FIGURE 9.1 Angiotensin II concentration-response curves in CHO-AEQ cells transfected with the human angiotensin AT1 receptor in the absence (open symbol) and presence (closed symbol) of 0.1 % BSA. The potency of Angiotensin II is higher in the presence of BSA.

et al., 2015). It appeared that the angiotensin II concentration-response curves were slightly leftward shifted in the presence of 0.1% BSA (Figure 9.1). The corresponding EC_{50} values were 0.31 and 4.54 nM, respectively, in the presence or absence of BSA, indicating that angiotensin II becomes more potent in the presence of BSA. Although this observation indicates that BSA is indeed capable of binding angiotensin II, the underlying mechanism needs to be further elucidated.

9.3.2 Influence on the Binding of Cholecystokinins

In contrast to the small effect on angiotensin II, a more pronounced effect of BSA is observed for the octopeptide cholecystokinin-8 (CCK-8). Cholecystokinins are a group of brain-gut peptides involved in several peripheral effects, including gall-bladder contraction and hepatic bile secretion as well as modulating (patho) physiological processes like panic disorders or anxiety and satiety. Cholecystokinin is a long precursor which is cleaved into CCK-8 and other active fragments CCK-33, CCK-39 and CCK-58. CCK-8, of which the tyrosine can be sulphated (CCK-8s) and the C-terminal carboxylic groups is amidated, exerts its actions via binding to two distinct G-protein-coupled receptors that is, CCK1R and CCK2R, which have a distinct localization and selectivity for sulphated versus non-sulphated CCK peptides. Conflicting data were published regarding the effect of BSA on CCK-8 binding. BSA was shown to prevent the binding of 125I-labeled CCK-8 to CCK receptors in pancreatic acini membranes (mainly expressing CCK1R) (Huang et al., 1995). On the other hand, BSA was reported to specifically inhibit 125I-labeled CCK-8 binding to cerebral cortex (predominantly CCK2R) but not pancreatic membranes (Wennogle et al., 1985). Surprisingly, BSA increased the potency of CCK-8 to stimulate amylase secretion in the pancreas as well as contraction of gastric smooth-muscle cells (Huang et al., 1995). While both studies indicate that CCK-8 can strongly bind to BSA, it appears that the complexed CCK-8 displays an enhanced potency to activate the CCK receptors. In agreement with these findings, we found that BSA profoundly enhanced the potency of CCK-8 to induce calcium responses as measured by the aequorin methodology in CHO-AEQ cells that were transfected with CCK1R (see Figure 9.2). The corresponding EC_{50} values for CCK-8s were 155 and 3.6 nM in the absence or presence of 0.1% BSA, respectively. BSA increased the potency of CCK-8 to inhibit the binding of [125-I]-CCK-8 to pancreatic acini, indicating the ability of BSA to convert low- to high-agonist affinity states of the CCK1R (Huang et al., 1995). Yet in this study, these authors proposed the ability of BSA to enhance the efficacy by which CCK-8 elicits the receptor responses, an alternative hypothesis that certainly merits further investigation.

9.3.3 Influence on the Binding of Free Fatty Acids to G-Protein-Coupled Receptors

The third example deals with the influence of BSA on the receptor binding of free fatty acids. Relatively recently, a small cluster of four rhodopsin-like GPCR were deorphanized by the identification of small to large free fatty acids as their ligands.

INFLUENCE OF SERUM ALBUMIN ON THE BIOLOGICAL ACTIVITY OF ENDOGENOUS AND SYNTHETIC LIGANDS OF G-PROTEIN-COUPLED RECEPTORS

Human Serum Albumin as a Typical Example of a Binder for Free Fatty Acids

Serum albumins display an immense capacity to bind diverse endogenous and exogenous compounds. This property makes them crucial carriers and/or depots, which not only affects the pharmacokinetics of many drugs but also can protect against toxins. Structural and biochemical studies have been able to reveal the multidomain organization of albumins and their different binding sites. Although most of the biochemical assays are carried out with BSA (as are those discussed in this chapter), most of the structural information is available for human serum albumin (HSA). Notwithstanding this, the structures of BSA and of rodent and equine serum albumin display substantial similarity to the structure of HSA. The sequence identity of BSA compared to HSA is 75.5%. Moreover, the root-mean-square deviation of atomic positions which is the measure of the average distance between the atoms (usually the backbone atoms) of superimposed proteins is only 1.1 Ångström when comparing HSA with BSA (Majorek et al., 2012). This, together with the lack of structural data on ligand-bound BSA, motivated us to provide a brief overview of information regarding the ligand-binding sites of HSA with a focus on the sites for fatty acids. As a full listing of the structural data on HSA and all the characterized binding sites is outside the scope of this chapter, we refer to the excellent review article of Fanali et al. (2012) for further reading.

One of the first albumin binding components was suspected to "lipoidal in nature" (Kendall, 1941). Subsequently, a huge number of studies characterized the presence of up to nine so-called fatty acid binding sites (FA1-FA9). This group of binding sites also accommodates the binding of a wide variety of other non-fatty acid compounds which may lead to improved plasma solubility, extension of the half-life of drugs and/or reduction of the free active concentration. Within Figure B9.1, the modular organization of HSA is depicted. HSA is organized in three homologous domains denoted as I (1-195), II (196-383) and III (384-585), which contain comparable amino acid sequences. Regarding the secondary structure, 68% of the amino acid residues are part of α-helices, while no β-sheet elements are present. Overall, HSA adopts a so-called globular heart-shaped conformation (see Figure B9.1).

The overall three-dimensional structure of HSA is maintained when determined in the absence or presence of endogenous or exogenous ligands (Fanali et al., 2012). Moreover, its overall structure is common among the vertebrate serum albumins. Since a vast number of liganded HSA structures have been elucidated, it is far beyond the scope of this chapter to discuss all these structures. Nevertheless, to illustrate the complexity of the fatty acid binding sites on serum albumin, the structure of HSA when bound to the C_{14} chain fatty acid myristic acid is depicted in Figure B9.1.

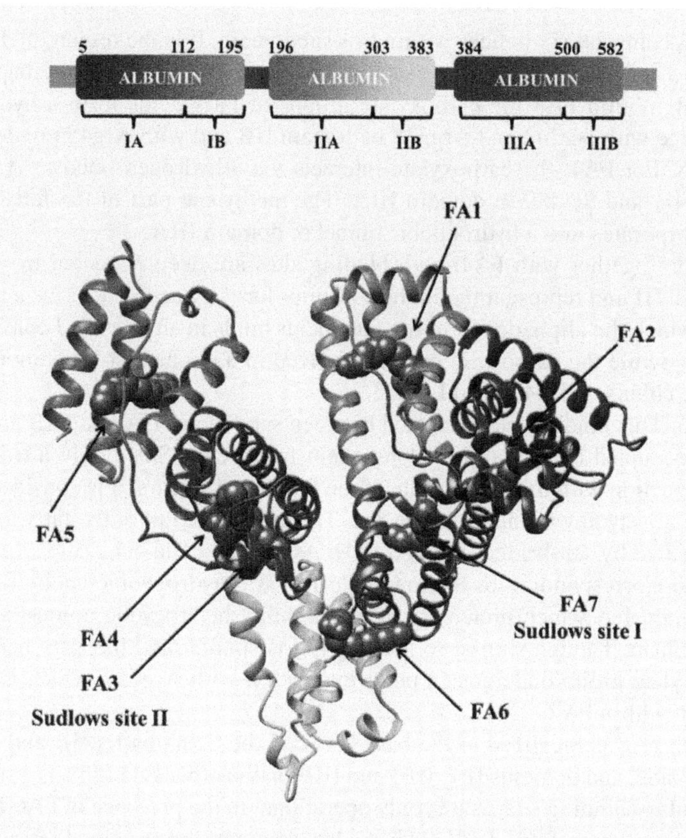

FIGURE B9.1 Domain organization of HSA bound with myristate. The upper part depicts the three conserved domains. The lower part shows the three-dimensional structure of these domains with the corresponding color. Myristate is shown as a space-filled structure in gray. Reproduced with permission from Fanali et al. (2012).

The major fatty acid binding sites of HSA are:

- FA1: The so-called heme binding pocket accommodates the binding of synthetic porphyrin, phthalocyanine and the pigment bilirubin, which is derived from heme-Fe degradation. All fatty acids (medium- to long-chain fatty acids) bind by forming a hydrogen bond between their carboxylate group and Arg117 in subdomain IB of HSA.
- FA2: This binding site is located and enclosed between subdomains IA and IIA and displays medium affinity for fatty acids. Binding of FAs to this site induces a conformational change of HSA in which domain I is rotated relatively to domain II. The carboxylate group of fatty acids forms hydrogen bonds with amino acid side chains in subdomain IIA (i.e. Tyr150, Arg257 and Ser287).
- FA3 and FA4: Both binding sites are located in a cavity within subdomain IIIA, and together they form the so-called Sudlow's site II. These binding

sites comprise six helices within this subdomain. It is the region of HSA to which aromatic carboxylates such as the anti-inflammatory drug ibuprofen bind. Within FA3 the carboxylate groups of fatty acids form a hydrogen bridge with Ser342 and Arg348 in domain IIB and with Arg485 in domain IIIA. For FA4, the carboxylate interacts via a hydrogen bond to Arg410, Tyr411 and Ser489 in domain IIIA. The methylene part of the fatty acids incorporates into a hydrophobic tunnel of domain IIIA.

- FA5: Together with FA4, both binding sites are deeply located in subdomain III and represent the highest affinity for fatty acids. It forms a tunnel in which the aliphatic tail of the fatty acids binds in an extended conformation, while the carboxylic head is involved in a strong salt-bridge with the side chains of Tyr401 and Lys525.
- FA6: This binding site is located between subdomain IIA and IIIB and can be occupied by medium- to long-chain fatty acids. Since only a transient interaction with the carboxylate of the fatty acids occurs, it is considered as a relatively low affinity binding site. The methylene tail of the fatty acids is attached by salt-bridges with Arg209, Asp324 and Glu354.
- FA7: Corresponding to Sudlow's site I, it is a hydrophobic cavity of subdomain IIA which preferentially binds bulky heterocyclic anions such as warfarin. Though similar to FA3/FA4, it is smaller and the fatty acid carboxylate group undergoes a polar interaction with Arg257, which is common within FA2.
- FA8: FA8 is localized at the base between the domains IA, IB and IC on one side, and domains IIA, IIIB and IIIB on the other side. This is a supplementary binding site, as it is only operational in the presence of FA6-bound medium chain fatty acids such as decanoic acid (also named capric acid with 11 C-atoms).
- FA9: This binding site is located at the upper part of the domains IA, IB and IC on one side, and domains IIA, IIIB and IIIB on the other side. Amino acids Glu187 of domain I and Lys432 of domain III form a salt-bridge with the fatty acids. Similar as FA8, this binding site is formed only by conformational changes induced by saturating concentrations of fatty acids.

In addition to the fatty acid binding sites, many other ligand-binding sites are described in HSA. They include binding sites for thyroxine (at least four distinct sites), for the growth and virulence-promoting bacterial surface protein PAB and for metal ions such as Mg^{2+}, Al^{2+}, Ca^{2+}, Mn^{2+}, Co^{2+} and Co^{3+}, Ni^{2+}, Cu^{+} and Cu^{2+}, Zn^{2+}, Cd^{2+}, Pt^{2+}, Au^{+} and Au^{2+}, Hg^{2+} and Tb^{3+} (for which three major binding sites are found). These sites overlap with those for the fatty acids but are not further discussed here.

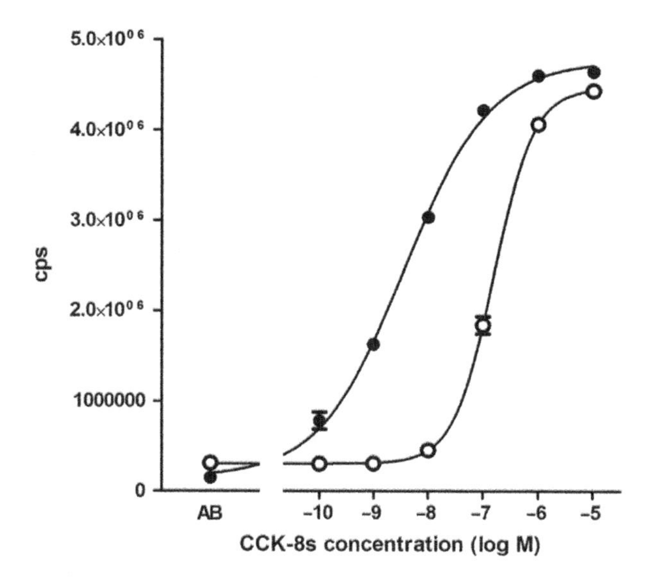

FIGURE 9.2 CCK-8s concentration-response curves in CHO-AEQ cells transfected with the human CCK receptor in the absence (empty symbol) and the presence (full symbol) of 0.1% BSA. The potency of CC8-8s is profoundly increased in the presence of BSA. (From Engelbeen S. and Vanderheyden P.M.L., unpublished data.)

Accordingly, they were baptized as free fatty acid receptors (FFA1-4). The FFA1 receptor was first deorphanized and is activated by medium (C7-C12) to long (C > 12) saturated and unsaturated chain fatty acids (Briscoe et al., 2003; Itoh et al., 2003; Kotarsky et al., 2003). The FFA2 and FFA3 receptors are activated by short-chain fatty acids (SCFAs, C2-C6) (Brown et al., 2003; Le Poul et al., 2003). FFA4 has a similar ligand profile as FFA1 (Hirasawa et al., 2008). Interest in the screening of ligands for FFA1 and FFA4 receptors came from their localization and their ability to enhance glucose-stimulated insulin secretion in pancreatic β-cells. With respect to the pharmacology, the FFA1 receptor was found to be selectively coupled to Gq/11 G-proteins, leading to elevation of intracellular calcium (Briscoe et al., 2003; Itoh et al., 2003; Kotarsky et al., 2003). Long before FFA1/4 receptors were deorphanized, their ligands were extensively described to bind to human and bovine serum albumin (Spector et al., 1969; Ashbrook et al., 1972, 1975; Kragh-Hansen, 1981). It is, therefore, not surprising that the presence of BSA modulates FFA1 ligand-mediated activation. Indeed, Stoddart et al. (2007) found that addition of 0.5% (w/v) BSA completely inhibited the receptor-mediated calcium response induced by 100 μM lauric acid (a typical LCFA FFA1 receptor agonist). In the same study, it was found that the FFA1 receptor displays full constitutive activation (in the absence of FFA1 ligands) by measuring [^{35}S]GTPγS binding in membranes and thereby makes it impossible to observe ligand-dependent activation. By making use of the ability to strongly bind free fatty acids, inclusion free fatty acid free BSA was able to reduce basal [^{35}S]GTPγS binding by complexing and removal of endogenous intracellular free fatty acids released during the membrane preparation (Stoddart et al., 2007). To confirm this notion, we have examined the absence and

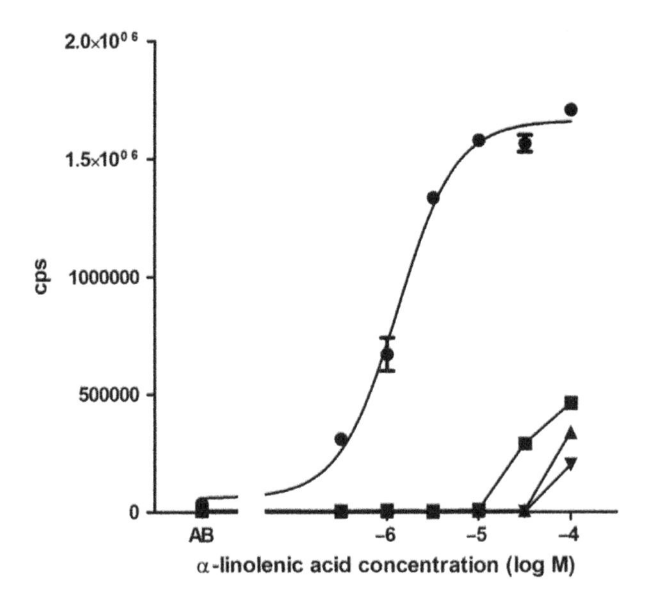

FIGURE 9.3 α-linolenic acid concentration-response curve in CHO-AEQ cells transfected with human FFA1 receptor calculated with the area under the curve of the transient luminescence response as a result of intracellular calcium rise in the absence of BSA (●). In the presence of 0.025% (■), 0.05% (▲) or 0.1% (▼) BSA the potency of α-linolenic acid is profoundly decreased. (From Engelbeen S. and Vanderheyden P.M.L., unpublished data.)

presence of BSA on α-linolenic acid-induced calcium responses in FFA1 receptor transfected CHO-AEQ cells. As shown in Figure 9.3, it was found that BSA as low as 0.025% (w/v) profoundly decreased the potency of this ligand. No concentration-response curves could be calculated in the presence of BSA since high concentrations of α-linolenic acid had presumably detergent effects on the cells. The EC_{50} values in the absence of BSA was 1.3 μM. Taken together, these data indicate that, in the case of free fatty acids, serum albumin can strongly influence the actions of endogenous ligands and possibly also that of synthetic agonists.

9.4 GENERAL CONSIDERATIONS

It is well known that serum albumin is one of the most abundant proteins present in blood plasma and that it serves as storage and transporter for both endogenously and exogenously administrated compounds. Accordingly, there is massive literature data describing the BSA binding of many compounds which might influence their transport, tissue distribution, metabolism and excretion. The intent of this chapter was to provide a few examples in which the influence of BSA on the ability of ligands to activate and bind to G-protein-coupled receptor was examined. The main message is that despite the wealth of data regarding the direct serum albumin binding of these ligands, the impact on the EC_{50} values (as determined by *in vitro* functional data) is quite variable. While the presence of BSA profoundly increases the potency of CCK-8-induced signaling, the opposite is seen for free fatty acid activation of FFA1 receptors, and this is despite the knowledge that both compounds are well described as binding to BSA. A basic knowledge of the underlying molecular mechanisms is of importance in the accurate determination of the potency of ligand-receptor activation and contributes to the further understanding of the action mechanism(s) of endogenous and synthetic ligands.

REFERENCES

Aires, V., Adote, S., Hichami, A., Moutairou, K., Boustani, E. S. & Khan, N. A. Modulation of intracellular calcium concentrations and T cell activation by prickly pear polyphenols. *Mol. Cell. Biochem.* **260**, 103–110 (2004).

Almajano, M. P., Delgado, M. E. & Gordon, M. H. Albumin causes a synergistic increase in the antioxidant activity of green tea catechins in oil-in-water emulsions. *Food Chem.* **102**, 1375–1382 (2007).

Ashbrook, J. D., Spector, A. A. & Fletcher, J. E. Medium chain fatty acid binding to human plasma albumin. *J. Biol. Chem.* **247**, 7038–7042 (1972).

Ashbrook, J. D., Spector, A. A., Santos, E. C. & Fletcher, J. E. L. Long chain fatty acid binding to human plasma albumin *J. Biol. Chem.* **250**, 2333–2338 (1975).

Babu, P. V. A. & Liu, D. Green tea catechins and cardiovascular health: An update. *Curr. Med. Chem.* **15**, 1840–1850 (2008).

Bahem, R., Hoffmann, A., Azonpi, A., Caballero-George, C. & Vanderheyden, P. Modulation of calcium signaling of angiotensin AT1, endothelin ETA, and ETB receptors by silibinin, quercetin, crocin, diallyl sulfides, and ginsenoside Rb1. *Planta Med.* **81**, 670–678 (2015).

Bonoli-Carbognin, M., Cerretani, L., Bendini, A., Almajano, M. P. & Gordon, M. H. Bovine serum albumin produces a synergistic increase in the antioxidant activity of virgin olive oil phenolic compounds in oil-in-water emulsions. *J. Agric. Food Chem.* **56**, 7076–7081 (2008).

Bourassa, P., Kanakis, C. D., Tarantilis, P., Pollissiou, M. G. & Tajmir-Riahi, H. A. Resveratrol, genistein, and curcumin bind bovine serum albumin. *J. Phys. Chem. B* **114**, 3348–3354 (2010).

Briscoe, C. P., Tadayyon, M., Andrews, J. L., Benson, W. G., Chambers, J. K., Eilert, M. M., Ellis, C., et al. The orphan G protein-coupled receptor GPR40 is activated by medium and long chain fatty acids. *J. Biol. Chem.* **278**, 11303–11311 (2003).

Brown, A. J., Goldsworthy, S. M., Barnes, A. A., Eilert, M. M., Tcheang, L., Daniels, D., Muir, A. I., et al. The orphan G protein-coupled receptors GPR41 and GPR43 are activated by propionate and other short chain carboxylic acids. *J. Biol. Chem.* **278**, 11312–11319 (2003).

Chitpan, M., Wang, X., Ho, C. T. & Huang, Q. Monitoring the binding processes of black tea thearubigin to the bovine serum albumin surface using quartz crystal microbalance with dissipation monitoring. *J. Agric. Food Chem.* **55**, 10110–10116 (2007).

Chiu, A. T., Carini, D. J., Duncia, J. V., Leung, K. H., McCall, D. E., Price, W. A., Wong, P. C., et al. DuP 532: A second generation of nonpeptide angiotensin II receptor antagonists. *Biochem. Biophys. Res. Commun.* **177**, 209–217 (1991).

Dawra, R. K., Makkar, H. P. S. & Singh, B. Protein-binding capacity of microquantities of tannins. *Anal. Biochem.* **170**, 50–53 (1988).

De Gasparo, M., Catt, K. J., Inagami, T., Wright, J. W. & Unger, T. International Union of Pharmacology. XIII. The angiotensin II receptors. *Pharmacol. Rev.* **52**, 415–472 (2000).

Del Rio, D., Rodriguez-Mateos, A., Spencer, J. P., Tognolini, M., Borges, G. & Crozier, A. Dietary (poly)phenolics in human health: Structures, bioavailability, and evidence of protective effects Against chronic diseases. *Antioxid. Redox Signal.* **18**, 1818–1892 (2013).

Dickinson, K. E. J., Cohen, R. B., Skwish, S., Delaney, C. L., Serafino, R. P., Poss, M. A., Gu, Z., et al. BMS-180560, an insurmountable inhibitor of angiotensin II-stimulated responses: Comparison with losartan and EXP3174. *Br. J. Pharmacol.* **113**, 179–189 (1994).

Engelbeen S. and Vanderheyden P.M.L., unpublished data. Functional characterization and influence of bovine serum albumin on α-linolenic acid mediated activation of free fatty acid 1 receptors in Chinese hamster ovary cells.

Fanali, G., di Masi, A., Trezza, V., Marino, M., Fasano, M. & Ascenzi, P. Human serum albumin: From bench to bedside. *Mol. Aspects Med.* **33**, 209–290 (2012).

Ferrer-Gallego, R., Gonçalves, R., Rivas-Gonzalo, J. C., Escribano-Bailón, M. T. & De Freitas, V. Interaction of phenolic compounds with bovine serum albumin (BSA) and α-amylase and their relationship to astringency perception. *Food Chem.* **135**, 651–658 (2012).

Frazier, R. A., Deaville, E. R., Green, R. J., Stringano, E., Willoughby, I., Plant, J. & Mueller-Harvey, I. Interactions of tea tannins and condensed tannins with proteins. *J. Pharm. Biomed. Anal.* **51**, 490–495 (2010).

Goebel-Stengel, M., Stengel, A., Taché, Y. & Reeve, J. R. The importance of using the optimal plasticware and glassware in studies involving peptides. *Anal. Biochem.* **414**, 38–46 (2011).

Hemingway, R. W. & Karchesy, J. J. *Chemistry and Significance of Condensed Tannins*. New York: Plenum Press (1989).

Hirasawa, A., Hara, T., Katsuma, S., Adachi, T. & Tsujimoto, G. Free fatty acid receptors and drug discovery. *Biol. Pharm. Bull.* **31**, 1847–1851 (2008).

Huang, S. C., Talkad, V. D., Fortune, K. P., Jonnalagadda, S., Severi, C., Delle Fave, G. & Gardner, J. D. Modulation of cholecystokinin activity by albumin. *Proc. Natl. Acad. Sci. U. S. A.* **92**, 10312–10316 (1995).

Ikeda, M., Ueda-Wakagi, M., Hayashibara, K., Kitano, R., Kawase, M., Kaihatsu, K., Kato, N., et al. Substitution at the C-3 position of catechins has an influence on the binding affinities against serum albumin. *Molecules* **22**, 1–12 (2017).

Itoh, Y., Kawamata, Y., Harada, M., Kobayashi, M., Fujii, R., Fukusumi, S., Ogi, K., et al. Free fatty acids regulate insulin secretion from pancreatic beta cells through GPR40. *Nature* **422**, 173–176 (2003).

Kotarsky, K., Nilsson, N. E., Flodgren, E., Owman, C. & Olde, B. A human cell surface receptor activated by free fatty acids and thiazolidinedione drugs. *Biochem. Biophys. Res. Commun.* **301**, 406–410 (2003).

Kragh-Hansen, U. Molecular aspects of ligand binding to serum albumin. *Pharmacol. Rev.* **33**, 17–53 (1981).

Kusuda, M., Hatano, T. & Yoshida, T. Water-soluble complexes formed by natural polyphenols and bovine serum albumin: Evidence from gel electrophoresis. *Biosci. Biotechnol. Biochem.* **70**, 152–160 (2006).

Le Poul, E., Loison, C., Struyf, S., Springael, J. Y., Lannoy, V., Decobecq, M. E., Brezillon, S., et al. Functional characterization of human receptors for short chain fatty acids and their role in polymorphonuclear cell activation. *J. Biol. Chem.* **278**, 25481–25489 (2003).

Li, J., Zhu, X., Yang, C. & Shi, R. Characterization of the binding of angiotensin II receptor blockers to human serum albumin using docking and molecular dynamics simulation. *J. Mol. Model.* **16**, 789–798 (2010).

Liu, L., Ma, Y., Chen, X., Xiong, X. & Shi, S. Screening and identification of BSA bound ligands from Puerariae lobata flower by BSA functionalized Fe3O4 magnetic nanoparticles coupled with HPLC-MS/MS. *J. Chromatogr. B* **887–888**, 55–60 (2012).

Maillard, M. P., Centeno, C., Frostell-Karlsson, Å., Brunner, H. R. & Burnier, M. Does protein binding modulate the effect of angiotensin II receptor antagonists? *J. Renin-Angiotensin-Aldosterone Syst.* **2**, S54–SS58 (2001).

Mather, J. P. Making informed choices: Medium, serum, and serum-free medium. How to choose the appropriate medium and culture system for the model you wish to create. *Methods Cell Biol.* **57**, 19–30 (1998).

Nicholson, J. P., Wolmarans, M. R. & Park, G. R. The role of albumin in critical illness. *Br. J. Anaesth.* **85**, 599–610 (2000).

Pal, S., Saha, C., Hossain, M., Dey, S. K. & Kumar, G. S. Influence of galloyl moiety in interaction of epicatechin with bovine serum albumin: A spectroscopic and thermodynamic characterization. *PLoS ONE* **7** (2012).

Quideau, S., Deffieux, D., Douat-Casassus, C. & Pouységu, L. Plant polyphenols: Chemical properties, biological activities, and synthesis. *Angew. Chem. Int. Ed. Engl.* **50**, 586–621 (2011).

Robertson, M. J., Barnes, J. C., Drew, G. M., Clark, K. L., Marshall, F. H., Michel, A., Middlemiss, D., et al. Pharmacological profile of GR117289 in vitro: A novel, potent and specific non-peptide angiotensin AT1 receptor antagonist. *Br. J. Pharmacol.* **107**, 1173–1180 (1992).

Soares, S., Mateus, N. & De Freitas, V. Interaction of different polyphenols with bovine serum albumin (BSA) and Human Salivary alpha-amylase (HSA) by fluorescence quenching. *J. Agric. Food Chem.* **55**, 6726–6735 (2007).

Spector, A. A., John, K. & Fletcher, J. E. Binding of long-chain fatty acids to bovine serum albumin. *J. Lipid Res.* **10**, 56–67 (1969).

Stoddart, L. A., Brown, A. J. & Milligan, G. Uncovering the pharmacology of the G protein-coupled receptor GPR40: High apparent constitutive activity in guanosine 5'-O-(3-[^{35}S]thio)triphosphate binding studies reflects binding of an endogenous agonist. *Mol. Pharmacol.* **71**, 994–1005 (2007).

Tolpygo, S. M., Pevtsova, E. I., Shoibonov, B. B. & Kotov, A. V. Free and protein-bound angiotensin II1-7 in the regulation of drinking behavior and hemodynamics in rats. *Bull. Exp. Biol. Med.* **153**, 623–626 (2012).

Tolpygo, S. M., Pevtsova, E. I., Shoibonov, B. B. & Kotov, A. V. Comparative study of the effects of free bound and carrier protein angiotensin II in experimental hypoglycemia and hyperglycemia **156**, 419–422 (2014).

Tolpygo, S. M., Pevtsova, E. I. & Kotov, A. V. Behavioral and hemodynamic effects of free and protein-bound angiotensin IV in rats in experimental hypo- and hyperglycemia: Comparative aspects. *Bull. Exp. Biol. Med.* **159**, 297–301 (2015).

Vazquez, E., Corchero, J. L. & Villaverde, A. Post-production protein stability: Trouble beyond the cell factory. *Microb. Cell Fact.* **10** (2011).

Vélez, A. J., Garcia, L. A. & de Rozo, M. P. In vitro interaction of polyphenols of coffee pulp and some proteins. *Arch. Latinoam. Nutr.* **35**, 297–305 (1985).

Vergote, V., Van Dorpe, S., Verbeken, M., Burvenich, C., Van de Wiele, C., Banks, W. A. & De Spiegeleer, B. Development of peptide receptor binding assays: Methods to avoid false negatives. *Regul. Pept.* **158**, 97–102 (2009).

Wennogle, L. P., Steel, D. J. & Petrack, B. Characterization of central cholecystokinin receptors using a radioiodinated octapeptide probe. *Life Sci.* **36**, 1485–1492 (1985).

Xiao, Y. & Isaacs, S. N. Enzyme-linked immunosorbent assay (ELISA) and blocking with bovine serum albumin (BSA)-not all BSAs are alike. *J. Immunol. Methods* **384**, 148–151 (2012).

Zinellu, A., Sotgia, S., Scanu, B., Pisanu, E., Giordo, R., Cossu, A., Posadino, A. M., Carru, C. & Pintus, G. Evaluation of non-covalent interactions between serum albumin and green tea catechins by affinity capillary electrophoresis. *J. Chromatogr. A* **1367**, 167–171 (2014).

Index

C

I

L

M

N

S

T

V